Insertion Devices for Synchrotron Radiation and Free Electron Laser

World Scientific Series on Synchrotron Radiation Techniques and Applications

Editors-in-Charge:
D. H. Bilderback *(CHESS, Cornell University, USA)*
K. O. Hodgson *(Dept of Chemistry and SSRL, Stanford University, USA)*
M. P. Kiskinova *(Sincrotrone Trieste, Italy)*
R. Rosei *(Sincrotrone Trieste, Italy)*

Published

Vol. 1 Synchrotron Radiation Sources — A Primer
        H. Winick

Vol. 2 X-ray Absorption Fine Structure (XAFS) for Catalysts & Surfaces
        ed. Y. Iwasawa

Vol. 3 Compact Synchrotron Light Sources
        E. Weihreter

Vol. 4 Novel Radiation Sources Using Relativistic Electrons
        P. Rullhusen, X. Artru & P. Dhez

Forthcoming

Photoelectron Spectroscopy of Solids and Surfaces
        K. C. Prince

X-ray Microscopy
        ed. G. R. Morrison

Medical Applications of Synchrotron Radiation
        eds. Al Thompson, B. Thomlinson & M. Ando

Synchrotron Radiation Studies of Non-Crystalline Biology Systems
        J. Bordas & A. Svensson

F. Ciocci
G. Dattoli
A. Torre
A. Renieri
ENEA INN-FIS, Frascati, Rome, Italy

Insertion Devices for Synchrotron Radiation and Free Electron Laser

World Scientific
Singapore • New Jersey • London • Hong Kong

Published by

World Scientific Publishing Co. Pte. Ltd.
P O Box 128, Farrer Road, Singapore 912805
USA office: Suite 1B, 1060 Main Street, River Edge, NJ 07661
UK office: 57 Shelton Street, Covent Garden, London WC2H 9HE

British Library Cataloguing-in-Publication Data
A catalogue record for this book is available from the British Library.

INSERTION DEVICES FOR SYNCHROTRON RADIATION AND FREE ELECTRON LASER

Copyright © 2000 by World Scientific Publishing Co. Pte. Ltd.

All rights reserved. This book, or parts thereof, may not be reproduced in any form or by any means, electronic or mechanical, including photocopying, recording or any information storage and retrieval system now known or to be invented, without written permission from the Publisher.

For photocopying of material in this volume, please pay a copying fee through the Copyright Clearance Center, Inc., 222 Rosewood Drive, Danvers, MA 01923, USA. In this case permission to photocopy is not required from the publisher.

ISBN 981-02-3832-0

This book is printed on acid-free paper.

Printed in Singapore by Uto-Print

Preface

Accelerated charged particles emit radiation. Albeit the work of Larmor (1897) and Lienard (1898), accomplished few years later the development of the Maxwell theory (1873), provided the theoretical foundations of this phenomenon, the first visual observation of what is now termed synchrotron radiation, dates back to about 50 years ago, with the first operations of the $70\,MeV$ synchrotron of General Electric (G.E.).

It is interesting to note that, after the pioneering work of Lienard and Larmor, this topic was largely unnoticed and gave rise to a brief flurry of interest with the work of Schott (1907,1912), who published a detailed analysis including the angular and frequency distribution of the emitted radiation as well as the polarization properties. There are at least two reasons, which may explain the initial lack of interest for this type of effect and for its theoretical implications

1) since its observation requires ultrarelativistic charged particles, it was regarded as a curiosity rather than a really observable phenomenon,

2) The use of the physical concepts, related to the synchrotron radiation, within the context of the microscopic theory of atomic phenomena was readily discouraged by the developing quantum mechanics and by its success in explaining the observed spectra.

During the second world war the problem was reconsidered by the Russian scientists Iwanenko and Pomeranchuk (1944), who argued that radiation, emitted by particles moving on circular paths and the consequent energy loss, may pose a limit on the maximum energy attainable with a circular accelerator.

The monumental work of Schwinger provided the final rigorous settlement of the theory and explained why the first experimental attempts by Blewett, to detect synchrotron radiation, failed. Although Blewett measured the particle energy loss in the $100\,MeV$ betatron of G.E. and found agreement with the theory, he looked for synchrotron radiation in the range of microwaves. The work of Schwinger pointed out that the spectrum is peaked at much higher frequencies and that the power emitted in the microwave region is negligible.

The subsequent discovery with the $70\,MeV$ G.E. synchrotron occurred by chance rather then by systematic investigation.

Before going further, let as underline that even-though any charged particle looses energy in the form of electromagnetic radiation, synchrotron radiation is the radiation emitted by charged particles travelling on curved paths. In the case of high energy electrons moving in the magnetic field of Storage Rings, the emitted radiation is extremely intense on a wide range of wavelengths, extending from the infra-red to the soft and hard X-ray region of the spectrum.

The historical development we have sketched shows that science proceeds gradually. Synchrotron radiation became a mature subject with the development of the technology of high energy accelerators. Synchrotron radiation was initially viewed as a limiting factor to accelerate particles at higher energies with circular machines, now its peculiar features of wide band, high spectral fluxes, directionality etc., make it a unique tool of investigations for basic and applied studies in biology, chemistry, medicine, physics and in applications to technology such as X-ray litography, micromechanics, material characterization and so on.

Machines dedicated to the production of synchrotron radiation are widespread all over the world and are not restricted to the most technologically advanced countries. Many synchrotron radiation facilities are located or are going to be realized in emerging or less developed countries, where they are seen as a way to develop new technologies and new tools in basic research and in industrial applications.

Synchrotron radiation sources are essentially provided by Storage Rings, with dimensions ranging from few meters in circumference and operating at a few hundred MeV to devices up to 1500 meters in circumference and operating in the GeV region.

The historical development of synchrotron radiation sources has proceeded through three steps which are known as the three generations of synchrotron radiation sources.

a) First generation

The Storage Rings of this generation were originally built for high energy research. The synchrotron radiation programs started in a parasitic mode and have grown to the point where facility is partially or fully dedicated to synchrotron radiation research.

b) Second generation

The sources of this generation are designed specifically to produce synchrotron light. These rings are characterized by a large number of beam lines, which may serve many users. The first example of second generation synchrotron radiation source is the 380 MeV SOR ring at the university of Tokyo.

c) Third generation

The rings of this generation have lower emittance, many straight sections for insertions and are designed to provide spectral photon fluxes much larger than those of the second generation.

In 1976 it was experimentally demonstrated that stimulated synchrotron radiation, namely the Free Electron Laser (FEL), can be obtained. Since then, various FEL facilities have been designed and operated. These new facts have opened the possibility of using Storage Rings or Linacs to realize fourth generation devices. Such sources would provide improvements of several orders of magnitude in the spectral fluxes, with respect to the performances of the third generation rings.

The amount of scientific and technical literature on synchrotron radiation and relevant sources and facilities is immense, it is therefore natural to ask why a further book on synchrotron radiation and insertion devices?.

The Authors have spent most of their scientific lives, working on the theory of synchrotron radiation, on the theory of FEL, on the design of wigglers, of undulators and of transport lines for electrons in high energy accelerators (Linacs and Storage-Rings), have participated to the design of FEL and proposed new schemes for the generation of X-ray sources. They witnessed the birth of FEL, contributed to its theoretical and technological developments, and contributed to the theory of synchrotron radiation, by presenting a comprehensive treatment of the spectroscopic details of the radiation emitted by relativistic electrons moving in magnetic undulators with different configurations. The above quoted experience can be helpful to provide a book which may be useful if it will give the correct idea of what is synchrotron radiation, how is produced, how is exploited and why, how will compete and cohabit with FEL devices, what will be its future.

The goal sounds ambitious!

We believe that a book is useful when it can be used by pedestrians and by scientists already expert in the field. We have therefore designed the book in such a way that the experts may use it as reference manual. We have indeed collected the main results by using practical formulae which can also be used for design purposes. The pedestrians will find concise results which privilege the physical and technological aspects and avoid as much as possible the mathematical details.

During the years of our activity we had the privilege of working with distinguished scientists in the field of synchrotron and Free Electron Laser radiation. They shared with us their points of view and provided the cultural background which made possible the completion of this book. We owe them our gratitude. Among them it is our pleasure to mention drs. G.K. Voykov and L. Giannessi who collaborated with us on various topics relevant to synchrotron radiation and FEL theory and in particular developed important methods to use the synchrotron radiation as a diagnostic tool for the characteristics of the emitting beam. It is also a pleasure to express our sincere appreciation to the generous and constant assistance of Dr. P.L. Ottaviani. Most of the ideas contained in chapter III are the result of ten years of joint research between him and one of the Authors (G.D.).

We have already mentioned that the scientific literature on synchrotron radiation and related topics is enormous, therefore, at the end of each chapter, the reader is not addressed to specific contributions, but mainly to books and review articles. Doing so we are not making justice to each contributors and for this we sincerely apologize in advance.

This book does not cover all the details of synchrotron radiation theory, of FEL theory and of the design and manufacturing aspects of the devices most commonly used to produce synchrotron and FEL radiation, its lecture should be therefore complemented by that of other books and review articles which are listed below, along with the seminal papers mentioned in this preface.

1. J. Larmor, Phil. Mag. 44, 503 (1897).
2. A. Lienard, L Eclairage Electrique 16, 5 (1898).

3. G.A. Schott, Ann. Phys. 24, 635 (1907).
4. G.A. Schott, Electromagnetic radiation, Cambrige University Press. (1912).
5. D. Ivanenko and I. Pomeranchuk, Phys. Rev. 65, 343 (1944).
6. J. Schwinger, Phys. Rev. 70, 798 (1946).
7. J.P. Blewett, Phys. Rev. 69, 87 (1946).
8. F.R. Elder, R.V. Langmnir and H.C. Pollock, Phys. Rev. 74, 52 (1948).
9. J.D. Jackson, Classical Electrodynamics, Wiley, New York (1962).
10. A.A. Sokolov and I.M. Ternov, Synchrotron Radiation, Akademic-Verlag, Berlin (1968).
11. S. Krinsky, M. Perlman and R.E. Watson, Characteristics of Synchrotron Radiation and of its Sources, in Handbook of Synchrotron Radiation, ed. by E.E. Koch, North-Holland, Amsterdam (1983).
12. G.Dattoli and A.Renieri, Experimental and Theoretical aspects of Free Electron Lasers, in Laser Handbook, Vol. IV, ed. by M.L. Stitch and M.S. Bass, North-Holland, Amsterdam (1984).
13. K.J. Kim, Characteristic of Synchrotron Radiation, AIP Conference proceedings, p.565, 184, Vol 1, AIP, New York (1989).
14. J.C. Marshall, Free Electron Lasers, Mac Millan Publishing Co, New York (1985).
15. A. Hofmann, Theory of Synchrotron Radiation, SSRL ACD note (SLAC), (Sept. 1986).
16. A. Hofmann, Characteristic of Sychrotron Radiation, Proceedings of CAS, CERN 90-3 (1990).
17. R.Barbini, F.Ciocci, G.Dattoli, L.Giannessi and A.Torre, Spectral Properties of the Undulator Magnet Radiation: Analytical and Numerical Treatment, La Rivista del Nuovo Cimento 6, (1990).
18. P.Luchini and H.Motz, Undulators and Free Electron Lasers, Claredon Press, Oxford (1990).
19. C.A. Brau, Free Electron Lasers, Academic Press, Boston (1990).
20. G.Dattoli, A.Renieri and A.Torre, Lectures on the Free Electron Laser Theory and Related Topics, World Scientific, Singapore (1995).

Foreword

Synchrotron radiation (S.R.) is the energy lost by a charged particle moving with relativistic speed on curved trajectories.

In particular high energy electrons traversing the field of bending, undulator or wiggler magnets, produce intense laser-like radiation on a broad range of frequencies not accessible with other sources.

To produce S.R. it is necessary to have

a) high-energy electrons

b) a magnetic field which constraints the electrons on a curved path.

It is evident that the most natural candidate as source of S.R. is a Storage Ring ($\mathcal{SR}$). In this device electrons (or positrons) circulate on a closed orbit. It is fairly natural that radiation be lost when the electron curve in the Storage Ring bending magnets. However they can also be forced to radiate by inserting wigglers or undulators (Fig. 1) in the straight sections of the Storage Ring.

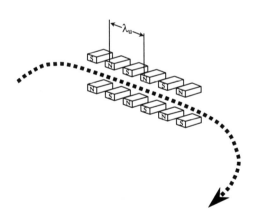

Figure 1   Undulator scheme

Wiggler and undulator magnets are the most commonly adopted insertion devices (ID) to produce S.R. . Since each one of these devices affects the electron motion in a different way, we can expect that the characteristics of the emitted radiation are not the same for any type of insertion.

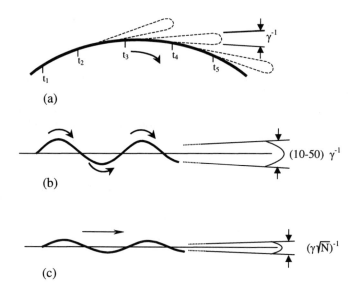

Figure 2   Schematic of emission characteristics for the three main types of synchrotron light sources: bending magnet (a), wiggler (b), undulator (c).

The bending magnets, or dipoles, bend the electron beam (e-beam) through short arcs, which close the orbit since add up to a total of 360 degrees. The electrons moving in bending magnets describe a circular orbit and produce a broad-band smooth spectrum.

Wigglers are essentially a sequence of bending magnets with alternate polarities. The emitted radiation has spectral characteristics analogous to those of bending magnets, with an intensity enhancement of a factor 2N, where 2N is the number of poles.

In the undulator the electrons propagate by executing smooth transverse oscillations, keeping into the emitted radiation cone in any point of the trajectory. As a result of the interference between radiation emitted from different points of the trajectory, the spectrum consists of a series of quasi-monochromatic peaks. The differences between the spectra of the three devices are, perhaps, better clarified by figs.2-4.

In the case of bending magnet (see fig. 2a), an observer placed on axis will receive a stream of photons by the radiating electron for a very short time, he will therefore measure a narrow electric field in the time domain, corresponding to a broad-band field in the frequency domain (see fig. 2b-c).

Being the wiggler an array of 2N bending magnets, the on axis observer will receive 2N distinct short time photon streams. The measured field will consist of a series of distinct sharp peaks in the time domain and of a broad-band spectrum, 2N time more intense than the single pole spectrum (see fig. ??a-c).

In the case of the undulator, the electrons do not execute large transverse oscillations. The on axis observer will receive an almost continuous signal in the time domain, in the frequency domain the spectrum is squeezed into a series of harmonics (see fig. 4a-c).

In the case of bending and wiggler magnets the spectrum is a smooth continuum characterized by the so called critical frequency

$$\varepsilon_c \, [KeV] = 0.665 \, B \, [T] \, E^2 \, [GeV] \qquad (1)$$

one half of the spectrum is radiated below ε_c, one half above ε_c. In eq. (1) B is the on axis magnetic peak field generated by the poles of the device, expressed in $Tesla$ ($10^4 \, Gauss$) and E is the electron energy in GeV ($10^9 \, eV$). To give an example for a 1-GeV Storage Ring with 1 $Tesla$ magnetic field, the critical frequency is $0.665 \, KeV$. By recalling that the wavelength is related to the energy of the radiation emitted by the practical formula

$$\lambda \, [cm] = \frac{1.2399 \cdot 10^{-7}}{\varepsilon_c \, [KeV]} \qquad (2)$$

we infer that one half of the spectrum is emitted above and below the UV region (see fig. 5).

As already remarked the undulator spectrum is provided by a series of nearly monochromatic peaks centered around (n is the order of the harmonics)

$$\varepsilon_n \, [KeV] = \frac{0.947 \, n \, E^2 \, [GeV]}{\lambda_u \, [cm] \, (1 + k^2/2)} \qquad (3)$$

where λ_u is the undulator period (see fig.1) and k is the undulator strength, linked to the on axis magnetic peak field and period by

$$k = 0.9362 \, B \, [T] \, \lambda_u \, [cm] \qquad (4)$$

For $E = 1 \, GeV$, $k = 1$ and $\lambda_u = 3 \, cm$, we find that the first harmonic is radiated around $\varepsilon_1 \cong 0.21 \, KeV$.

To give a further feeling on the numbers involved in, we note that the energy lost per machine turn by an electron in a bending magnet is

$$\delta E \, [KeV] = 26.6 \, B \, [T] \, E^3 \, [GeV] \qquad (5a)$$

while the corresponding power, lost by a stored current, is

$$P_{S.R.} \, [KW] = I \, [A] \, \delta E \, [KeV] \qquad (5b)$$

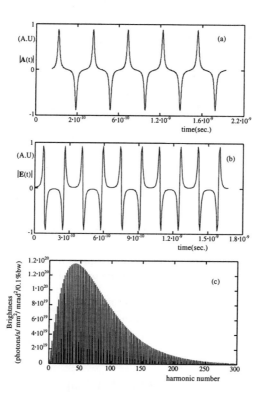

Figure 3 Characteristics of the e.m. radiation emitted in a wiggler with $\lambda_u = 10\,cm$ and strength parameter $k = 4.5$: (a) vector potential and (b) electric field modulus vs time, (c) brightness vs the harmonic number.

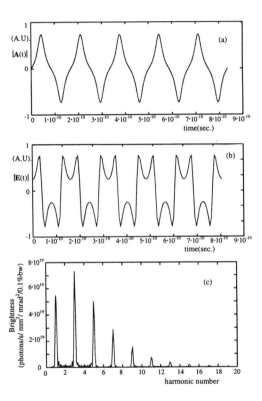

Figure 4 Characteristics of the e.m. radiation emitted in an undulator with $\lambda_u = 5\,cm$ and strength parameter $k = \sqrt{2}$: (a) vector potential and (b) electric field modulus vs time, (c) brightness vs the harmonic number.

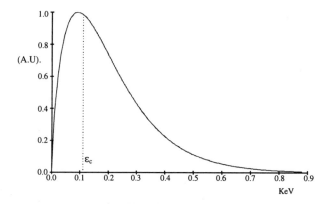

Figure 5 Synchrotron radiation spectrum

where I is the beam current expressed in Ampers.

According to eqs. (5a), the energy lost by a $1\,GeV$ electron in the field of $1\,Tesla$, is $26.6\,KeV$, and a stored current of $1\,A$ radiates $26.6\,KW$ per machine turn. Only a fraction of this power can be exploited for research purposes but all the lost power must be replaced by the Storage Ring accelerating system, to guarantee the stability of the device.

In the case of wigglers or undulators eq. (5a) should be replaced by

$$\delta E\,[KeV] = 0.633 \cdot E^2\,[GeV] \cdot \left\langle B^2\,[T] \right\rangle \cdot L\,[m] \tag{5c}$$

where $< B^2[T] >$ represents the average square value of the magnetic field of the device over its length L. The energy lost by one GeV electrons, moving in a $1\,m$ undulator with $< B^2 > \cong 1\,T^2$, is $0.633\,KeV$ and the corresponding power, radiated by a $1\,A$ beam, is $0.633\,KW$. The difference with the bending magnet case is that all the radiation emitted in the undulator or wiggler device can be delivered into the experimental stations.

We have mentioned at the beginning of these general remarks that S.R. exhibits laser- like properties. This is a very important point which should be clarified before entering into the main body of the book. In some experiments it is important that the sample receives a high photon flux, which is defined as

$$F = \#\,photons\,/\,\sec \cdot mrad \cdot \Delta\omega_B \tag{6a}$$

where $\Delta\omega_B$ is the unit bandwidth.

Other experiments demand for a very high brightness beam, this latter quantity being defined as

$$B = \# \ photons \ / \ \Delta\Omega \cdot S \cdot \Delta\omega_B \qquad (6b)$$

where $\Delta\Omega$ is the unit solid angle and S is the unit source area.

Along with brightness, high coherent power is required. Coherence and brightness are strictly related to the transverse quality of the e-beam, namely to its transverse phase-space area. To smaller emittances correspond larger brightness The emittances are in turn determined by the structure of the transport system inside the ring. Emittances of about 100-200 $nmrad$ are typical of second generation rings, insertion devices on these rings may produce photon beam with brightness 10^{16}- 10^{17} ($photons/sec/mm^2/mrad^2/0.1\%\Delta\omega_B$). Third generation rings are designed to produce one order of magnitude lower emittances and photon beam with brightness up to 10^{19} ($photons/sec/mm^2/mrad^2/0.1\%\Delta\omega_B$).

It is important to keep in mind that the properties of the emitted radiation depend on the characteristics of the radiating beam, which are in turn determined by the structure of the transport system, or more technically by the lattice. The analysis of insertion devices for S.R. cannot be viewed by itself, but as a part of a whole carefully integrated in the rest of the system. A part of this book will be, therefore, dedicated to the design of lattices and to their interplay with the characteristics of the emitted radiation. The design and construction of IDs usually concern a) the effects on the e-beam b) the spectral characteristics of the emitted radiation.

Both points are important for the optimization of the devices. The first, crucial point for devices operating in Storage Rings, will be discussed in chap. 3. A further important issue is linked to the characterization of the magnetic field of the device. Wigglers and undulators are most commonly realized by means of permanent magnet materials. The technological aspects relevant to the construction of these devices will be extensively discussed in the book as well as the methods relevant to the field measurement.

Free Electron Lasers (FEL) devices, have been independently developed since the last twenty years. In a FEL a beam of relativistic electrons, not necessarily from a Storage Ring, is injected into a magnetic undulator, which provides a transverse component to the electron motion, thus allowing the coupling of the beam with a copropagating TE wave, nearly resonant with one of the harmonics (normally the first) of the undulator spectrum. The coupling induces an energy modulation of the beam, which is followed by a density modulation and by a coherent emission, occurring when the density modulation is of the same order of the wave-length of the TE-wave.

In a FEL oscillator the emitted radiation, stored in an optical cavity, reinteracts with the e-beam and becomes amplified until gain saturation occurs. In fig. 6 we have reported the schematic layout of an ID in a Storage Ring. The laser optical cavity and the undulator are inserted on the straight section of the ring. The maximum attainable laser power is linked to the total power emitted in one machine turn via synchrotron radiation, namely

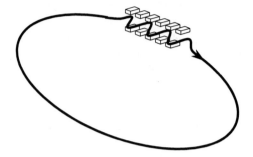

Figure 6 Schematic layout of an ID on a straight section of a Storage Ring.

$$P_L \cong P_{S.R.} \,/\, 2N$$

where N is the number of periods of the undulator.

Experiments requiring higher brightness, higher coherence, higher peak power are benefitting from the peculiar characteristics of a Storage Ring FEL.

We must, finally, underline that oscillator FELs are limited to the region in which mirrors to confine the produced radiation are available. We can safely assume a lower limit for wavelengths around $180\,nm$, i.e. not above VUV region (see fig. 5).

FELs operating in the X-region require a mirrorless configuration, known as SASE (self amplified spontaneous emission) which employs high current and high energy Linacs. In this case the output laser power is proportional to a fraction of the e-beam power and may reach power densities of the order of $10^5\,MW/cm^2$.

The preliminary remarks we have given, yield, perhaps, an idea of how the field of S.R. is close to a significant evolution and that the interplay with FEL may provide significant improvements and enlarge the number of possible applications.

This book contains a description of conventional $\mathcal{SR}$ insertion devices and includes some comments on FEL and on its use.

Contents

Preface	v
Foreword	ix

1 An introduction to the theory of charged particle transport — 1
- 1.1 Summary of the properties of relativistic particles. — 1
- 1.2 Introducing the basic elements of a transport system. — 6
  - 1.2.1 Dipole or bending magnet — 6
  - 1.2.2 The quadrupole magnet — 6
- 1.3 Transport system and phase-space evolution. — 10
- 1.4 The synchrotron magnet and the Dispersion function. — 15
- 1.5 Transverse motion coupling and sextupole magnets. — 19
- 1.6 Periodicity conditions on circular transport and motion stability conditions. — 20
- 1.7 Chromaticity, resonances and concluding remarks. — 24
- 1.8 Problems — 28

A Bending magnets and magnetic lenses — 35

B Treatment of charged beam transport using Lie algebra — 39

2 Generalities on synchrotron radiation — 45
- 2.1 Introduction — 45
- 2.2 Lienard-Wiechert potentials — 50
- 2.3 Spectral properties of the synchrotron radiation — 53
- 2.4 Damping and beam equilibrium conditions. — 57
- 2.5 The Fokker-Planck treatment of the longitudinal equilibrium problem — 64
- 2.6 Synchrotron radiation from magnetic undulators — 66
- 2.7 Concluding remarks — 71
- 2.8 Problems — 74

3 Generalities on Free Electron Lasers — 81
- 3.1 Introduction and low gain equations. — 81
- 3.2 High-gain regimes — 88
- 3.3 FEL strong signal regime — 95
- 3.4 Storage Ring FEL dynamics — 100
- 3.5 FEL Optical Klystron devices — 103
- 3.6 Concluding remarks — 107
- 3.7 Problems — 112

4 Optical systems in the geometrical and wave optics framework — 119
- 4.1 Introduction — 119
- 4.2 The optical phase space — 122
- 4.3 Quadratic Hamiltonian and optical ray matrices — 132
- 4.4 Quadratic Hamiltonians and optical propagators — 139
- 4.5 Problems — 145

5 Wigner distribution and synchrotron radiation sources — 149
- 5.1 Introduction — 149
- 5.2 Optical Wigner distribution function — 151
- 5.3 Moments of Wigner distribution function and phase space ellipse — 157
- 5.4 Optical characteristics of synchrotron radiation sources — 167
- 5.5 Problems — 176

6 Synchrotron radiation sources, insertion devices and beam current limitations — 179
- 6.1 Introduction — 179
- 6.2 Undulator brightness — 182
- 6.3 Wiggler and bending magnets — 183
- 6.4 Polarization characteristics of insertion devices — 186
- 6.5 Emittance optimization — 191
- 6.6 Effect of the insertion device on the beam parameters — 197
- 6.7 Current limitations — 200
  - 6.7.1 Beam-gas interaction — 201
  - 6.7.2 Ion trapping processes — 202
  - 6.7.3 Intra-beam scattering — 204

7 Constructing and measuring insertion devices — 217
- 7.1 Introduction — 217
- 7.2 Rare hearth permanent magnets — 219
- 7.3 Measuring techniques for PM blocks — 222
- 7.4 Magnetic field correcting techniques — 228
- 7.5 Magnetic field measurements — 229

8 Free Electron Lasers as insertion devices 237
8.1 Introduction 237
8.2 FEL oscillator brightness 238
8.3 FEL Storage Ring brightness 243
8.4 SASE FEL brightness 246
8.5 Emission by a prebunched e-beam 248
8.6 Self-induced higher order generation 255
8.7 Self-induced harmonic generation in Storage Rings 264
8.8 Storage-Ring FELs and longitudinal instabilities. 269
8.9 Harmonic generation and FEL devices operating at short wave-lengths 275
8.10 Fourth generation synchrotron radiation sources: Linacs or Storage Rings. 281

9 Synchrotron radiation beam lines: X-ray optics 287
9.1 Introduction 287
9.2 Role of X-ray optics: preservation of the source brightness and emittance. 295
9.3 X-ray mirrors. 296
9.4 X-ray focusing mirrors 308
9.5 Multilayers 319
9.6 Crystal monochromator 325
9.7 X-rays from synchrotron radiation sources: some applications. 346
9.8 Conclusions 351
9.9 Problems 351

Index 353

Chapter 1
AN INTRODUCTION TO THE THEORY OF CHARGED PARTICLE TRANSPORT

Summary

This chapter contains a brief outline of the properties of relativistic particles and of the methods of charged beam transport, with particular reference to electrons. Technical details are treated in two Appendices relevant to the magnetic multipole expansion and to the Lie algebraic theory of charged beam transport.

1.1 Summary of the properties of relativistic particles.

The peculiar nature of synchrotron radiation (S.R.) is due to the fact that it is emitted by charged particles moving at relativistic speed on a curved path. Before entering into more technical details, it is convenient to clarify, from a quantitative point of view, what relativistic means, thus devoting this introductory section to the properties of relativistic particles. These preliminary remarks, albeit very well known to the reader, have the purpose of providing familiarity with the notation and the terminology we will use in the rest of the book.

A charged particle moving in a uniform electric field is subjected to a constant acceleration provided by

$$\mathbf{a} = \frac{q \cdot \mathbf{E}}{m_q} \qquad (1.1.1)$$

where E is the electric field vector, q and m_q are the charge and the mass of the particle.

The final velocity acquired by the particle after crossing a region with potential difference V, is

$$v_p = \sqrt{\frac{2qV}{m_q}} \qquad (1.1.2)$$

If we express V in volts and normalize q and m_q to the electron charge and mass* (e, m_0) we obtain

---
*Recall that $e = 1.602 \cdot 10^{-19} C$, $m_0 = .511 \cdot 10^6 eV$, $10^6 eV = 1 MeV$

$$v_p\,[cm/\sec] = 5.97 \cdot 10^7 \alpha_n \sqrt{V\,[Volt]} \tag{1.1.3}$$

$$\alpha_n = \sqrt{\frac{m_e}{m_q}}$$

In the case of the proton we obtain $\alpha_n \cong 2.33 \cdot 10^{-2}$. The above equations hold for non relativistic particle only. By recalling, indeed, that the light velocity is $2.997 \cdot 10^{10} cm/\sec$, equation (1.1.3), if assumed valid for any velocity, would state that an electron can reach c, after crossing a region having a potential difference of $0.25 \cdot 10^3 Volts$.

By non relativistic particle we mean a particle which is moving at low velocity compared to that of light. It is therefore convenient to introduce the ratio

$$\beta = \frac{v}{c} \tag{1.1.4}$$

which is usually called "reduced relativistic velocity".

According to the above definition we will say that a particle is moving at non-relativistic speed whenever

$$\beta \ll 1 \tag{1.1.5}$$

The above inequality does not mean too much from the quantitative point of view. Let us therefore recall that in accordance with the theory of relativity the mass of an electron moving at speed v is

$$m = \frac{m_e}{\sqrt{1-\beta^2}} \tag{1.1.6}$$

By recalling that the mass is equivalent to energy, we also obtain that the total equivalent mass of an electron which has crossed a region of potential difference V is

$$m = m_0 \cdot \left(1 + \frac{eV}{m_0 c^2}\right) = m_0 \cdot \left(1 + 1.94 \cdot 10^{-6} V\right) \tag{1.1.7}$$

By comparing equations (1.1.6) and (1.1.7) we end up with

$$v = c \cdot \sqrt{1 - \frac{1}{(1 + 1.96 \cdot 10^{-6}\,V)^2}} \tag{1.1.8}$$

which can be fairly helpful to establish more quantitative conclusions.

When the electron kinetic energy is around $100 keV$ it has reached a velocity which is about 54% of the light velocity, at $1\,MeV$ the reduced relativistic velocity is around 0.94, while at $1 GeV$ ($10^3\,MeV$) β is 0.99999987. Figs. 1,2 are perhaps more explicative, they contain indeed the dependence of β and of the relative mass

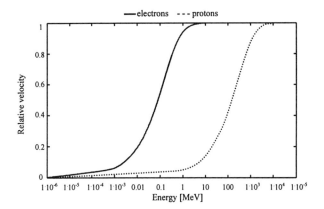

Figure 1 Relative velocity vs. energy. Continuous line electrons, dotted line protons.

vs. the particle kinetic energy and a comparison between the electron and proton cases[†]. It is evident that a proton can be considered relativistic when it acquires kinetic energies larger than $1 GeV$.

It is interesting to note (see fig. 2) that, with increasing kinetic energy, the relative electron mass become equivalent to that of much heavier particles.

An other important quantity which is usually exploited to define the degree of relativisticity of a particle is the ratio of the kinetic energy to the rest mass

$$\chi = \frac{E_k}{m_q \cdot c^2} \qquad (1.1.9)$$

since

$$mc^2 = m_q c + E_k \qquad (1.1.10a)$$

we can also state that

$$m = m_q (1 + \chi) \qquad (1.1.10b)$$

and

[†]The analogous of eq.(1.1.8) for the proton reads

$$v = c \cdot \sqrt{1 - \frac{1}{\left(1 + 1.94 \cdot 10^{-6} \frac{m_0}{m_p} V\right)^2}}$$

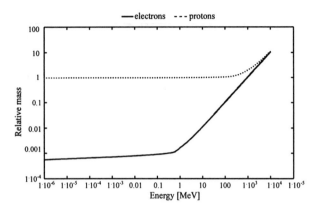

Figure 2 Relative mass, normalized to proton rest mass, vs. energy. Continuous line electrons, dotted line protons.

$$v = \frac{c}{1+\chi} \cdot \sqrt{\chi \cdot (2+\chi)} \qquad (1.1.10c)$$

A more commonly adopted quantity is the relativistic factor γ provided by the relation

$$m_q \gamma c^2 = m_q c^2 + E_k \qquad (1.1.11a)$$

i.e.

$$\gamma = \chi + 1 \qquad (1.1.11b)$$

the relativistic factor can also be viewed as the ratio between the total energy and the rest mass. It also clear that if $E_k \gg m_q c^2$ then $\gamma \simeq \chi$.

The momentum of a relativistic particle is the product of its relativistic mass by its velocity, namely

$$p = m_q \gamma v \qquad (1.1.12a)$$

In the case of highly relativistic speed it is expressed in GeV/c, in fact since $v \simeq c$ equation (1.1.12a) yields

$$p \simeq \frac{E}{c} \qquad (1.1.12b)$$

Before concluding this section, we introduce the concept of magnetic rigidity of a particle which may be exploited as a further measure of the relativistic nature of a particle.

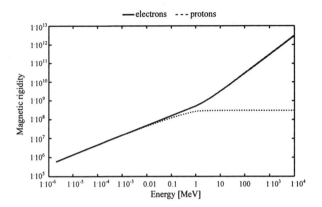

Figure 3 Behaviour of the electron magnetic rigidity (continuous line) and velocity (dotted line) vs. energy.

It is well known that a charged particle moving in a uniform magnetic field B describes a circle of radius r. According to the Lorentz force we find

$$qrB = m_q \gamma v \tag{1.1.13}$$

By using equations (1.1.10c) and (1.1.11b) we find

$$r \cdot B = \frac{1}{qc}\sqrt{E_k^2 + 2E_k m_q c^2} \tag{1.1.14}$$

The above quantity is usually called magnetic rigidity of the particle. It is evident that for non-relativistic energies the magnetic rigidity is proportional to $\sqrt{E_k}$, while for increasing speeds it goes like E_k. The transition between non-relativistic and relativistic regions is provided by the region exhibiting a change in the slope of rB vs. E_k. This region is clearly visible in fig. 3 where we have reported the electron magnetic rigidity and velocity.

It is finally worth stressing that, by normalizing everything to the electron charge and mass, we can write the particle magnetic rigidity in practical units as

$$r\,[cm] \cdot B\,[kG] = \frac{10.7}{2\pi} \cdot \frac{e}{q} \cdot \sqrt{\chi_e^2 + 2\chi_e \cdot \frac{m_q}{m_0}} \tag{1.1.15}$$

After these preliminary remarks we will come into more technical details concerning the charged beam transport systems.

1.2 Introducing the basic elements of a transport system.

In the previous section we have tried to clarify the meaning of the attribute relativistic for particles moving with high speed. We have noted that this notion can be quantitatively expressed by means of suitable parameters, linking the mass, the energy and the velocity of a particle.

We have also introduced a parameter of noticeable importance, namely the magnetic rigidity, which may also be viewed as a measure of the relative difficulty of bending relativistic particles with respect to their low energy counterparts.

Albeit trivial, we have reached an important conclusion, a magnetic field can be used to bend charged particles moving on a linear path. We can, therefore, argue that suitable combinations of magnetic elements can be exploited to guide and focus an electron beam in a Storage Ring or in any transport device.

Lattice is the term which is usually adopted to indicate the system of magnetic lenses which realizes the transport device. It is also clearly understood that the elements in which S.R. is produced: bending magnets, undulators and wigglers are themselves the components of a lattice, whose design is the first most important choice in the realization of a project dedicated to a S.R. device. In the case of a Storage Ring the lattice determines the emittance of the electron beam, the brightness of the photon beam, the beam life time, the number of insertion devices that can be placed into the straight sections, the cost of the entire project etc.

In the following we will give the basic ideas of a transport system and of the relevant components.

1.2.1 Dipole or bending magnet

A homogeneous magnetic field B_0 oriented along the vertical direction (see fig.4) is generated by a magnet with flat pole shoes (see Appendix A for details) and the radius of curvature of a ultrarelativistic particle of charge e and momentum p moving in the magnet, is given in practical units by

$$\frac{1}{\rho}\left[m^{-1}\right] = 0.2998 \cdot \frac{B_0\,[T]}{p\,[GeV/c]} \qquad (1.2.1)$$

1.2.2 The quadrupole magnet

This type of magnet is used to focus the e-beam (see fig.5),

we demand therefore that its field has a linear deviation from the axis (see Appendix A), namely

$$B_y = -g \cdot x \quad , \quad B_x = -g \cdot y \qquad (1.2.2)$$

where g is the gradient of the quadrupole. It is convenient to relate g to the momentum of the particle thus defining the quadrupole strength,

$$k = \frac{eg}{p} \qquad (1.2.3a)$$

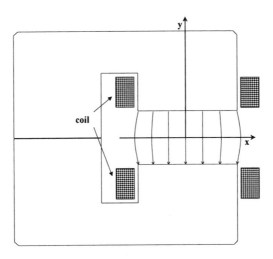

Figure 4   Schematic view of a dipole magnet.

which in practical units reads

$$k\left[m^{-2}\right] = 0.2998 \cdot \frac{g\,[T/m]}{p\,[GeV/c]} \qquad (1.2.3b)$$

An interesting property of the quadrupole is that the horizontal force component depends on the horizontal position only and not on vertical position of the particle trajectory. In fact the Lorentz equations of motion write

$$\mathbf{F} = m_0 \gamma \frac{d\mathbf{v}}{dt} = -e\mathbf{v} \times \mathbf{B} \qquad (1.2.4a)$$

and the relevant components read

$$\begin{aligned} F_x &= evgx \\ F_y &= -evgy \end{aligned} \qquad (1.2.4b)$$

where (see fig.6 for the axis orientation) v is the particle motion in the s-direction.

Being $\mathbf{v}$ perpendicular to $\mathbf{v} \times \mathbf{B}$ the energy and the relativistic mass of the particle are not changed, therefore, since $v \simeq c$ and $s \simeq ct$, equation (1.2.4b) can be cast in the form of the following second order linear equation ($k > 0$)

$$\frac{d^2x}{ds^2} = k_x \cdot x \quad ; \quad k_x = +k \qquad (1.2.5)$$

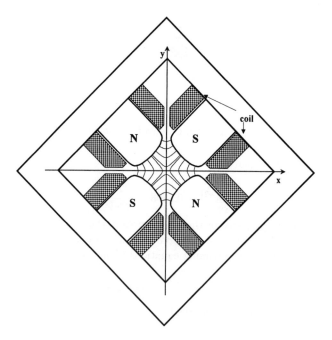

Figure 5 Cross section of a quadrupole magnet.

Introducing the basic elements of a transport system

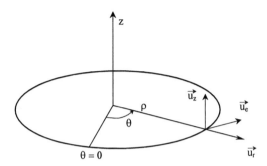

Figure 6   Coordinate system for a particle motion in a circular accelerator.

$$\frac{d^2y}{ds^2} = k_y \cdot y \quad ; \quad k_y = -k$$

The transverse motions are decoupled and the motion along the y direction is a simple harmonic motion, while in the x-direction the solutions are not of the oscillatory type. Accordingly we will say that the quadrupole magnet is focusing or defocusing according to wether $k_\xi < 0$ or $k_\xi > 0$ ($\xi = x, y$).

Before considering more complicated situations, we can summarize the above discussion as follows: bending and quadrupole magnets are the basic elements of a lattice.

If the bending magnets were the only elements of the lattice, particles with spatial coordinates different from those of the ideal orbit, i.e. that of a reference electron moving e.g. on a trajectory with zero transverse coordinates, would move progressively away from this orbit. A beam of electrons is provided by an ensemble of particles with different coordinates, angles and energies. The whole beam would therefore spread out and eventually be lost. For this reason the quadrupole magnets provide the elements which counteract the spreading of the particles with different coordinates. Fig.7 yields a pictorial illustration of the concepts discussed so far. The motion around the reference trajectory will be called betatron motion.

We did not mention that along with quadrupole and bending magnets the free space or drift section should be considered as one of the main element of the lattice. The next section will be devoted to the description of general methods to treat the transport of electrons through dipole, quadrupoles and drift sections.

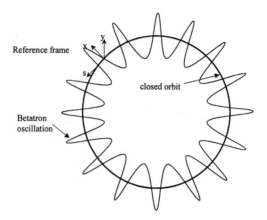

Figure 7 Descriptive view of the closed orbit and a betatron oscillation.

Since we will be mainly interested to electrons, the term electron and particle will be interchanged.

It is worth to emphasize that the longitudinal coordinate of propagation s plays the role of time and therefore the chosen frame of reference moves with the particles.

1.3 Transport system and phase-space evolution.

We have seen that if we limit ourselves to the case of transverse uncoupled motion, the problem of electron evolution through a quadrupole magnet is reduced to a second order differential equation of the type 1.2.5. The evolution through a drift section can be recovered by setting $k_y = 0$.

The above mentioned equation of motion can be derived from the Hamiltonian

$$H = \frac{1}{2}p_\xi^2 + \frac{1}{2}k_\xi \xi^2 \quad , \quad \xi = x, y \qquad (1.3.1)$$

where p_ξ and ξ are canonical coordinates. The characteristics of the whole beam motion can be derived from the analysis of the evolution of the relevant particle distribution. An effective tool of analysis is offered by the study of the Liouville phase-space evolution. Denoting by $f(p_x, x)$ such a distribution, from the Liouville theorem we obtain

$$\frac{d}{dt}f(p_\xi, \xi) = p'_\xi \frac{\partial f}{\partial p_\xi} + \xi' \frac{\partial f}{\partial \xi} + \frac{\partial f}{\partial s} \qquad (1.3.2)$$

Transport system and phase-space evolution

Since s plays the role of the time the primes denote derivatives with respect to s. By exploiting the Hamilton equations of motion we get from equation (1.3.2)

$$\frac{\partial f}{\partial s} = -p_\xi \frac{\partial f}{\partial \xi} + k_\xi \xi \frac{\partial f}{\partial p_\xi} \qquad (1.3.3)$$

Before going further it is worth clarifying the physical meaning of p_x which within the present context is just provided by

$$p_\xi = \frac{d\xi}{ds} = \xi' \qquad (1.3.4)$$

Since we assume that electrons do not deviate significantly from the axis motion, ξ' can be just viewed as the angle formed by the particle with the axis of motion. The $f(p_\xi, \xi)$ describes therefore the evolution of the distribution of the particles divergence and coordinates along the lattice.

Equation (1.3.3) can be solved exactly. We assume indeed that the general form of $f(p_\xi, \xi)$ is of the type

$$f(p_\xi, \xi) = \frac{1}{2\pi\varepsilon_\xi} \cdot \exp\left[-\frac{1}{2} \frac{\beta_\xi p_\xi^2 + 2\alpha_\xi p_\xi \xi + \gamma_\xi \xi^2}{\varepsilon_\xi}\right] \qquad (1.3.5)$$

where (β, α, γ) are called the beam Twiss parameters and satisfy the relation

$$\beta_\xi \cdot \gamma_\xi - \alpha_\xi^2 = 1 \qquad (1.3.6)$$

to ensure the normalization of (1.3.6). From equation (1.3.5) we infer that

$$\begin{aligned}
\sigma_\xi^2 &= <\xi^2> - <\xi>^2 = \varepsilon_\xi \beta_\xi \\
\sigma_{\xi'}^2 &= <p_\xi^2> - <p_\xi>^2 = \varepsilon_\xi \gamma_\xi \\
\sigma_{\xi,p_\xi} &= <\xi p_\xi> - <\xi><p_\xi> = -\varepsilon_\xi \alpha_\xi
\end{aligned} \qquad (1.3.7a)$$

where the brackets denote average on the distribution (1.3.5). It is therefore clear that β_ξ is linked to the beam r.m.s. dimension, γ_ξ its r.m.s. divergence and α_ξ is linked value of the second mixed momentum. According to equation (1.3.5) and (1.3.6) we obtain the relation

$$\sigma_{p_\xi}\sigma_\xi = \sqrt{1 + \alpha_\xi^2}\varepsilon_\xi \qquad (1.3.7b)$$

which indicates that when $\alpha_\xi \neq 0$ the product of the beam r.m.s. divergence and section is not proportional to the constant ε_ξ, which, as we will see below, plays a role of paramount importance.

According to the Liouville theorem the phase-space area is a conserved quantity. Accordingly the shape and the orientation of the locus containing positions and angles of the particle distributions change as a function of s, but its area remain

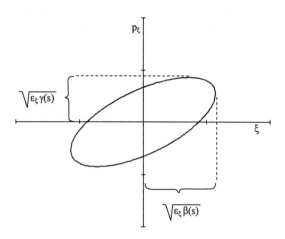

Figure 8 Phase-space representation of the e-beam

constant. The quadratic form argument of the distribution function (1.3.5) may be used as a geometric representation of phase-space locus, namely (see fig. 8)

$$\gamma_\xi \xi^2 + 2\alpha_\xi \xi p_\xi + \beta_\xi p_\xi^2 = \varepsilon_\xi \qquad (1.3.7c)$$

It is evident that it is an ellipse and ε_ξ linked to the ellipse area by $A_\xi = \pi\varepsilon_\xi$[†].

The physical meaning of the Twiss parameters should be clear. In particular the β function together with ε determine the maximum amplitude of the betatron oscillations. The units of γ are meter·radian and meter/radian for the β functions.

The evolution of f is provided by the variation of the Twiss parameters as function of s. By inserting (1.3.5) into (1.3.3) we obtain that the (β, α, γ) are governed by the equations of motion

$$\beta'_\xi = -2\alpha_\xi \quad , \quad \alpha'_\xi = k_\xi \beta_\xi - \gamma_\xi \quad , \quad \gamma'_\xi = 2k_\xi \alpha_\xi \qquad (1.3.8)$$

The evolution of the phase space locus is just provided by the s-dependence of the Twiss parameters and a convenient matrix form of the quadratic form 1.3.7c is provided below

$$\begin{pmatrix} \xi & p_\xi \end{pmatrix} \begin{pmatrix} \gamma_\xi & \alpha_\xi \\ \alpha_\xi & \beta_\xi \end{pmatrix} \begin{pmatrix} \xi \\ p_\xi \end{pmatrix} = \varepsilon_\xi \qquad (1.3.9)$$

the quadrupole strength k may be an s-dependent function, but it can be easily shown that the solution (1.3.5) holds in general.

[†]Some times A_ξ instead of ϵ_ξ is called the beam emittance.

Even though (1.3.5) are currently exploited in accelerator physics, we prefer to follow a different procedure which can be easily generalized in the case of coupled transverse motion. We set indeed

$$f(p_\xi, \xi) = \frac{1}{2\pi |\hat{\Sigma}_\xi|} \cdot \exp\left[-\frac{1}{2} \mathbf{z}^T \hat{\Sigma}_\xi^{-1} \mathbf{z}\right] \tag{1.3.10}$$

where

$$\mathbf{z} = \begin{pmatrix} \xi \\ p_\xi \end{pmatrix} \quad , \quad \hat{\Sigma}_\xi = \begin{pmatrix} \beta_\xi \varepsilon_\xi & -\alpha_\xi \varepsilon_\xi \\ -\alpha_\xi \varepsilon_\xi & \gamma_\xi \varepsilon_\xi \end{pmatrix} \quad , \quad |\hat{\Sigma}_\xi| = \det \hat{\Sigma}_\xi$$

According to (1.3.7) the matrix $\hat{\Sigma}_\xi$ can be viewed as the fluctuation tensor.

Furthermore according to (1.3.8) the evolution of the phase space distribution is essentially due to the evolution of $\hat{\Sigma}_\xi$, which, according to the method of characteristics, is provided by

$$\hat{\Sigma}_\xi(s) = \hat{U}(s) \hat{\Sigma}_\xi(0) \hat{U}^T(s) \tag{1.3.11a}$$

where $\hat{U}$ denotes the 2x2 matrix

$$\hat{U} = \begin{pmatrix} C & S \\ C' & S' \end{pmatrix} \tag{1.3.11b}$$

with S, C being the sin- and cos-like solution of the oscillatory equation (1.2.5), note that (see problem 1) the determinant of $\hat{U}$ is 1, namely

$$S'C - SC' = 1 \tag{1.3.11c}$$

The evolution of the Twiss parameters can be directly inferred from equation (1.3.11a), thus getting

$$\begin{pmatrix} \beta \\ \alpha \\ \gamma \end{pmatrix} = \begin{pmatrix} C^2 & -2CS & S^2 \\ -CC' & CS' - SC' & -SS' \\ C'^2 & -2C'S' & S'^2 \end{pmatrix} \cdot \begin{pmatrix} \beta_0 \\ \alpha_0 \\ \gamma_0 \end{pmatrix} \tag{1.3.12}$$

As already noted the beam evolution through a straight section of length l, can be obtained as a particular case of (1.3.12). By setting $k = 0$ in equation (1.3.12), we obtain for the evolution matrix

$$\hat{U} = \begin{pmatrix} 1 & l \\ 0 & 1 \end{pmatrix} \tag{1.3.13}$$

and, accordingly, the Twiss parameters after a drift l propagate as

$$\begin{pmatrix} \beta \\ \alpha \\ \gamma \end{pmatrix} = \begin{pmatrix} 1 & -2l & l^2 \\ 0 & 1 & -l \\ 0 & 0 & 1 \end{pmatrix} \cdot \begin{pmatrix} \beta_0 \\ \alpha_0 \\ \gamma_0 \end{pmatrix} \tag{1.3.14}$$

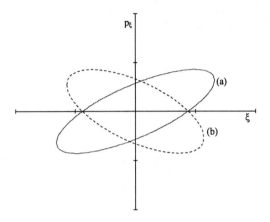

Figure 9   Phase-space evolution for the propagation through a drift section. The plot shows the ellipse slope before (a) and after (b) the passage

The corresponding evolution of the phase-space ellipse is shown in fig.9
In the case of focusing or defocusing quadrupole we have (for k constant)

$$\begin{pmatrix} C & S \\ C' & S' \end{pmatrix} = \begin{pmatrix} \cos\psi & \frac{1}{\sqrt{|k|}}\sin\psi \\ -\sqrt{|k|}\sin\psi & \cos\psi \end{pmatrix} \qquad (1.3.15a)$$

and

$$\begin{pmatrix} C & S \\ C' & S' \end{pmatrix} = \begin{pmatrix} \cosh\psi & \frac{1}{\sqrt{|k|}}\sinh\psi \\ \sqrt{|k|}\sinh\psi & \cosh\psi \end{pmatrix} \qquad (1.3.15b)$$

with

$$\psi = \sqrt{|k|}\cdot s \qquad (1.3.16)$$

The matrices (1.3.15a,1.3.15b) can be exploited to write down the evolution of the Twiss parameters. However, in the case in which the following condition holds

$$\sqrt{|k|}\cdot s \ll 1 \qquad (1.3.17)$$

it is easily verified that the quadrupole magnet acts on the beam as a thin lens with focus[§]

$$F = k \cdot s \tag{1.3.18}$$

very often $|f| = 1/F$ instead of F is used.
The evolution of the Twiss coefficients is then specified by

$$\begin{pmatrix} \beta \\ \alpha \\ \gamma \end{pmatrix} = \begin{pmatrix} 1 & 0 & 0 \\ -F & 1 & 0 \\ F^2 & -2F & 1 \end{pmatrix} \cdot \begin{pmatrix} \beta_0 \\ \alpha_0 \\ \gamma_0 \end{pmatrix} \tag{1.3.19}$$

The simple considerations we have developed so far, provide the basic elements of the design transport "technology", the exercises at the end of the chapter contain the necessary complement for a deeper understanding.

1.4 The synchrotron magnet and the Dispersion function.

The synchrotron magnet combines the effect of a dipole and of a quadrupole magnet. Its field is specified by the components (see Appendix A)

$$\begin{aligned} B_y &= B_0 - g \cdot x \\ B_x &= -g \cdot y \end{aligned} \tag{1.4.1}$$

The equation of motion of an electron traversing this field and moving on a circular path are specified by

$$\begin{aligned} \frac{d^2x}{ds^2} &= \frac{1}{r} - \frac{eB}{m_0 \gamma v} + \frac{eg}{m_0 \gamma v} \cdot x \\ \frac{d^2y}{ds^2} &= -\frac{eg}{m_0 \gamma v} \cdot y \end{aligned} \tag{1.4.2}$$

where

$$r = x + \rho \tag{1.4.3}$$

Furthermore we set

$$m_0 \gamma v = p = p_0 \cdot \left(1 + \frac{\Delta p}{p_0}\right) \tag{1.4.4}$$

[§]from eq.(1.3.14) in the case of thin lens approximation we get

$$\begin{pmatrix} C & S \\ C' & S' \end{pmatrix} = \begin{pmatrix} 1 & 0 \\ -\frac{1}{f} & 1 \end{pmatrix}$$

further comments on this approximation can be found in Appendix B

where p_0 is the momentum of the reference particle. If $x << r$ and $\Delta p << p_0$ the first of equations (1.4.2) can be cast in the form[¶]

$$\frac{d^2x}{ds^2} - \left(k - \frac{1}{\rho^2}\right) \cdot x = \frac{1}{\rho} \cdot \frac{\Delta p}{p_0} \tag{1.4.5}$$

The term $\frac{1}{\rho} \cdot x$ accounts for the weak focusing of the bending magnet, but what makes (1.4.5) different from (1.2.5) is the presence of the inhomogeneous term.

It is evident that equation (1.4.5) can be derived from the Hamiltonian

$$H = \frac{1}{2} p_x^2 - \frac{1}{2} \left(k - \frac{1}{\rho^2}\right) \cdot x^2 - \frac{1}{\rho} \cdot \frac{\Delta p}{p_0} \cdot x \tag{1.4.6}$$

and the relevant Liouville equation writes

$$\frac{\partial f}{\partial s} = -p_x \frac{\partial f}{\partial x} - \left(\bar{k} \cdot x + \frac{1}{\rho} \cdot \frac{\Delta p}{p_0}\right) \cdot \frac{\partial f}{\partial p_x} \tag{1.4.7}$$

$$\bar{k} = k - \frac{1}{\rho}$$

To better appreciate the role of the non-homogeneous term in equation 1.4.5, study the evolution of the function f by using the (see Appendix B) and simplify the problem by neglecting the quadratic part of the potential in (1.4.6) [‖].

$$f = \exp\left[s \cdot \left(-p_x \frac{\partial}{\partial x} - \frac{1}{\rho} \cdot \frac{\Delta p}{p_0} \frac{\partial}{\partial p_x}\right)\right] \cdot f_0 \tag{1.4.8}$$

where f_0 is the initial distribution.

By noting that (see Appendix B)

$$\exp\left[s \cdot \left(-p_x \frac{\partial}{\partial x} - \frac{1}{\rho} \cdot \frac{\Delta p}{p_0} \frac{\partial}{\partial p_x}\right)\right] \tag{1.4.9}$$

$$= \exp\left(-s\, p_x \frac{\partial}{\partial x}\right) \cdot \exp\left(-\frac{s}{\rho} \cdot \frac{\Delta p}{p_0} \frac{\partial}{\partial p_x}\right) \cdot \exp\left(\frac{1}{2} \frac{s^2}{\rho} \cdot \frac{\Delta p}{p_0} \frac{\partial}{\partial x}\right)$$

and by assuming that f_0 is provided by equation(1.3.5) we and up with (see Appendix B)

[¶] Eq. (1.4.5) can be derived after noting that $\frac{1}{r} \approx \frac{1}{\rho} \cdot \left(1 - \frac{x}{\rho}\right)$ and $\frac{1}{\rho} = \frac{eB_0}{p_0}$, $k = \frac{eg}{p_0}$ and by neglecting second order terms, like $x \cdot \frac{\Delta p}{p_0}$.

[‖] We assume that ρ does not depends on s.

The synchrotron magnet and the Dispersion function

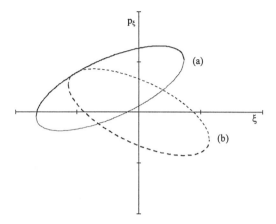

Figure 10 Phase-space evolution including the inhomogeneous term contribution. The figure shows the ellipse before (a) and after (b) the passage.

$$f = \exp\left(-sp_x\frac{\partial}{\partial x}\right) \cdot \frac{1}{2\pi\varepsilon} \exp\left[-\frac{1}{2\varepsilon}\left(\gamma\bar{x}^2 + 2\alpha\bar{x}\bar{p}_x + \beta\bar{p}_x^2\right)\right] \quad (1.4.10)$$

where
$$\bar{x} = x + \frac{1}{2}\frac{s^2}{\rho} \cdot \frac{\Delta p}{p_0}$$
$$\bar{p}_x = p_x - \frac{s}{\rho} \cdot \frac{\Delta p}{p_0}$$

The effect of the inhomogeneous term is that of modifying the phase-space evolution, The center of the ellipse is shifted off the position $x = 0$, $p_x = 0$ and then the evolution of the Twiss parameters occurs as in the standard free propagation evolution (see fig.10).

More in general the evolution of the phase-space distribution can be studied by defining the $\mathbf{z}$ column vector of equation (1.4.10) as follows

$$\mathbf{z} = \begin{pmatrix} x \\ p_x \\ \frac{\Delta p}{p_0} \end{pmatrix} \quad (1.4.11)$$

and the f function as

$$f(p_x, x, \frac{\Delta p}{p_0}) = \frac{1}{(2\pi)^{3/2}|\Sigma_x|^{1/2}} \cdot \exp\left[-\frac{1}{2}\mathbf{z}^T\hat{\Sigma}_x^{-1}\mathbf{z}\right] \quad (1.4.12)$$

where

$$\hat{\Sigma}_x = \begin{pmatrix} \beta_x \varepsilon_x & -\alpha_x \varepsilon_x & 0 \\ -\alpha_x \varepsilon_x & \gamma_x \varepsilon_x & 0 \\ 0 & 0 & \sigma_p^2 \end{pmatrix} \quad , \quad |\Sigma_x|^{1/2} = \varepsilon \cdot \sigma_p \tag{1.4.13}$$

The distribution (1.4.12) is slightly more general than the case considered so far. It includes indeed also the particle relative momentum distribution and σ_p is the relevant r.m.s. value.

The evolution of the $\hat{\Sigma}_x$ matrix is obtained from equation (1.3.11a) with the only difference that the matrix $\hat{U}$ should be replaced by**

$$\hat{U} = \begin{pmatrix} C & S & D \\ S' & C' & D' \\ 0 & 0 & 1 \end{pmatrix} \tag{1.4.14}$$

where D is called dispersion function and is associated with the homogeneous part of the solution of the second order differential equation (1.4.5)

$$D(s) = S(s) \int_{s_0}^{s} \frac{1}{\rho(s')} C(s') \, ds' - C(s) \int_{s_0}^{s} \frac{1}{\rho(s')} S(s') \, ds' \tag{1.4.15a}$$

and (see problem 6)

$$D'(s) = S'(s) \int_{s_0}^{s} \frac{1}{\rho(s')} C(s') \, ds' - C'(s) \int_{s_0}^{s} \frac{1}{\rho(s')} S(s') \, ds' \tag{1.4.15b}$$

It is also important to emphasize that, according to (1.4.12) we find that the first two relations of equation (1.3.7a) should be modified as

$$\begin{align} \sigma_x &= \sqrt{\varepsilon_x \beta_x + \sigma_p^2 D^2} \\ \sigma_{x'} &= \sqrt{\varepsilon_x \gamma_x + \sigma_p^2 D'^2} \end{align} \tag{1.4.16}$$

The reader may complement the previous considerations with the further readings suggested at the and of the chapter. He may also prove his degree of understanding of electronic transport by solving the proposed exercise.

**The matrix $\hat{U}$ is now exploited to evaluate the evolution of the beam fluctuation tensor and not that of the Twiss parameters.

1.5 Transverse motion coupling and sextupole magnets.

In the previous section we have considered linear optical elements. The sextupole magnet generates a non linear field specified by the components (see Appendix A)

$$B_y = \frac{1}{2}g' \cdot (x^2 - y^2) \qquad (1.5.1)$$
$$B_x = g' \cdot xy$$

The equations of motion of an electron experiencing the field (1.5.1) are provided by

$$\frac{d^2x}{ds^2} = -\frac{1}{2}m_s \cdot (x^2 - y^2) \qquad (1.5.2)$$
$$\frac{d^2y}{ds^2} = m_s \cdot xy$$

where m_s is the sextupole strength which in practical units is defined as

$$m_s \left[m^{-3}\right] = 0.2998 \frac{g' \left[T/m^2\right]}{p_0 \left[GeV/c\right]} \qquad (1.5.3)$$

Equations (1.5.2) are significantly different from those of the previous sections for two reasons: they couple transverse motions and are non-linear. Coupling and non linearity do not come necessarily together.

The field of a 45^0 rotated quadrupole (see Appendix A) is indeed provided by

$$B_y = -g \cdot y \qquad (1.5.4)$$
$$B_x = g \cdot x$$

and the Lorentz equations of motion yield the following system of second order equation

$$\frac{d^2x}{ds^2} = -k \cdot y \qquad (1.5.5)$$
$$\frac{d^2y}{ds^2} = -k \cdot x$$

which are coupled and linear.
Equations (1.5.5) can be derived from the Hamiltonian

$$H = \frac{1}{2}\left(p_x^2 + p_y^2\right) + k \cdot xy \qquad (1.5.6)$$

and the relevant Liouville equation writes

$$\frac{\partial f}{\partial s} = \left[-p_x \frac{\partial}{\partial x} - p_y \frac{\partial}{\partial y} + V_x(x,y) \frac{\partial}{\partial p_x} + V_y(x,y) \frac{\partial}{\partial p_y} \right] \cdot f \qquad (1.5.7)$$

where the potential $V(x,y)$ is specified by

$$V(x,y) = k \cdot xy \qquad (1.5.8)$$

and the subscripts x and y denote first derivative with respect to x or y.

The phase-space evolution is now characterized by four variables and the function f can be in general cast in the form

$$f(x, p_x, y, p_y) = \frac{1}{(2\pi)^2 |\Sigma_x|^{1/2}} \cdot \exp\left[-\frac{1}{2} \mathbf{z}^T \hat{\Sigma}^{-1} \mathbf{z}\right] \qquad (1.5.9)$$

$$\mathbf{z} = \begin{pmatrix} x \\ p_x \\ y \\ p_y \end{pmatrix}$$

and $\hat{\Sigma}$ is the fluctuation tensor. The solution of (1.5.9) is still provided by (1.3.11a) with $\hat{U}$ being the evolution matrix associated to the coupled system of differential equations (1.5.5) (see Appendix B for further comments).

Let us now come back to the sextupole, whose relevant "equation of motion" can be derived from the Hamiltonian

$$H = \frac{1}{2} p_x^2 + \frac{1}{2} p_y^2 + \frac{1}{6} m_s \cdot \left(x^3 - 3xy^2 \right) \qquad (1.5.10)$$

The non linearity of the problem does not allow the use of matrix-type operators to characterize the distribution evolution. Different methods can be exploited and are outlined in Appendix B.

Before closing this section we believe worth reminding that the magnetic field can be expanded in multipoles. The relevant practical formulae will be discussed in Appendix A.

1.6 Periodicity conditions on circular transport and motion stability conditions.

In the previous sections we have described the basic elements of electron beam transport through a magnetic structure. The considerations we have developed apply to the case of single passage through the magnetic elements and should be complemented by adding the conditions of periodicity and of motion stability.

To clarify the first condition we note that a circular accelerator may have a certain degree of symmetry. If we assume that it is provided by a certain number of identical cells with periodicity $\mathcal{L}$ [††], we must also assume that the characteristic parameters of the cell should exhibit the same periodicity. For example the "elastic constant" of the quadrupole should satisfy the condition

$$k(s+\mathcal{L}) = k(s) \qquad (1.6.1)$$

According to the Floquet-Bloch theorem the sin and cos-like solutions appearing in the evolution matrix $\hat{U}$ may be rewritten as

$$\begin{pmatrix} C(s) & S(s) \\ C'(s) & S'(s) \end{pmatrix} = \begin{pmatrix} \tilde{C}(s) & \tilde{S}(s) \\ \tilde{C}'(s) & \tilde{S}'(s) \end{pmatrix} \cdot \begin{pmatrix} \cos\left(\frac{\mu s}{\mathcal{L}}\right) & \sin\left(\frac{\mu s}{\mathcal{L}}\right) \\ -\sin\left(\frac{\mu s}{\mathcal{L}}\right) & \cos\left(\frac{\mu s}{\mathcal{L}}\right) \end{pmatrix} \qquad (1.6.2)$$

where the tilded functions have also periodicity $\mathcal{L}$. from equation(1.6.2) it follows that

$$\begin{pmatrix} C(s+m\mathcal{L}) & S(s+m\mathcal{L}) \\ C'(s+m\mathcal{L}) & S'(s+m\mathcal{L}) \end{pmatrix} = \begin{pmatrix} \tilde{C}(s) & \tilde{S}(s) \\ \tilde{C}'(s) & \tilde{S}'(s) \end{pmatrix} \cdot \begin{pmatrix} \cos(m\mu) & \sin(m\mu) \\ -\sin(m\mu) & \cos(m\mu) \end{pmatrix} \qquad (1.6.3)$$

with m being an integer and representing the number of passages through the cell.

The role played by μ, which in the present context can be just viewed as a phase advance per period $\mathcal{L}$, will be more deeply discussed below.

According to the previous discussion the effect of the periodicity is mainly contained in the phase advance μ. Before discussing the motion stability condition, we note that the parametric equation of the phase-space ellipse(see fig. 8 and eq. (1.3.7c)) can be written in the form (see problem 8) ($\xi' = p_\xi$)

$$\xi(s) = \sqrt{\varepsilon_\xi \beta_\xi(s)} \cos(\phi_\xi(s)) \qquad (1.6.4)$$

$$\xi'(s) = -\alpha_\xi(s) \sqrt{\frac{\varepsilon_\xi}{\beta_\xi(s)}} \cos(\phi_\xi(s)) - \sqrt{\frac{\varepsilon_\xi}{\beta_\xi(s)}} \sin(\phi_\xi(s))$$

If we demand that evolution through a periodic section be just that of producing a phase advance μ we must require that

$$\beta_\xi(s+\mathcal{L}) = \beta_\xi(s) \quad , \quad \alpha_\xi(s+\mathcal{L}) = \alpha_\xi(s) \quad , \quad \phi_\xi(s+\mathcal{L}) = \phi_\xi(s) + \mu_\xi \qquad (1.6.5)$$

It is also easily checked that the transfer operator

[††] $\mathcal{L}$ may also be the length of the ring.

$$\hat{U}(s+\mathcal{L} \mid s) = \hat{1}\cos\mu_\xi + \hat{S}\cdot\hat{T}_\xi \sin\mu_\xi, \qquad (1.6.6)$$

$$\hat{T}_\xi = \begin{pmatrix} \beta_\xi & \alpha_\xi \\ \alpha_\xi & \gamma_\xi \end{pmatrix}, \qquad \hat{S} = \begin{pmatrix} 0 & -1 \\ 1 & 0 \end{pmatrix}$$

and $\hat{1}$ is the unity matrix, the matrix $\hat{S}$ is usually called the symplectic matrix, since

$$\left(\hat{S}\cdot\hat{T}_\xi\right)^2 = -\hat{1} \qquad (1.6.7)$$

it is easy to prove that

$$\hat{U}(s+\mathcal{L} \mid s) = \exp\left[\mu_\xi \hat{S}\cdot\hat{T}_\xi\right] \qquad (1.6.8)$$

Furthermore it is worth noting that (see problem 9)

$$\hat{U}(s+m\mathcal{L} \mid s) = \exp\left(m\mu_\xi \hat{S}\cdot\hat{T}_\xi\right) = \hat{1}\cos(m\mu_\xi) + \hat{S}\cdot\hat{T}_\xi \sin(m\mu_\xi) \qquad (1.6.9)$$

A necessary and sufficient condition to guarantee the stability of the motion is that the transfer matrix (1.6.8) remains bounded for $m \to \infty$ and this is true for μ real only.

Let us now come back to equation (1.6.4), which along with equations 1.3.8 ensures that (see problem 5)

$$\phi_\xi(s) = \int_0^s \frac{ds'}{\beta_\xi(s)} \qquad (1.6.10)$$

from which it also follows

$$\mu_\xi = \int_s^{s+\mathcal{L}} \frac{ds'}{\beta_\xi(s)} \qquad (1.6.11)$$

an other quantity of noticeable importance is the tune linked to μ by the relation

$$\nu_\xi = \frac{\mu_\xi}{2\pi} \qquad (1.6.12)$$

and measuring the number of betatron oscillations per periodicity length.

It is worth closing this section by discussing a specific example. According to the problem 14 a lattice consisting only of equally spaced focusing and defocusing lenses, arranged as in fig. 11

namely first a focusing lens (F), then a drift section (O), third a defocusing lens (D) and finally another drift section (O), has a transfer matrix (see problem 14)

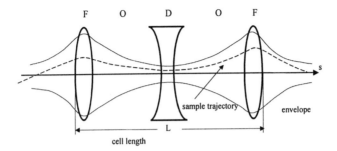

Figure 11   FODOF cell

$$\hat{U}(s+\mathcal{L} \mid s) = \begin{pmatrix} 1 - \frac{L}{2f} - \frac{L^2}{(2f)^2} & L + \frac{L^2}{4f} \\ -\frac{L}{2f^2} & 1 + \frac{L}{2f} \end{pmatrix} \qquad (1.6.13)$$

Any matrix

$$\hat{M} = \begin{pmatrix} A & B \\ C & D \end{pmatrix} \qquad (1.6.14)$$

with unit determinant can always be cast in the Courant-Snyder form (1.6.6), by putting

$$\mu = \cos^{-1}\left(\frac{A+D}{2}\right) \qquad (1.6.15)$$

$$\alpha = \frac{A-D}{2\sin\mu} \quad , \quad \beta = \frac{B}{\sin\mu} \quad , \quad \gamma = -\frac{C}{\sin\mu}$$

We can therefore conclude that the phase-advance per cell is

$$\mu = \cos^{-1}\left(1 - \frac{L^2}{8f^2}\right) \qquad (1.6.16a)$$

and thus stability is achieved when

$$\frac{L}{4f} < 1 \qquad (1.6.16b)$$

or what is the same

$$f > \frac{L}{4} \qquad (1.6.16c)$$

We have obtained a remarkably simple result: the motion is stable provided that the focal length is greater than $\frac{1}{2}$ the lens spacing. (for further comments see problem 15).

1.7 Chromaticity, resonances and concluding remarks.

We have stressed that the focussing or defocussing effect of a quadrupole is inversely proportional to the particle energy.

Electrons with different energies therefore are focussed (or defocussed) in different ways. This fact, in analogy to optical lenses, is called chromatic effect. We may therefore argue that this effect will provide a perturbation of the betatron motion by modifying the betatron tune, such a modification is called "chromaticity" is denoted by c_ξ and is linked to the tune shift by

$$\delta v_\xi = c_\xi \frac{\Delta p}{p_0} \tag{1.7.1}$$

an idea of how c_ξ should be evaluated is offered by the following simple considerations.

In the case of a constant k quadrupole the tune variation writes

$$\Delta v_\xi = \frac{1}{2\pi} \sqrt{k_\xi} L_q \tag{1.7.2}$$

where L_q is the length of the quadrupole. Any (Δv) shift due to the deviation from the nominal value reads

$$\delta(v_\xi) \simeq \frac{1}{4\pi} \frac{k_q}{\sqrt{k_\xi}} L_q \tag{1.7.3}$$

now since

$$\beta_\xi = \frac{1}{\sqrt{k_\xi}} \quad , \quad \delta k_\xi = -k_\xi \frac{\Delta p}{p_0} \tag{1.7.4}$$

we end up with

$$\delta(v_\xi) \simeq -\frac{1}{4\pi} \beta_\xi \, k_\xi \, L_q \frac{\Delta p}{p_0} \tag{1.7.5}$$

in the case of non constant β_ξ and k_ξ, by extending (1.7.5) to the whole circumference, we get

$$c_\xi = -\frac{1}{4\pi} \oint \beta_\xi \, k_\xi \, ds \tag{1.7.6}$$

Furthermore by using the second of eqs.(1.3.9) we end up with

$$c_\xi = -\frac{1}{4\pi} \oint \left(\alpha'_\xi + \gamma_\xi\right) ds \tag{1.7.7}$$

the first term vanishes, when integrated along the ring, so that

$$c_\xi = -\frac{1}{4\pi} \oint \gamma_\xi \, ds \tag{1.7.8}$$

There are several reasons to correct chromaticities. In particular, since they modify the machine tunes, their correction is necessary to prevent unwanted instabilities. The magnetic field of the sextupoles is usually exploited to compensate these effects.

The nature of the correction mechanism can be easily understood. We note indeed that the motion in the x-direction can be divided in two parts, the ordinary betatron contribution (x_β) and the dispersion dependent term (x_D), so that

$$x = x_\beta + x_D \tag{1.7.9}$$

Inserting (1.7.9) into the field components (1.5.1), we end up with

$$\begin{aligned} B_y &= (g' \, x_D) \, x_\beta + \frac{1}{2} g' \left(x_\beta^2 + x_D^2 - y_\beta^2\right) \\ B_x &= (g' \, x_D) \, y_\beta + \frac{1}{2} g' \, x_\beta \, y_\beta \end{aligned} \tag{1.7.10}$$

If we redefine $g' \, x_D = g_D$, i.e. as an amplitude dependent gradient, we can recognize the first contributions on the r.h.s. as a quadrupole term. This fact is utilized to offset the linear energy dependence of the focusing strength.

A better idea of how the sextupole works in the chromaticity correction is offered by fig.12

The sextupole is placed at $x = x_D$ in the horizontal plane $y = 0$. Particles having $\frac{\Delta p}{p_0} > 0$ are deflected towards the central orbit and particles with $\frac{\Delta p}{p_0} < 0$ move away from it (see fig.12).

Since the horizontal and vertical chromaticities are both negative, but the equivalent quadrupole contributions of (1.7.10) have opposite focussing and defocussing effects, two families of sextupoles are required, both placed in the dispersive region. From eqs. (1.7.6) and (1.7.10) it follows that the compensated chromaticity is just provided by[‡‡]

[‡‡]Note that (see eq.1.5.2)

$$k_D = \frac{e g' x_D}{p_0} = m_s D(s) \cdot \frac{\Delta p}{p_0}$$

To reduce the sextupole strength the sextupoles are located where the dispersion is large.

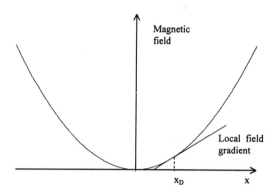

Figure 12   Sextupole field and local field gradient for an orbit displaced by x_D

$$\bar{c}_\xi = -\frac{1}{4\pi} \oint \left[k_\xi(s) - m_s D_\xi(s) \frac{\Delta p}{p_0} \right] \beta_\xi(s) \ ds \qquad (1.7.11)$$

The horizontally correcting sextupoles are located in regions in which the horizontal β function is large and the vertical β function is low. The converse is true for the vertical corrections.

The sextupole compensation is unavoidable, but along with the quadrupole like contributions they provide the further non linear terms, contained in the square brackets of (1.7.10) and responsible for transverse motion coupling.

In the analysis of the motion stability we have imposed that μ must be positive, on account of eq.(1.6.9) we should impose that

$$\mu_{x,y} \neq 2\pi \cdot m_{x,y} \qquad (1.7.12)$$

where $m_{x,y}$ are integers. The analysis of the motion shows that there are certain values of the tunes that are potentially dangerous for the motion stability, in particular those satisfying the condition

$$m \ v_x \pm n \ v_y = p \qquad (1.7.13)$$

where m, n, p are integers. Fig.13 shows a typical tune diagram, the machine working point is chosen at a reasonable distance from the resonance lines.

It is not difficult to realize that, besides chromaticity effects, unavoidable magnet imperfections may provide further tune shifts, which, following a procedure analogous to that yielding eq.(1.7.6), write

Chromaticity, resonances and concluding remarks

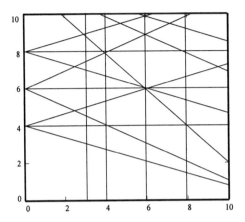

Figure 13   Example of stability diagram.

$$\Delta \nu_\xi = -\frac{1}{4\pi} \oint \beta_\xi(s) \, \Delta k_\xi(s) \, ds \qquad (1.7.14)$$

where $\Delta k_\xi(s)$ is the quadrupole strength error.

A gradient error may change the β function itself, in fact

$$\Delta \beta_\xi(s) = -\frac{\beta_\xi(s)}{2\sin(2\pi\nu_\xi)} \oint \beta_\xi(s') \, \Delta k_\xi(s') \, \cos\left(\frac{2}{\phi_\xi(s')} - \phi_\xi(s) - 2\pi\nu_\xi\right) ds' \quad (1.7.15)$$

The above result shows that the amplitude of the motion increases indefinitely for integer values of $2\nu_\xi$.

The origin of (1.7.13) can be explained as follows : dipole excite first order resonances, quadrupoles second order resonances, sextupoles third order ...

In the case of the sextupole there are four unstable resonances

$$\begin{aligned} 3\nu_x &= p \quad, \quad 2\nu_x + \nu_y = p \\ \nu_x + 2\nu_y &= p \quad, \quad 3\nu_y = p \end{aligned} \qquad (1.7.16)$$

This section concludes the chapter on electron transport, most of the concept we have developed here will be exploited in chapter III where we will discuss the longitudinal motion dynamics. The reader is addressed to the exercises and the bibliography at the end of the chapter to complete and complement the matter treated in the previous sections.

The order of the resonance is defined as $|m| + |n|$.

1.8 Problems

1) Show that

$$CS' - SC' = 1$$

(hint: use the notion of wronskian and the relevant Abel theorem, which states that the wronskian of a pair of interdependent solutions of

$$y'' + a(x) y' + y = 0$$

is $w(x) = c \cdot \exp\left[-\int_0^x a(x') \, dx'\right]$).

2) Use eqs.(1.3.8) to prove that the amplitude function β satisfies the differential equations

$$\frac{1}{2}\beta\beta'' - \frac{1}{4}\beta'^2 + k\beta^2 = 1$$
$$\beta''' + 4\beta' + 2k'\beta = 0$$

3) Define $\sigma = \sqrt{\varepsilon}\beta$ and prove that

$$\sigma'' + k\sigma = \frac{\varepsilon^2}{\sigma^3}$$

4) Show that the evolution matrix

$$\hat{U} = \begin{pmatrix} C & S \\ C' & S' \end{pmatrix}$$

can be cast in the form

$$\begin{pmatrix} C & S \\ C' & S' \end{pmatrix} = \begin{pmatrix} \sqrt{\beta}\cos\psi & \sqrt{\beta}\sin\psi \\ \sqrt{\gamma}\cos\chi & \sqrt{\gamma}\sin\chi \end{pmatrix}$$

and that

$$\sin(\chi - \psi) = \frac{1}{\sqrt{1 + \alpha^2}}$$

5) Obtain the eqs. (1.3.8) from the parametrization of the $\hat{U}$ matrix of problem 4 and show that

$$\psi = \int_0^s \frac{1}{\beta(s')} ds'$$

6) Use eq.(1.4.15) to prove that

$$D''(s) + kD = \frac{1}{\rho}$$

7) Prove that the quantity

$$J = \gamma\xi^2 + 2\alpha\xi\xi' + \beta\xi'^2$$

is an invariant of motion for a dynamical system ruled by a quadratic Hamiltonian of the type

$$H = \frac{1}{2}\xi'^2 + \frac{1}{2}k_\xi\xi^2$$

8) Show that eqs. (1.4.6) provide a parametric representation of the phase-space ellipse.

9) Prove identity 1.6.9 and find an appropriate expression for $e^{\hat{A}}$ where

$$\hat{A} = \begin{pmatrix} a & b \\ c & d \end{pmatrix}$$

10) Show that the phase space ellipse can be written in the alternative form

$$\frac{1}{\beta}\left(\xi^2 + (\alpha\xi + \beta\xi')^2\right) = \varepsilon$$

represent the locus in the plane

$$Y = \frac{\xi}{\sqrt{\beta}} \quad , \quad T = \frac{\alpha\xi + \beta\xi'}{\sqrt{\beta}}$$

and show that $Y' = \frac{T}{\beta}$.

11) Consider a system made up of two thin lenses each of focal length f (see fig.14), one focussing one defocussing, separated by a distance L. Show that the system is focussing if $|f| > L$.

$$\begin{pmatrix} 1 & 0 \\ \frac{1}{f} & 1 \end{pmatrix} \begin{pmatrix} 1 & L \\ 0 & 1 \end{pmatrix} \begin{pmatrix} 1 & 0 \\ -\frac{1}{f} & 1 \end{pmatrix} = \begin{pmatrix} 1 - \frac{L}{f} & L \\ -\frac{L}{f^2} & 1 + \frac{L}{f} \end{pmatrix}$$

12) Clarify the concept of thin lens approximation and note that this approximation can not be deduced from the Taylor expansion.

Furthermore explain why the application of the Taylor expansion would violate the Liouville theorem (see also problem 17-19). (hint. evaluate the determinant of the $\hat{U}$ matrix).

13) Evaluate the matrix of a quadrupole of length $10\ m$ with a gradient of $80\ T/m$

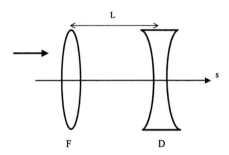

Figure 14

traversed by a particle of energy 20 TeV. Discuss the validity of thin lens approximation.

14) Suppose that a particle traverses first a focussing lens (F) with a focal length f; second a drift (O) of length $\frac{L}{2}$, third a defocussing lens (D) with focal length f and fourth an other drift (O) of length $\frac{L}{2}$. Show that the matrix for this cell is

$$\hat{U} = \hat{U}_O \hat{U}_D \hat{U}_O \hat{U}_F = \begin{pmatrix} 1 - \frac{L}{2f} - \frac{L^2}{(2f)^2} & L + \frac{L^2}{4f} \\ -\frac{L}{2f^2} & 1 + \frac{L}{2f} \end{pmatrix}$$

15) Evaluate α, β, γ for the previous problem discuss the motion stability by considering the case $f = \frac{L}{2}$.

16) Establish where are located the maximum and minimum values of the amplitude function for a simple FODO cell, show that they are given by

$$\beta_{max} = 2f \cdot \left[\frac{1 + \sin(\mu/2)}{1 - \sin(\mu/2)}\right]^{\frac{1}{2}}$$

$$\beta_{min} = 2f \cdot \left[\frac{1 - \sin(\mu/2)}{1 + \sin(\mu/2)}\right]$$

provide a graphic representation of the Twiss parameters through the FODO cell using $f = 3\ m$ and $L = 5\ cm$. Show the evolution of the phase-space ellipse and illustrate graphically the concept of phase advance for the individual particle trajectory. (hint see fig.8).

17) Use the approximation

$$\hat{U} = e^{-sV'(\xi)\frac{\partial}{\partial p_\xi}} e^{sp_\xi \frac{\partial}{\partial \xi}}$$

to study the (ξ, p_ξ) coordinate evolution, discuss the error of the approximation compare with (10) and prove the symplecticity of the transformation.

18) prove that (14) is a symplectic transformation. Consider the approximation

$$\hat{U} = e^{-\frac{s}{2}V'(\xi)\frac{\partial}{\partial p_\xi}} e^{sp_\xi \frac{\partial}{\partial \xi}} e^{-\frac{s}{2}V'(\xi)\frac{\partial}{\partial p_\xi}}$$

study the relevant transformation.

19) Specialize $V(\xi)$ to the case of a quadratic potential and provide a matrix representation for the approximations in problems 17-18.

20) Discuss the thin lens approximation for a 45^0 rotated quadrupole.

21) Use (16) to study the motion coupling induced by a sextupole magnet.

Assuming that the initial distribution of an e-beam traversing the magnet is provided by

$$f(x, p_x, y, p_y) = f(x, p_x) \, f(y, p_y)$$

with $f(\xi, p_\xi)$ given by (1.3.5) evaluate the x and y emittances.

(hint: remind that $\varepsilon_\xi = \begin{vmatrix} \sigma_\xi^2 & \sigma_{\xi, p_\xi} \\ \sigma_{\xi, p_\xi} & \sigma_{p_\xi}^2 \end{vmatrix}$).

Discuss the emittance conservation after showing that (16) yields symplectic transformation.

22) Use the Lie operator method to study the evolution of a dynamical system ruled by the Hamiltonian

$$H = \frac{p^2}{2m} - mgx$$

by using the Lie operator method and without applying any approximation.

(hint: note that : $H : x = -p/m$, $(: H :)^2 \, x = -g$, $(: H :)^3 \, x = 0$).

23) Apply the Lie operator method to the study of the Harmonic oscillator problem.

(hint: if $H = \frac{p^2}{2m} + \frac{1}{2}m\omega^2 x^2$, note that

$$(-t : H :)^{2n} \, x = (-1)^{2n} \, (\omega t)^{2n} \, x$$
$$(-t : H :)^{2n+1} \, x = (-1)^n \, (\omega t)^{2n+1} \, \frac{p}{m\omega}).$$

24) By using the notion of symplectic matrix prove that the Jacobian matrix of a symplectic transformation satisfies the condition

$$\hat{J} \, \hat{S} \, \hat{J}^T = \hat{S}$$

25) Prove the fallacy of the following argument: the linearity of $: H :$ can be used to state that

$$\frac{d}{dt}(\mathbf{z}_1 + \mathbf{z}_2) = \dot{\mathbf{z}}_1 + \dot{\mathbf{z}}_2 = -:H:\mathbf{z}_1 - :H:\mathbf{z}_2 = -:H:(\mathbf{z}_1 + \mathbf{z}_2)$$

Accordingly $\mathbf{z}_1 + \mathbf{z}_2$ satisfies the same Hamilton equations of $\mathbf{z}_1$ and $\mathbf{z}_2$, we have proved therefore that the Hamiltonian is always linear.

26) Consider the phase-space distribution (1.3.5), show that the spatial and momentum distributions are provided by

$$q(\xi) = \frac{1}{\sqrt{2\pi}\sigma_\xi} e^{-\frac{x^2}{2\sigma_\xi^2}}$$

$$g(\xi) = \frac{1}{\sqrt{2\pi}\sigma_{p_\xi}} e^{-\frac{p_\xi^2}{2\sigma_{p_\xi}^2}}$$

respectively.

27) Show that (1.3.5) can also be cast in the form

$$f(\xi, p_\xi) = \frac{\beta_\xi}{2\pi\sigma_\xi^2} e^{-\left[\xi^2 + \left(\beta_\xi \xi' + \alpha_\xi \xi\right)^2\right]/2\sigma_\xi^2}$$

and that the area in the phase-space containing a fraction f of the beam is

$$r_\xi = -\frac{2\pi\sigma_\xi^2}{\beta_\xi} \ln(1-f)$$

28) As already noted chromaticity is the variation of the focusing strength of the magnetic lenses with the particle momentum.

It is an example of coupling between momentum and betatron motion that leads to emittance growth.

show that if we denote by

$$k(\delta) = k(0) \cdot (1 - \delta)$$

the quadrupole strength dependence on the momentum deviation

$$1 - \delta = \frac{1}{1+\eta} \quad , \quad \eta = \frac{\Delta p}{p_0}$$

we obtain the following emittance variation

$$\frac{\Delta\varepsilon}{\varepsilon} = \frac{1}{2}\xi^2 \varepsilon^2 \quad , \quad \xi = k(0)\,\beta$$

for an e-beam traversing the magnet with an initial distribution provided by (1.3.5).

29) Consider a beam passing through a thin foil where the particles are scattered at random. When a particle is scattered its position x is not changed, but its horizontal slope is changed from $x' + \delta x'$ and averages to zero. Show that the emittance growth is simply provided by

$$\Delta \varepsilon^2 = 16 \left\langle x^2 \cdot \delta x'^2 \right\rangle .$$

30) In the following table we have reported the Cartesian solutions of magnetic vector potentials (see Appendix A)

	n	Regular	Skew
Q	2	$x^2 - y^2$	$2xy$
S	3	$x^3 - 3xy^2$	$3x^2y - y^3$
O	4	$x^4 - 6x^2y^2 + y^4$	$4x^3y - 4xy^3$
D	5	$x^5 - 10x^3y^2 + 5xy^4$	$5x^4y - 10x^2y^3 + y^5$

Write the relevant Hamiltonian for a charged particle crossing the field. Study the system evolution by using the thin-lens approximation and the approximations discussed in Appendix B.

Appendix A
BENDING MAGNETS AND MAGNETIC LENSES

In this Appendix the reader can be found some details relevant to the analytical treatment of the main magnetic elements of a particle beam transport line or of the Storage Ring lattice.

Bending magnets

Bending magnets generate a dipolar field normal to plane of the reference orbit.
The radius of curvature of a charge q and mass m_q traversing with speed v a constant dipolar field of intensity B can be deduced from the Lorentz equation (1.2.4a) and reads

$$\frac{1}{\rho} = \frac{qB}{m_q \gamma v}$$

where γ is the relativistic factor.
In order to obtain both deflection and focalization of a particle-beam, traversing the bending magnet, one can induce a field gradient, shaping the magnetic field by canting the pole surfaces(see fig 4). The field gradient index is given by the following expression

$$n = -\frac{\rho}{B} \cdot g$$

where

$$g = \frac{\partial B}{\partial \rho}$$

Magnetic lenses

This section is devoted to study field distribution generated by an ideal magnetic lens.
The theoretical field has only components in the transverse plane with respect the direction of motion of the particles, this is true only assuming an infinite length of the lens in this direction . Because the interior of a magnetic lens is current-free

one can define a scalar magnetic potential $V(x, y)$ which can be calculated from the two-dimensional Laplace equation

$$\frac{\partial^2 V(x,y)}{\partial x^2} + \frac{\partial^2 V(x,y)}{\partial y^2} = 0 \qquad (A.0.1)$$

The axial symmetry of a magnetic lens allows the use of polar coordinates in order to find an analytical solution for the scalar potential $V(r, \theta)$. In polar coordinates the two-dimensional Laplace equation reads

$$r\frac{\partial}{\partial r}\left(r\frac{\partial}{\partial r}V(r,\theta)\right) + \frac{\partial^2 V(r,\theta)}{\partial \theta^2} = 0 \qquad (A.0.2)$$

Assuming

$$V(r,\theta) = R(r) \cdot \Theta(\theta) \qquad (A.0.3)$$

eq.(A.0.2) reads

$$\frac{r}{R}\frac{\partial}{\partial r}\left(r\frac{\partial R}{\partial r}\right) = -\frac{1}{\Theta}\frac{\partial^2 \Theta}{\partial \theta^2} \qquad (A.0.4)$$

The general solution of this equation reads

$$V(r,\theta) = \sum_{n=1}^{\infty} [A_n \cos(n\theta) + B_n \sin(n\theta)] \cdot \left(C_n \; r^n + D_n \; r^{-n}\right) + (E + F\theta)(G + H \ln r) \qquad (A.0.5)$$

Using the following boundary conditions for a quadrupolar magnet geometry

$$\begin{array}{ll} 1) & V(r,\theta) = V(r,\theta + 2\pi) \\ 2) & V(r,\theta) \quad finite \; at \; the \; origin \\ 3) & -V(r,-\theta) = V(r,\theta) = -V(r,\theta + \frac{\pi}{2}) \end{array} \qquad (A.0.6)$$

one finally obtain for the scalar potential

$$V(r,\theta) = \sum_{n=2,6,10,\ldots} B_n \sin(n\theta) \cdot r^n \qquad (A.0.7)$$

An ideal quadrupole with constant gradient has $B_2 \neq 0$, $B_n = 0$ for $n = 6, 10, \ldots$

returning to Cartesian coordinates

$$V(x,y) = 2B_2 \; xy + 6B_6 \; xy \cdot \left(x^4 - \frac{10}{3}x^2 y^2 + y^4\right) + \ldots \qquad (A.0.8)$$

The expression of the scalar potential for a 45^0 rotating quadrupole is

$$V(x,y) = B_2 \left(x^2 - y^2\right) + B_6 \left[-x^6 + 15\, x^2 y^2 \left(x^2 - y^2\right) + y^6\right] + \ldots \quad (A.0.9)$$

One can follows a similar procedure for the sextupole magnet geometry, using as third of the boundary conditions (A.0.6) the sextupole symmetry properties

$$-V(r,-\theta) = V(r,\theta) = -V(r,\theta + \tfrac{\pi}{3})$$

Assuming $B_3 \neq 0$, $B_n = 0$ for $n = 9, 15, \ldots$, from (A.0.5) one obtains

$$V(r,\theta) = B_3 \sin(3\theta)\, r^3 \quad (A.0.10)$$

that, in Cartesian coordinates, reads

$$V(x,y) = 3B_3\, y \left(x^2 - \frac{y^2}{3}\right) \quad (A.0.11)$$

Appendix B
TREATMENT OF CHARGED BEAM TRANSPORT USING LIE ALGEBRA

The treatment of charged beam transport developed in the main body of the chapter is general enough to be extended to non linear elements like sextupoles, whose role has been shown to be crucial to correct the quadrupole chromaticities. Unfortunately the non linear case does not allow the use of the matrix method which is both efficient and physically transparent. Other methods, possibly of approximate nature, should be used. The main point is therefore that of choosing the most appropriate approximation. To clarify this point we have to go back to the concept of thin lens approximation, according to which the effect of a quadrupole satisfying the condition $\sqrt{ks} \ll 1$ can be treated by using the matrix

$$\hat{M}_L = \begin{pmatrix} 1 & 0 \\ F & 1 \end{pmatrix} \quad , \quad F = \frac{1}{ks} \tag{B.0.1}$$

which cannot be derived from the evolution matrix by using a Taylor approximation, which yields

$$\hat{M}_T = \begin{pmatrix} 1 & s \\ F & 1 \end{pmatrix} \tag{B.0.2}$$

Albeit (B.0.2) is more accurate it has the unpleasant feature that its determinant is not 1.

From the physical point of view, the fact that $\det \hat{M} \neq 1$ implies the non conservation of the phase-space area and this may bud to unphysical particle losses. This effect becomes significant in circular accelerators, since cumulates turn after turn.

The previous point is easily understood, the condition $\det\left(\hat{U}(s)\right) \neq 1$ implies indeed that $\det\left(\hat{\Sigma}(s)\right) \neq \det\left(\hat{\Sigma}(0)\right)$ (see eqs.1.3.10).

The distribution f is therefore not correctly normalized after the propagation.

According to the previous discussion we will consider appropriate any approximation which preserves the phase-space area conservation. In a technical language we say that we will privilege "symplectic" approximations. The concept of symplecticity is fairly delicate and will be clarified in the following.

We have noted that most of the problems we have treated in the previous sections can be treated by exploiting the Hamiltonian formalism.

The Hamilton equations of motion can be solved by using the classical evolution operator or Lie operator. The Hamilton equations of motion for a system with Hamiltonian H writes in the form

$$\frac{d}{ds}q = -\{H, q\} \tag{B.0.3}$$
$$\frac{d}{ds}p = -\{H, p\}$$

The braces denotes Poisson brackets and we remind that

$$\{F(q,p), G(q,p)\} = \frac{\partial F}{\partial q}\frac{\partial G}{\partial p} - \frac{\partial F}{\partial p}\frac{\partial G}{\partial q} \tag{B.0.4}$$

Accordingly we can formally write the solution of eqs.(B.0.3) as

$$\mathbf{z}(s) = e^{-:H:\,s}\mathbf{z} \tag{B.0.5}$$

if H is not explicitly s-dependent and we have denoted with $:H:$ the operator

$$:H: = \frac{\partial H}{\partial q}\frac{\partial}{\partial p} - \frac{\partial H}{\partial p}\frac{\partial}{\partial q} \tag{B.0.6}$$

Although (B.0.5) is formally very simple, the detailed evaluation may provide a formidable or even impossible problem. As already remarked the possibility of obtaining exact solutions is limited to particular cases only, we are in the majority of cases forced to perform approximations.

We can cast the evolution operator in eq.(B.0.5) as

$$\hat{U} = e^{s\left(p_\xi \frac{\partial}{\partial \xi} - V'(\xi)\frac{\partial}{\partial p_\xi}\right)} \tag{B.0.7}$$

It is evident that the problems in finding an exact solution arise because the kinetic and potential parts of the operator $:H:$ do not commute and, except for linear and quadratic potentials, we cannot obtain straightforward disentanglements.

In spite of the non-commutativity, we approximate the evolution operator as

$$\hat{U} \simeq e^{s p_\xi \frac{\partial}{\partial \xi}}\, e^{-sV'(\xi)\frac{\partial}{\partial p_\xi}} \tag{B.0.8}$$

and by using the operational rule

$$e^{x\frac{\partial}{\partial y}}f(y) = f(y + x) \tag{B.0.9}$$

we obtain

$$\hat{U}\begin{pmatrix}\xi \\ p_\xi\end{pmatrix} = \begin{pmatrix}\xi + s\,p_\xi \\ p_\xi - V'(\xi + s\,p_\xi)\end{pmatrix} \tag{B.0.10}$$

Denoting by $\xi^{(1)}$ and $p_\xi^{(1)}$ the new coordinates, it is easy to verify that the determinant of the Jacobian matrix

$$\hat{J} = \begin{pmatrix} \frac{\partial \xi^{(1)}}{\partial \xi} & \frac{\partial \xi^{(1)}}{\partial p_\xi} \\ \frac{\partial p_\xi^{(1)}}{\partial \xi} & \frac{\partial p_\xi^{(1)}}{\partial p_\xi} \end{pmatrix} \qquad (B.0.11)$$

is 1. This ensures that the transformation induced by the approximation (B.0.8) is symplectic*.

If $V(\xi)$ is e quadratic potential, we can rewrite (B.0.10) as

$$\hat{U}\begin{pmatrix} \xi \\ p_\xi \end{pmatrix} = \begin{pmatrix} 1 & 0 \\ -ks & 1 \end{pmatrix}\begin{pmatrix} 1 & s \\ 0 & 1 \end{pmatrix}\begin{pmatrix} \xi \\ p_\xi \end{pmatrix} \qquad (B.0.12)$$

It is evident that the approximation (B.0.8) reduces the effect of a quadrupole to that of a thin lens and a straight section. It is also evident that thin-lens approximation can be viewed as a kind of impulsive approximation.

We must however underline that (B.0.8) is accurate to the second order only, a later approximation (to third order) is provided by the symplectic split decoupling, namely

$$\hat{U} \cong e^{\frac{s}{2} p_\xi \frac{\partial}{\partial \xi}} e^{-sV'(\xi)\frac{\partial}{\partial p_\xi}} e^{\frac{s}{2} p_\xi \frac{\partial}{\partial \xi}} \qquad (B.0.13)$$

which induces transformations of the type

$$\hat{U}\begin{pmatrix} \xi \\ p_\xi \end{pmatrix} = \begin{pmatrix} \xi + s\, p_\xi - \frac{s^2}{2} V'(\xi + \frac{s}{2}\, p_\xi) \\ p_\xi - sV'(\xi + \frac{s}{2}\, p_\xi) \end{pmatrix} \qquad (B.0.14)$$

whose symplectic nature can be detected by checking the unitary of the determinant of the associated Jacobian matrix. In the case of the quadratic potential eq.(B.0.12) should be modified as

$$\hat{U}\begin{pmatrix} \xi \\ p_\xi \end{pmatrix} \cong \begin{pmatrix} 1 & \frac{s}{2} \\ 0 & 1 \end{pmatrix}\begin{pmatrix} 1 & 0 \\ -ks & 1 \end{pmatrix}\begin{pmatrix} 1 & \frac{s}{2} \\ 0 & 1 \end{pmatrix}\begin{pmatrix} \xi \\ p_\xi \end{pmatrix} \qquad (B.0.15)$$

which states that the action of a quadrupole can be reduced to an OFO.

Further elements, clarifying the role of symplectic approximation can be found in problems 17-25

Let us now discuss the thin lens approximation associated to sextupole. According to the previous discussion we apply the impulsive approximation and write (see eq.1.5.10)

---

*Strictly speaking this condition is necessary and sufficient for one dimensional motions, a more general condition is provided in problem 24.

$$\hat{U} \simeq e^{-\frac{s}{2}m_s(x^2-3y^2)\frac{\partial}{\partial p_x}} e^{s\, m_s xy \frac{\partial}{\partial p_y}} \begin{pmatrix} x \\ p_x \\ y \\ p_y \end{pmatrix} = \begin{pmatrix} x \\ p_x - \frac{s}{2}m_s(x^2-3y^2) \\ y \\ p_y + s\, m_s xy \end{pmatrix} \quad \text{(B.0.16)}$$

The above relation yields a clear idea of the transverse motion coupling and offers the possibility of evaluating (see problem 21) the effect on the beam emittance.

Suggested bibliography

For further details on charged beam transport systems the reader is addressed to:

1. E.D. Courant and H.S. Snyder, *Theory of the Alternating Gradient Synchrotron*, Ann. Phys., $\underline{3}$, 1 (1958).
2. K. Steffen, *Fundamentals of Accelerator Optics*, Proc. of CERN-Accelerator School (CAS) on *Synchrotron Radiation and free Electron Lasers*, S. Turner ed. CERN 90-03 (1990).
3. D.A. Edwards and M.J. Syphers, *An Introduction to the Physics of Particle Accelerators*, AIP Conference Proceedings, 184, p. 2 *Physics of Particle Accelerators* Vol. 1 New York. (1989).
4. R. Talman, *Transverse Motion of Single Particles in Accelerators*, ibidem p. 190.
5. F. Willeke and G. Ripken, *Methods of Beam Optics*, ibidem p. 758.

For the Lie algebraic treatment of beam transport optics the reader is addressed to:

6. G. Dattoli and A. Torre, *A General View to Lie Algebraic Methods in Applied Mathematics, Optics and Transport Systems for Charged Beam Accelerators*, in *Dynamical Symmetries and Chaotic Behaviour in Physical Systems*, World Publ. Co., Singapore (1991).
7. E. Forest, *Canonical Integrators as Tracking codes*, AIP Conference Proceedings, 1106, p. 2 *Physics of Particle Accelerators* Vol. 1 New York. (1989).
8. A.J. Dragt and J.M. Finn, *Lie Series and Invariants for Analytic Symplectic Maps*, J. Math. Phys., $\underline{17}$, 2215 (1976).
9. F. Neri, *Lie Algebras and Canonical Integration*, unpublished, University of Maryland (1987).

Chapter 2
GENERALITIES ON SYNCHROTRON RADIATION

Summary

The properties of synchrotron radiation and its effect on electron beam dynamics are described. We discuss the three kinds of possible sources: bending magnets, wiggler and undulator.

In view of its importance for Free Electron Laser devices, we will devote particular cure to the properties of undulator radiation and discuss its dependence on the e-beam characteristics.

2.1 Introduction

In this chapter we will provide a heuristic description of the properties of synchrotron radiation (S.R.) and address the reader to the bibliography at the end of the chapter for a deeper and more rigorous treatment.

We have already remarked that an accelerated charged particle emits radiation. Electrons accelerated in a few hundred MeV accelerator emit a wide spectrum stretching from visible to frequencies of the order of $10^{15} Hz$, which, according to the well known relations

$$\varepsilon [eV] = 0.41356 \cdot 10^{-14} \cdot \nu [Hz] \qquad (2.1.1)$$
$$\varepsilon [eV] = 1.2398 \cdot 10^4 \cdot \frac{1}{\lambda \left[\mathring{A}\right]}$$

corresponds to photon energies ranging in the region of few eV and wavelength of the order of tens of $\mathring{A}ngstrom$.

The emitted radiation is concentrated in a cone whose apex decreases as the electron energy increases and is so intense that it is visible to the naked eye.

The points we will try to clarify in these introductory remarks are
1) The angular properties
2) The spectral width
3) The intensity dependence on the electron maximum energy.

Consider an electron moving on an arbitrary trajectory and a stationary observer as in fig. 1.

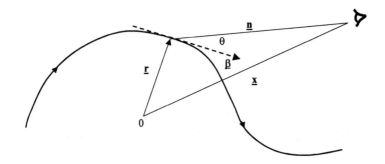

Figure 1 Electron trajectory and stationary observer.

If $\mathbf{r}(t')$ is the position of the electron at the time t' and $\mathbf{x}$ the position of the observer, the distance-particle observer will be provided by $\mathbf{R}(t') = \mathbf{x} - \mathbf{r}(t')$. An electromagnetic signal emitted by an electron at the time t' will be received by the observer at the time

$$t = t' + \frac{R(t')}{c} \qquad (2.1.2)$$

The above relation shows the existence of two times, namely
a) the time of the emitter
b) the time of the observer.
The importance of eq.(2.1.2) is readily understood; by keeping indeed the derivative of both sides with respect to t' we get

$$\frac{dt}{dt'} = 1 - \mathbf{n} \cdot \boldsymbol{\beta} \qquad (2.1.3)$$

where $\mathbf{n}$ is the versor in the $\mathbf{R}$ direction.

Eq. (2.1.3) yields the rate of change of the observer time with respect to that of the emitter. Any small time $\Delta t'$ will be related to the corresponding interval of the observer by

$$\Delta t \cong \left(\frac{dt}{dt'}\right) \Delta t' \qquad (2.1.4)$$

Since we are considering relativistic particles, their motion will appear, to a stationary observer, to occur on a time scale much shorter than that of the real motion. Denoting by ϑ the angle between $\mathbf{R}$ and $\boldsymbol{\beta}$ we get

$$\frac{dt}{dt'} = 1 - \beta \cos \vartheta \qquad (2.1.5)$$

By expanding for small ϑ, by keeping the relativistic limit of β ($\sim 1 - \frac{1}{2\gamma^2}$) we obtain

$$\frac{dt}{dt'} = \frac{1}{2\gamma^2}\left(1 + \gamma^2\vartheta^2\right) \qquad (2.1.6)$$

The difference between emitter and observer time is large, and for $\vartheta \sim \gamma^{-1}$ we get

$$\Delta t \cong \frac{\Delta t'}{\gamma^2} \qquad (2.1.7)$$

Eq. (2.1.7) is a consequence of the relativistic nature of the electron motion, which follows the emitted photon with approximately the same velocity. This is the reason why the motion appears to the observer to occur on a shorter time scale. The strength of the electric field at the observer is proportional to the apparent acceleration of motion seen by the observer. The apparent acceleration is larger for larger time squeezing, as a consequence a large amount of radiation will be observed for small angles, i.e. when the time-squeezing effect is large.

A first important consequence of the above heuristic considerations is that the emitted S.R. is essentially concentrated in a cone with maximum opening angle

$$\vartheta \leq \frac{1}{\gamma} \qquad (2.1.8)$$

A further consequence we may draw is relevant to the spectrum broadness. In fig. 2 we show the trajectory of an electron moving in a bending magnet, more precisely it describes the limited portion of the trajectory contributing to the radiation seen by an observer in a given direction. The radiation, detected by the observer, corresponds to an arc of length $2\rho/\gamma$. The length of the radiation pulse is the difference between the time the electron takes to travel around the arc from A to B and the time for the photon to travel directly from A to B.

According to the figure we obtain

$$\Delta t \cong \frac{2\rho}{\gamma\beta c} - \frac{2\rho \cdot \sin\left(1/\gamma\right)}{c} \simeq \frac{\rho}{\gamma^3 c} \qquad (2.1.9)$$

The typical frequency associated with the time interval (2.1.9) will be quite large, namely*

$$\omega_{typ} \simeq \frac{\gamma^3 c}{\rho} \left(\sim \frac{1}{\Delta t}\right) \qquad (2.1.10)$$

this last relation explains the wide band of S.R. spectrum.

---

*an other important quantity is usually associated to ω_{typ} and is called critical frequency and is provided by $\omega_c = \frac{3}{2}\omega_{typ}$.

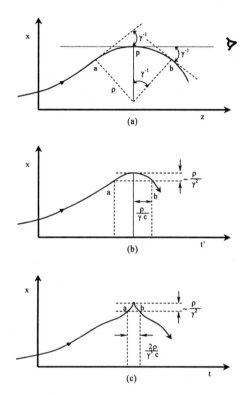

Figure 2 Electron trajectory in a bending magnet.

Introduction

The final point we will discuss, is the dependence of the radiated intensity on the particle energy.

Let us consider a particle subject to an acceleration a' in a frame where its motion can be considered non relativistic. According to the Larmor formula the power radiated by the particle will be provided by

$$P = \frac{2}{3} \cdot \frac{e^2}{c^3} a'^2 \qquad (2.1.11)$$

By considering transverse acceleration only with respect to the direction of motion, a' will be linked to the acceleration in the laboratory frame by

$$a' = \gamma^2 a \qquad (2.1.12a)$$

see problem II.1, being the radiated power a Lorentz invariant, we end up with

$$P = \frac{2}{3} \cdot \frac{e^2}{c^3} a^2 \gamma^4 \qquad (2.1.12b)$$

which can be rewritten as

$$P = \frac{2}{3} \cdot \frac{e^2 c}{\rho^2} \gamma^4 \qquad (2.1.12c)$$

after replacing a with the centripetal acceleration ($a = c^2/\rho$), and

$$P = \frac{2}{3} \cdot \frac{e^4}{m^4 c^5} \cdot B^2 E^2 \qquad (2.1.12d)$$

where the following identity between bending radius ρ, magnetic field B and particle energy E, holds

$$\rho = \frac{c}{e} \cdot \frac{E}{B} \qquad (2.1.13a)$$

which, in practical units, reads

$$\rho\,[m] = \frac{E\,[GeV]}{0.3\,B\,[T]} \qquad (2.1.13b)$$

the total energy lost per turn in a circular accelerator, will be therefore given by

$$\mathcal{U}_0 = \frac{2}{3} e^2 \beta^3 \gamma^4 \oint \frac{ds}{\rho^2} \qquad (2.1.14a)$$

which for an isomagnetic lattice (uniform bending radius in bending magnets), yields

$$\mathcal{U}_0 = \frac{4}{3} \pi \frac{e^2}{\rho} \gamma^4 \qquad (2.1.14b)$$

or in practical units

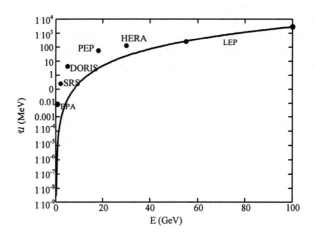

Figure 3 Energy loss per turn vs. electron energy

$$\mathcal{U}_0\,[eV] \cong 8.85 \cdot 10^4 \frac{E^4\,[GeV]}{\rho\,[m]} \cong 2.65 \cdot 10^4 E^3\,[GeV]\ B\,[T] \qquad (2.1.14c)$$

In fig.3 (see also tab. II.1) we have reported the energy loss per turn and the critical energy in various electron Storage Rings.

In this section we have provided a few introductory elements relevant to the main aspects of S.R.. The first five exercises at the end of the chapter are suggested to the reader to prove his degree of understanding on the subject and to get further physical insight.

Table II.1

	$E\,[GeV]$	$\rho\,[m]$	$\mathcal{U}_0\,[MeV]$	$\varepsilon_c\,[KeV]$
EPA	0.6	1.43	$8 \cdot 10^{-3}$	0.34
SRS	2	5.6	0.25	3.2
DORIS	5	12.5	4.5	22.5
PEP	18	166	56.1	78.1
HERA	30	550	140	109
LEP	55	3100	261	119
	100	3100	2855	715

2.2 Lienard-Wiechert potentials

In the previous section we have given a heuristic description of the properties of synchrotron radiation and we have shown that its peculiar features are due to the

Lienard-Wiechert potentials

existence of two distinct times, namely that of the observer and of the emitter linked by eq. (2.1.2).

this section is dedicated to a more rigorous derivation which employs the so called Lienard- Wiechert field, which have been deeply discussed in the bibliography quoted at the end of the chapter.

The scalar and vector potentials associated with a non relativistic-accelerated particle are provided by the standard relation

$$\Phi = \frac{e}{R} \quad , \quad \mathbf{A} = \frac{e\mathbf{v}}{R} \tag{2.2.1}$$

where R is the distance of the particle from the observer. To evaluate the fields associated to a relativistic particle, we must keep in mind that

a) the potential experienced by the observer is created by the particle at a previous time;

b) charges moving towards the observer contributes for a larger time and with larger potential values.

According to the above points it is understandable that the relativistic counterparts of (2.2.1) reads

$$\Phi(t) = e \left[\frac{1}{R(1 - \mathbf{n} \cdot \boldsymbol{\beta})} \right]_{ret} \tag{2.2.2}$$

$$\mathbf{A}(t) = e \left[\frac{\mathbf{v}}{R(1 - \mathbf{n} \cdot \boldsymbol{\beta})} \right]_{ret}$$

where the subscript ret means that the quantities in the square brackets are evaluated at the time t' (see eq. 2.1.2).

The electric and magnetic fields associated with (2.2.2) are evaluated by means of the standard expressions and writes

$$\mathbf{E}(t) = e \left[\frac{\mathbf{n} - \boldsymbol{\beta}}{R^2 \gamma^2 (1 - \mathbf{n} \cdot \boldsymbol{\beta})^3} \right]_{ret} + \frac{e}{c} \left[\frac{\mathbf{n} \times (\mathbf{n} - \boldsymbol{\beta}) \times \dot{\boldsymbol{\beta}}}{R(1 - \mathbf{n} \cdot \boldsymbol{\beta})^3} \right]_{ret} \tag{2.2.3}$$

$$\mathbf{B}(t) = \frac{\mathbf{n} \times \mathbf{E}(t)}{c}$$

(see problem II.6).

From the above relation we can conclude that

a) Electric and magnetic fields are perpendicular;

b) for $\dot{\beta} = \beta = 0$ we recover the Coulomb law;

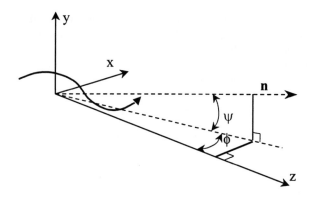

Figure 4 Coordinate system and angles definition in the istantaneous radiation emission.

c) the first term in the electric field, independent of the acceleration, falls off as $\frac{1}{R^2}$ and dominates at short distances;

d) the acceleration dependent term dominates in the "far field limit" i.e. at larger distances. For this reason this term is usually considered.

The Pointing vector ($\mathbf{S} = \mathbf{E} \times \mathbf{B}$) describes the energy flow associated with the fields (2.2.3), accordingly we get for the energy radiated per unit solid angle

$$\frac{dP}{d\Omega} = R^2 (\mathbf{S} \cdot \mathbf{n}) = \frac{1}{c} |RE|^2 \qquad (2.2.4a)$$

and in unit time of the emitter

$$\frac{dP(t')}{d\Omega} = \frac{dP}{d\Omega} \frac{dt}{dt'} = \frac{dP}{d\Omega} (1 - \mathbf{n} \cdot \boldsymbol{\beta}) \qquad (2.2.4b)$$

By keeping the far field contribution only, we eventually end up with

$$\frac{dP(t')}{d\Omega} = \frac{e^2}{4\pi c} \frac{\left|\mathbf{n} \times (\mathbf{n} - \boldsymbol{\beta}) \times \dot{\boldsymbol{\beta}}\right|^2}{(1 - \mathbf{n} \cdot \boldsymbol{\beta})^5} \qquad (2.2.5)$$

where $\mathbf{n}$ is the observation versor ($\mathbf{n} = (\sin\vartheta \cos\varphi, \sin\vartheta \sin\varphi, \cos\vartheta)$).

In the case of circular motion ($\dot{\boldsymbol{\beta}} = \beta^2 \frac{c}{\rho}(-1,0,0)$) we obtain (see problem II.7 and fig. 4 for the angles definition)

$$\frac{dP(t')}{d\Omega} = \frac{2e^2}{4\pi c} \frac{\gamma^6}{\rho^2} \left[\frac{1 + 2\gamma^2 \vartheta^2 (1 - 2\cos^2\varphi) + \gamma^4 \vartheta^4}{(1 + \gamma^2 \vartheta^2)^5} \right] \qquad (2.2.6)$$

Spectral properties of the synchrotron radiation

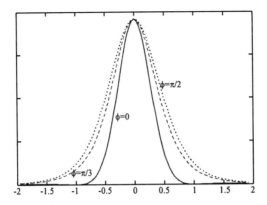

Figure 5   Istantaneous power vs. $\gamma\vartheta$ for different values of φ.

According to fig.5 we may conclude that the results of the previous section where qualitatively correct and in fact the radiation is peaked in the forward direction within a cone of aperture $1/\gamma$.

2.3   Spectral properties of the synchrotron radiation

In the previous section we have evaluated the Lienard-Wiechert fields, associated with a relativistic particle, in the time domain. It is therefore evident that the relevant frequency dependence will be specified by the Fourier-transform

$$\tilde{\mathbf{E}}(\omega) = \frac{1}{\sqrt{2\pi}} \int_{-\infty}^{+\infty} \mathbf{E}(t)\, e^{i\omega t} dt \tag{2.3.1a}$$

furthermore

$$\mathbf{E}(t) = \frac{1}{\sqrt{2\pi}} \int_{-\infty}^{+\infty} \tilde{\mathbf{E}}(\omega)\, e^{-i\omega t} d\omega \tag{2.3.1b}$$

The total energy detected by the observer during one particle passage is provided by

$$\frac{d\mathcal{I}}{d\Omega} = \int \frac{dP}{d\Omega} dt = \frac{1}{c}\int_{-\infty}^{+\infty} (RE)^2\, dt \tag{2.3.2}$$

by inserting (2.3.1b) in (2.3.2) we end up with

$$\frac{d\mathcal{I}}{d\Omega} = \int \frac{dP}{d\Omega} dt = \frac{1}{c}\int_{-\infty}^{+\infty} 2\left|R\tilde{E}(\omega)\right|^2 d\omega \tag{2.3.3}$$

and, according to (2.3.1a),

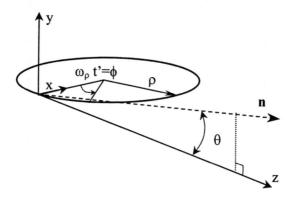

Figure 6 Geometry used to describe the motion in bending magnet

$$\frac{d^2 I}{d\Omega d\omega} = \frac{1}{2\pi c} \left| \int_{-\infty}^{+\infty} (RE)\, e^{i\omega t} dt \right|^2 \tag{2.3.4}$$

Using again the acceleration dependent part of the electric field, we obtain

$$\frac{d^2 I}{d\Omega d\omega} = \frac{e^2}{4\pi^2 c} \left| \int_{-\infty}^{+\infty} \left[\frac{\mathbf{n} \times (\mathbf{n} - \boldsymbol{\beta}) \times \dot{\boldsymbol{\beta}}}{(1 - \mathbf{n} \cdot \boldsymbol{\beta})^3} \right]_{ret} e^{i\omega t} dt \right|^2 \tag{2.3.5a}$$

or as a function of the observer time

$$\frac{d^2 I}{d\Omega d\omega} = \frac{e^2}{4\pi^2 c} \left| \int_{-\infty}^{+\infty} \left[\frac{\mathbf{n} \times (\mathbf{n} - \boldsymbol{\beta}) \times \dot{\boldsymbol{\beta}}}{(1 - \mathbf{n} \cdot \boldsymbol{\beta})^2} \right]_{ret} e^{i\omega \left(t' - \mathbf{n} \cdot \frac{\mathbf{r}(t')}{c} \right)} dt' \right|^2 \tag{2.3.5b}$$

finally if **n** is not a time-dependent function (2.3.5b) can be cast in the form (see problem II.9)

$$\frac{d^2 I}{d\Omega d\omega} = \frac{e^2 \omega^2}{4\pi^2 c} \left| \int_{-\infty}^{+\infty} \mathbf{n} \times (\mathbf{n} \times \boldsymbol{\beta})\, e^{i\omega \left(t' - \mathbf{n} \cdot \frac{\mathbf{r}(t')}{c} \right)} dt' \right|^2 \tag{2.3.5c}$$

we have now all the elements to evaluate the spectral characteristics of the radiation emitted by charged particles moving on an arbitrary path.

The simplest case is the emission from an electron moving on a bending magnet, according to the result of problem II.10 and referring to fig. 6

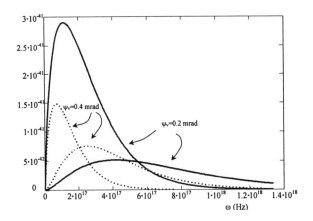

Figure 7 Horizontal and vertical polarization components of the peak intensity vs ω for two different values of ψ_v

we get[†]

$$\frac{d^2\mathcal{I}}{d\Omega d\omega} = \frac{3\,e^2}{4\pi^2 c}\gamma^2\left(\frac{\omega}{\omega_c}\right)^2\left(1+\gamma^2\psi_v^2\right)\left[K_{2/3}^2\left(\xi\right)+\frac{\gamma^2\psi_v^2}{1+\gamma^2\psi_v^2}K_{1/3}^2\left(\xi\right)\right] \quad (2.3.6)$$

$$\xi = \frac{\omega}{2\omega_c}\left(1+\gamma^2\psi_v^2\right)^{3/2}\quad,\quad \omega_c=\frac{3}{2}\frac{\gamma^3 c}{\rho}$$

where $K_{2/3,1/3}^2$ are specified by the integrals

$$K_{2/3}(\eta) = \sqrt{3}\int_0^\infty \tau\sin\left[\frac{3}{2}\eta\left(\tau+\frac{1}{3}\tau^3\right)\right]d\tau \quad (2.3.7)$$

$$K_{1/3}(\eta) = \sqrt{3}\int_0^\infty \cos\left[\frac{3}{2}\eta\left(\tau+\frac{1}{3}\tau^3\right)\right]d\tau$$

The two contributions in eq.(2.3.6) correspond to radiation polarized in the horizontal and vertical plane. When ψ_v increases the vertical polarization increases as indicated in fig.7 (for further comments see also problem II.11)

It is clear that, by setting $\psi_v = 0$ in eq.(2.3.6), we get the dependence on $\frac{\omega}{\omega_c}$ of the on axis peak intensity, namely

$$\left[\frac{d^2\mathcal{I}}{d\Omega d\omega}\right]_{\psi_v=0} = \frac{3\,e^2}{4\pi^2 c}\gamma^2\left(\frac{\omega}{\omega_c}\right)^2 K_{2/3}^2\left(\frac{\omega}{\omega_c}\right) \quad (2.3.8)$$

[†]For the definition of the observation angles see fig. 6 ψ_v represents the observation angle in the vertical plane.

The energy emitted per unit frequency can be obtained by integrating (2.3.6) over the angles thus getting

$$\frac{dI}{d\omega} = \sqrt{3}\,\frac{e^2}{c}\,\frac{\omega}{\omega_c}\int_{\frac{\omega}{\omega_c}}^{\infty} K_{5/3}\left(\frac{\omega'}{\omega_c}\right) d\left(\frac{\omega'}{\omega_c}\right) \qquad (2.3.9)$$

The above equation can be further manipulated to get

$$\frac{dI}{d\omega} = \mathcal{U}_0 \cdot S(x) \quad , \quad x = \frac{\omega}{\omega_c} \qquad (2.3.10)$$

where

$$S(x) = \frac{9\sqrt{3}}{8\pi} \cdot x \int_x^{\infty} K_{5/3}(x')\, dx' \qquad (2.3.11)$$

and must be normalized to unity, to get

$$\int_0^{\infty}\left(\frac{dI}{dx}\right) dx = \mathcal{U}_0 \qquad (2.3.12a)$$

It is worth noting that

$$\int_0^1 \left(\frac{dI}{dx}\right) dx = \frac{\mathcal{U}_0}{2} \qquad (2.3.12b)$$

i.e. half of the power is emitted below the critical frequency and half above.

Fig.8 is very helpful for a better understanding of the properties of synchrotron radiation and it is peaked at $\omega/\omega_c = 0.29$. For further comments see problems (II.12-16).

The formulae we have given in this section refers to the action of a single particle, the extension to the case of a beam of electrons with average current I_b writes

$$\frac{d^2 I}{d\Omega d\omega} = \frac{d^2 \mathcal{I}}{d\Omega d\omega} \cdot \frac{I_b}{e} \qquad (2.3.13)$$

By expressing I_b in $Amps$ and the photon energy in eV, we obtain 2.3.13 expressed in $Watts/mrad^2/eV$

$$\frac{d^2 I}{d\Omega d\varepsilon} = 2.124 \cdot 10^{-3} E^2\,[GeV] \cdot I_b\,[A] \cdot \frac{\varepsilon}{\varepsilon_c} \cdot \left(1 + \gamma^2 \psi_v^2\right) \left[K_{2/3}^2(\xi) + \frac{\gamma^2 \psi_v^2}{1 + \gamma^2 \psi_v^2} K_{1/3}^2(\xi)\right] \qquad (2.3.14)$$

where ε is the photon energy and ε_c the corresponding critical energy.

Synchrotron radiation intensity are some times expressed as number of photons per second obtained by dividing the power in a given frequency interval by the appropriate photon energy $\hbar\omega$. On the other site the power divided by $\hbar$ gives the number of photons per second per unit relative bandwidth

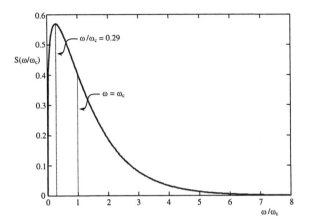

Figure 8 The function $S(x)$ vs $x = \omega/\omega_c$.

$$\frac{d^2\mathcal{I}}{d\Omega\,(d\omega/\omega)} = \frac{d^2I}{d\Omega d\omega} \cdot \frac{1}{\hbar} \qquad (2.3.15a)$$

thus getting in practical units ($photons/s/0.1\%\,bandwidth/mrad^2$) the so called "photon-Flux" (see fig.9)

$$\frac{d^2\mathcal{I}}{d\Omega\,(d\omega/\omega)} = 1.32 \cdot 10^{13} E^2\,[GeV] \cdot I_b\,[A] \cdot \left(\frac{\varepsilon}{\varepsilon_c}\right)^2 \cdot \qquad (2.3.15b)$$

$$\cdot \left(1 + \gamma^2\psi_v^2\right) \cdot \left[K_{2/3}^2(\xi) + \frac{\gamma^2\psi_v^2}{1+\gamma^2\psi_v^2} K_{1/3}^2(\xi)\right]$$

2.4 Damping and beam equilibrium conditions.

In circular accelerators the energy lost by synchrotron radiation is restored by the radio-frequency accelerating system, namely by one or more R.F. cavities located around the orbit. The cavity field has a sinusoidal dependence and the energy gained by an electron arriving at time t, writes as

$$\mathcal{U}_{R.F.} = e\,\hat{V}\,\sin\left(\omega_{R.F.}\,t + \pi - \varphi\right) \qquad (2.4.1)$$

where $\hat{V}$ is the peak voltage and $\omega_{R.F.}$ the relevant frequency which is chosen to be an integer multiple of the machine revolution frequency.

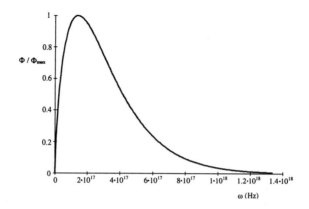

Figure 9 photon-Flux normalized to the maximum vs ω for $\psi_v = 0$.

We will define synchronous those particles gaining exactly $\mathcal{U}_0$ once crossing the cavity. A non-synchronous particle, i.e. one with energy $E = E_0 + \varepsilon$, will experience an energy variation provided by

$$\frac{d\varepsilon}{dt} = \frac{e\,V(\tau) - \mathcal{U}_0(\varepsilon)}{T_0} \qquad (2.4.2)$$

where τ is the arrival interval time at the R.F. . Furthermore since more energetic particles execute a larger closed orbit whose period is linked to the synchronous period by

$$\frac{T(\varepsilon) - T_0}{T_0} = \frac{\alpha_c\,\varepsilon}{E_0} \qquad (2.4.3)$$

where α_c is the so called momentum compaction, we find that after one revolution the time interval τ will be reduced by

$$\frac{d\tau}{dt} = -\frac{\alpha_c\,\varepsilon}{E_0} \qquad (2.4.4)$$

Even though eqs. (2.4.4) and (2.4.2) are sufficient to describe the dynamics of the longitudinal motion, we can get a deeper insight by linearizing eq. (2.4.2) by setting

$$eV(\tau) = \mathcal{U}_0 + \left[e\,\hat{V}\,\omega_{R.F.}\,\cos\varphi_s\right]\tau \qquad (2.4.5)$$

$$\mathcal{U}_0(\varepsilon) = \mathcal{U}_0 + \left(\frac{\partial \mathcal{U}_0(\varepsilon)}{\partial \varepsilon}\right)_{\varepsilon=0}\varepsilon$$

thus finally getting

$$\frac{d}{dt}\begin{pmatrix}\tau\\ \varepsilon\end{pmatrix} = \begin{pmatrix} 0 & -\frac{\alpha_c}{E_0}\\ \hat{\Lambda} & -2\alpha_\varepsilon\end{pmatrix}\begin{pmatrix}\tau\\ \varepsilon\end{pmatrix} \quad (2.4.6)$$

$$\hat{\Lambda} = \frac{e\,V_{R.F.}\,\cos\varphi_s}{T_0}, \qquad \alpha_\varepsilon = \frac{1}{2T_0}\left(\frac{\partial\mathcal{U}_0(\varepsilon)}{\partial\varepsilon}\right)_{\varepsilon=0}$$

which yields

$$\frac{d^2\tau}{dt^2} + 2\alpha_\varepsilon\frac{d\tau}{dt} + \Omega_s^2\,\tau = 0 \quad (2.4.7)$$

$$\frac{d^2\varepsilon}{dt^2} + 2\alpha_\varepsilon\frac{d\varepsilon}{dt} + \Omega_s^2\,\tau = 0$$

$$\Omega_s^2 = \frac{e\,\alpha_c\,\hat{V}\,\omega_{R.F.}\,\cos\varphi_s}{E_0\,T_0}$$

It is therefore convenient to introduce the longitudinal damping time τ_ε defined as

$$\tau_\varepsilon = T_0\cdot\left(\frac{\partial\mathcal{U}_0(\varepsilon)}{\partial\varepsilon}\right)^{-1} = \frac{T_0\cdot\varepsilon}{\mathcal{U}_0}$$

We could now exploit eqs. (2.4.7) to describe the longitudinal phase-space evolution, we must however underline that the above results represents only a part of the problem, we should indeed complete the analysis by including the effects due to the quantum nature of the emission process.

Just to provide an idea of what we should expect we give a simple heuristic argument which can be exploited to get a feeling of the amount of energy spread induced by the quantum fluctuations. By assuming that the emission of quanta is statistically independent, the process can be considered Poissonian. The fluctuation of the emitted number of quanta in a damping time will be therefore proportional to the square root of photons emitted in this time. By denoting with N the number of emitted photons per second and assuming that each photon carries a characteristic energy $u_c = \hbar\omega_c$ we get $\sigma_\varepsilon \sim \sqrt{N\tau_\varepsilon}\cdot u_c$.

Before going further we emphasize again the quantum nature of the emitted radiation. Within this context a quantity of paramount importance is the number $n(u)$ of photons per second radiated (on average) by a single electron per unit energy interval

$$n(u)\Delta u = \frac{d\mathcal{I}}{d\omega}\cdot\frac{c}{2\pi\rho}\cdot\frac{1}{\hbar\omega}\cdot\Delta\omega \quad (2.4.8)$$

it is therefore evident that

$$n(u) = \frac{P}{u_c^2} \cdot \frac{S(u/u_c)}{(u/u_c)} = \frac{9\sqrt{3}P}{8\pi u_c^2} \cdot \int_{u/u_c}^{\infty} K_{5/3}(x)\, dx \qquad (2.4.9)$$

From the above expression we can easily evaluate (see problems II.17, 18) the total number of emitted photons

$$N = \int_0^\infty n(u)\, du = \frac{15\sqrt{3}}{8} \cdot \frac{P}{u_c} \qquad (2.4.10a)$$

the mean square photon energy

$$<u> = \frac{\int_0^\infty u \cdot n(u)\, du}{N} = \frac{8}{15\sqrt{3}} \cdot u_c \qquad (2.4.10b)$$

and the mean-square energy

$$<u^2> = \frac{\int_0^\infty u^2 \cdot n(u)\, du}{N} = \frac{11}{27} \cdot u_c^2 \qquad (2.4.10c)$$

After these remarks, let us go back to eq.(2.4.7) and note that in presence of quantum fluctuation it can be replaced by

$$\frac{d^2\varepsilon}{dt^2} + 2\alpha_\varepsilon \frac{d\varepsilon}{dt} + \Omega_s^2\, \varepsilon = -\frac{d}{dt}\left[\tilde{P}(t) - P\right] \qquad (2.4.11)$$

where (see problem II.19)

$$\tilde{P}(t) = \sum_i u_i \delta(t - t_i) \qquad (2.4.12)$$

Equation (2.4.11) is a stochastic differential equation since u_i and t_i are stochastic variables. The solution of (2.4.11) can be written as (see problem II.20)

$$\varepsilon(t) = -\frac{1}{\Omega_s} \int_{-\infty}^t dt' \frac{d\tilde{P}(t')}{dt'} \cdot \exp\left[-\alpha_\varepsilon(t - t')\right] \cdot \sin\left[\Omega_s(t - t')\right] \qquad (2.4.13)$$

thus getting after integration by part for $\alpha_\varepsilon \ll \Omega_s$ (see problem II.21)

$$\varepsilon(t) = -\sum_{t_i = -\infty}^t n_i \cdot \exp\left[-\alpha_\varepsilon(t - t_i)\right] \cdot \cos\left[\Omega_s(t - t_i)\right] \qquad (2.4.14)$$

Since $\varepsilon(t)$ is usually much larger than the typical photon energies involved in the emission process and since ε_i are randomly distributed, we can conclude that $\varepsilon(t)$ consists of a large sum of statistically independent positive or negative small values. The central limit theorem ensures therefore that the probability that $\varepsilon(t)$ lies within ε and $\varepsilon + \delta\varepsilon$ is

$$w(\varepsilon)d\varepsilon = \frac{1}{\sqrt{2\pi}\sigma_\varepsilon} \exp\left(-\frac{\varepsilon^2}{2\sigma_\varepsilon^2}\right) \qquad (2.4.15)$$

The problem is now that of evaluating the standard deviation σ_ε.

We introduce the quantity

$$A^2 = \varepsilon^2 + \frac{1}{\Omega_s^2} \cdot \left(\frac{d\varepsilon}{dt}\right)^2 \qquad (2.4.16)$$

which in absence of damping and quantum fluctuation is an invariant, being linked to the "Hamiltonian" of the first of equation (2.4.11) (for $\alpha_\varepsilon = 0$ and without the fluctuation contribution). The emission of a photon implies that

$$A^2 + \delta A^2 = (\varepsilon - u)^2 + \frac{1}{\Omega_s^2} \cdot \left(\frac{d\varepsilon}{dt}\right)^2 \qquad (2.4.17)$$

from which it follows that

$$\delta A^2 = -2u\varepsilon + u^2 \qquad (2.4.18)$$

by comparing equations (2.4.18) and (2.4.11) we can associate the linear term in ε to the damping part and u^2 to the diffusive contribution. Accordingly we can write, after an average on the photon distribution,

$$\frac{d}{dt}\left\langle A^2 \right\rangle = -\frac{2}{\tau_\varepsilon}\left\langle A^2 \right\rangle + N\left\langle u^2 \right\rangle \qquad (2.4.19a)$$

which at equilibrium yields

$$\left\langle A^2 \right\rangle = \frac{N\tau_\varepsilon \left\langle u^2 \right\rangle}{2} \qquad (2.4.19b)$$

since

$$\left\langle A^2 \right\rangle = \frac{\left\langle \varepsilon^2 \right\rangle}{2} \qquad (2.4.20)$$

we find

$$\sigma_\varepsilon^2 = \left\langle \varepsilon^2 \right\rangle = \frac{N\tau_\varepsilon \left\langle u^2 \right\rangle}{4} \propto \sqrt{N\tau_\varepsilon}\, u_c \qquad (2.4.21)$$

As expected from the previous heuristic considerations.

We must stress that the averaging procedure in the above identities should be extended to one machine turn, we find therefore that the last term in equation (2.4.19a) should be replaced by

$$\left\langle N\left\langle u^2 \right\rangle \right\rangle = \frac{55}{24\sqrt{3}} \left\langle P \cdot u_c \right\rangle \qquad (2.4.22)$$

where

$$P = \frac{2}{3}e^2c\frac{\gamma^4}{\rho^2} \quad , \quad u_c = \frac{3}{2}\frac{\hbar\, c}{\gamma^3\, \rho} \qquad (2.4.23)$$

By noting that

$$\tau_\varepsilon = \frac{2\, E_0}{J_\varepsilon <P>} \qquad (2.4.24)$$

where the role and the meaning of J_ε will be discussed below, we end up with

$$\left(\frac{\sigma_\varepsilon}{E_0}\right)^2 = \frac{55}{32\sqrt{3}}\frac{\hbar}{m\, c}\frac{\gamma^2 <1/\rho^3>}{J_\varepsilon <1/\rho^2>} \qquad (2.4.25)$$

which for an isomagnetic machine yields (see problem II.22)

$$\left(\frac{\sigma_\varepsilon}{E_0}\right)^2 = C_q \frac{\gamma^2}{J_\varepsilon\, \rho} \quad , \quad C_q = \frac{55}{32\sqrt{3}}\frac{\hbar}{m\, c} = 3.84 \cdot 10^{-13} m \qquad (2.4.26)$$

We can now evaluate σ_τ relevant to longitudinal spatial distribution by noting that, in absence of damping, equations (2.4.14) can be calculated from the Hamiltonian

$$H = \frac{1}{2}(\Omega_s \tau^2 + \frac{\alpha_c}{E_0}\varepsilon^2) \qquad (2.4.27)$$

according to the equipartition theorem we find (see problem II.23)

$$\Omega_s \sigma_\tau = \frac{\alpha_c}{E_0}\sigma_\varepsilon \qquad (2.4.28)$$

thus getting, for an isomagnetic machine,

$$\sigma_\tau^2 = \left(\frac{\alpha_c}{\Omega_s}\right)^2 C_q \frac{\gamma^2}{\sigma_\varepsilon\, \rho} \qquad (2.4.29)$$

The transverse beam dimension can be obtained by using an analogous treatment.

We proceed by noting that the emission at a point with zero dispersion gives rise to a change in the off energy function and, therefore, introduces a change in the betatron motion amplitude, for the radial components we get indeed

$$\delta x_\beta = -D(s)\frac{u}{E_0} \quad , \quad \delta x'_\beta = -D'(s)\frac{u}{E_0} \qquad (2.4.30)$$

to the usual Courant-Snyder invariant we should therefore add the quantity

$$\delta A^2 = \left(\gamma\, D^2 + 2\alpha\, D\, D' + \beta\, D'^2\right)\frac{u^2}{E_0^2} = \mathcal{H}\frac{u^2}{E_0^2} \qquad (2.4.31)$$

Damping and beam equilibrium conditions

By following the same procedure as for the longitudinal case, we can conclude that the linear term in u are associated to the damping, so that in conclusion we get the following equilibrium equation (see problem II.24)

$$\frac{<A^2>}{2} = \frac{\tau_x}{4}\frac{<N<u^2>\mathcal{H}>}{E_0^2} \qquad (2.4.32)$$

which defines the equilibrium horizontal emittance through the relation

$$\varepsilon_x = \frac{<A^2>}{2} = C_q \frac{\gamma^2}{J_x}\frac{<\mathcal{H}/\rho^3>}{<1/\rho^2>} \qquad (2.4.33)$$

the horizontal beam dimension and divergence are therefore linked to the Twiss parameters and equilibrium emittance through

$$\sigma_x = \left[\varepsilon_x \beta_x + D^2 \left(\frac{\sigma_\varepsilon}{E_0}\right)^2\right] \qquad (2.4.34)$$

$$\sigma'_x = \left[\varepsilon_x \gamma_x + D'^2 \left(\frac{\sigma_\varepsilon}{E_0}\right)^2\right]$$

As to the vertical dimension let us note that the vertical emittance arises from processes like

a) coupling of the vertical and horizontal betatron motion due to skew-quadrupole field errors

b) vertical dispersion errors associated to vertical bending fields produced by angular positioning errors of dipoles and vertical positioning errors of quadrupoles.

The vertical emittance depends only on errors and it can therefore be estimated statistically. The effect can be described through a coefficient κ, defined in such a way that the sum of the horizontal and vertical emittances is constant, we therefore find

$$\varepsilon_x = \frac{1}{1+\kappa}\varepsilon_{x,0} \qquad \varepsilon_y = \frac{\kappa}{1+\kappa}\varepsilon_{x,0} \qquad (2.4.35)$$

the quantity $\varepsilon_{x,0}$ is evaluated through equation (2.4.33) and is usually called the natural emittance.

Before closing this section let us note that we have denoted by $<1/\rho^2>$ quantities of the type

$$I_2 = \oint \frac{1}{\rho^2}ds \qquad (2.4.36)$$

which provide one of the synchrotron radiation integrals, the other being expressed as

$$I_3 = \oint \frac{1}{|\rho^3|} ds \quad , \quad I_4 = \oint \frac{D}{\rho}\left(\frac{1}{\rho^2} - 2\kappa\right) ds \quad , \quad I_5 = \oint \frac{\mathcal{H}}{|\rho^3|} ds \quad (2.4.37)$$

It is therefore evident that the above integrals play a role of paramount importance for the determination of the beam properties in addition they enter in the definition of the damping times. According to the Robinson theorem the damping times can be expressed through the relation

$$\tau_i = \frac{2\, E_0\, T_0}{J_i\, \mathcal{U}_0} \quad , \quad i = x, y, \varepsilon \quad (2.4.38)$$

and J_i are the partition numbers satisfying the relation

$$\sum_i J_i = 4 \quad (2.4.39)$$

and in particular

$$J_x = 1 - \frac{I_4}{I_1} \quad , \quad J_y = 1 \quad , \quad J_\varepsilon = 1 + \frac{I_4}{I_1} \quad (2.4.40)$$

For further comments see problem II.26.

2.5 The Fokker-Planck treatment of the longitudinal equilibrium problem

In the previous section we have presented an analysis of the longitudinal electron motion based on heuristic considerations, in this section we will employ more rigorous argument based on the use of the Fokker-Planck equation.

The photon emission in a magnet can be viewed as a random process, affecting the electron energy variable. This process will be assumed to be "Markovian", which amounts to saying that the probability that at the time $t + \delta t$ the electron energy assumes a value $\varepsilon + \delta\varepsilon$ depends only on the value ε taken at the time t and on the variation $\delta\varepsilon$ and not on the previous history.

Let us denote the afore-mentioned probability by $W(\varepsilon, \delta\varepsilon)$ and the distribution function of ε at the time t as $f(\varepsilon, t)$, the distribution at the time $t + \delta t$ will be provided by the Master equation

$$f(\varepsilon, t + \delta t) = \int f(\varepsilon_0, t) \cdot W(\varepsilon_0, \varepsilon - \varepsilon_0)\, d\varepsilon_0 \quad (2.5.1)$$

By defining $\delta\varepsilon = \varepsilon - \varepsilon_0$ and by expanding with respect to this quantity assumed to be small we get

$$f(\varepsilon, t+\delta t) = \int f(\varepsilon - \delta\varepsilon, t) \cdot W(\varepsilon - \delta\varepsilon, \delta\varepsilon) \, d(\delta\varepsilon) \quad (2.5.2)$$

$$\simeq \int d(\delta\varepsilon) \left\{ \begin{array}{l} f(\varepsilon,t) \cdot W(\varepsilon,\delta\varepsilon) - \delta\varepsilon \frac{\partial}{\partial \varepsilon}[f(\varepsilon,t) \cdot W(\varepsilon,\delta\varepsilon)] + \\ +\frac{1}{2}(\delta\varepsilon)^2 \frac{\partial^2}{\partial \varepsilon^2}[f(\varepsilon,t) \cdot W(\varepsilon,\delta\varepsilon) + \ldots] \end{array} \right\}$$

By expanding the l.h.s. too

$$f(\varepsilon, t+\delta t) = f(\varepsilon, t) + \delta t \frac{\partial}{\partial t} f(\varepsilon, t) \quad (2.5.3)$$

and by noting that

$$\int W(\varepsilon, \delta\varepsilon) \, d(\delta\varepsilon) = 1$$
$$\int W(\varepsilon, \delta\varepsilon) \, \delta\varepsilon \, d(\delta\varepsilon) = <\delta\varepsilon> \quad (2.5.4)$$
$$\int W(\varepsilon, \delta\varepsilon) \, \delta\varepsilon^2 \, d(\delta\varepsilon) = <(\delta\varepsilon)^2>$$

we end up with the following one-dimensional Fokker-Planck diffusion equation

$$\frac{\partial}{\partial t} f(\varepsilon, t) = -\frac{\partial}{\partial \varepsilon}\left(\frac{<\delta\varepsilon>}{\delta t} f(\varepsilon, t)\right) + \frac{1}{2}\frac{\partial^2}{\partial \varepsilon^2}\left(\frac{<\delta\varepsilon^2>}{\delta t} f(\varepsilon, t)\right) \quad (2.5.5)$$

The extension to the case of two variables, accounting for electron energy and longitudinal position, is straightforward, we find indeed

$$\frac{\partial}{\partial t} f(\varepsilon, t) = -\frac{\partial}{\partial z}\left(\frac{<\delta z>}{\delta t} f(\varepsilon, t)\right) - \frac{\partial}{\partial \varepsilon}\left(\frac{<\delta\varepsilon>}{\delta t} f(\varepsilon, t)\right) + \quad (2.5.6)$$

$$+\frac{1}{2}\left\{ \begin{array}{l} \frac{\partial^2}{\partial z^2}\left(\frac{<\delta z^2>}{\delta t} f(\varepsilon, t)\right) + \frac{\partial^2}{\partial \varepsilon^2}\left(\frac{<\delta\varepsilon^2>}{\delta t} f(\varepsilon, t)\right) + \\ +2\frac{\partial^2}{\partial z \partial \varepsilon}\left(\frac{<\delta\varepsilon \, \delta z>}{\delta t} f(\varepsilon, t)\right) \end{array} \right\}$$

By using the equations derived in the previous section we get up to the first order in δt (i.e. in the machine turn T)

$$<\delta z> = -(\alpha_c \, \varepsilon_c) \cdot T \quad , \quad <\delta\varepsilon> = \left(\frac{\omega_s^2}{c\alpha_c} \cdot z - \frac{2}{T_s}\varepsilon\right) \cdot T \quad (2.5.7)$$

$$<\delta z^2> = 0 \quad , \quad <\delta z \, \delta\varepsilon> = 0 \quad , \quad <\delta\varepsilon^2> = \frac{<u^2>}{E_0}$$

thus finding in conclusion

$$\frac{\partial}{\partial t}f(\varepsilon,t) = (c\alpha_c)\left\{\varepsilon\frac{\partial f(\varepsilon,t)}{\partial z} - \omega_s^2\frac{z}{c\alpha_c}\frac{\partial f(\varepsilon,t)}{\partial \varepsilon} + \frac{\partial}{\partial \varepsilon}\left(\frac{2\varepsilon}{T_\varepsilon}f(\varepsilon,t) + \mathcal{D}\frac{\partial f(\varepsilon,t)}{\partial \varepsilon}\right)\right\}$$
(2.5.8)

where $\mathcal{D}$ is

$$\mathcal{D} = \frac{55}{16\sqrt{3}}\frac{\hbar}{mc}\frac{\gamma^2}{J_\varepsilon}\frac{<1/\rho^3>}{<1/\rho^2>}$$
(2.5.9)

By introducing the r.m.s. values, coinciding with those given in the previous section,

$$\sigma_\varepsilon = \sqrt{\frac{\mathcal{D}\,\tau}{2}} \quad , \quad \sigma_z^0 = \frac{c\alpha_c}{\omega_s}\sigma_\varepsilon$$
(2.5.10a)

and the rescaled variables

$$v = \frac{\varepsilon}{\sigma_\varepsilon} \quad , \quad x = \frac{z}{\sigma_z} \quad , \quad \xi = \omega_s T_\varepsilon \quad , \quad \bar{t} = \omega_s t$$
(2.5.10b)

we can recast equation (2.5.8) in the dimensionless form

$$\frac{\partial}{\partial t}f(x,v) = \left\{\left(v\frac{\partial}{\partial x} - x\frac{\partial}{\partial v}\right) + \frac{2}{\xi}\left(v\frac{\partial}{\partial v} + \frac{\partial^2}{\partial v^2}\right) + \frac{2}{\xi}\right\}\cdot f(x,v)$$
(2.5.11)

According to problem II.27 the distribution

$$f_0(x,v) = \frac{1}{2\pi\sigma_x\sigma_v}\exp\left[-\frac{1}{2}\left(\frac{v^2}{\sigma_v^2} + \frac{x^2}{\sigma_x^2}\right)\right]$$
(2.5.12)

is the stationary solution of (2.5.11) only for $\sigma_v = \sigma_x = 1$ and it is interesting to note that if $\sigma_v, \sigma_x \neq 1$, the system tends to the equilibrium condition and according to wether σ_η, $\eta = v, x$ is larger or smaller than unity the damping process is dominating until the equilibrium is reached and viceversa.

For further comments see problem II.27.

2.6 Synchrotron radiation from magnetic undulators

The previous sections have been devoted to the properties of synchrotron radiation emitted by charged particles moving on circular path. we have explored essentially the properties of bending magnet radiation and we have shown that one of its peculiar features is the large bandwidth of the spectral distribution which extends up to the critical frequency. As is well known the undulator magnet was initially proposed to reduce the bandwidth and to enhance the power emitted in a specific range of frequencies.

Synchrotron radiation from magnetic undulators

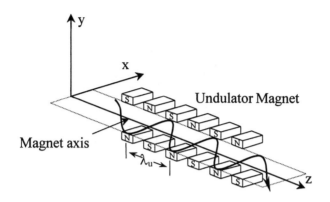

Figure 10 Sketch of the electron trajectory in an undulator magnet.

Albeit we will more carefully discuss the properties and the constructive aspects of an undulator in the following, we remind that an undulator is a spatially periodic structure (see fig.10) in which the electrons execute transverse oscillations and emits radiation centered at the wavelength

$$\lambda_1 = \frac{\lambda_u}{2\gamma^2}\left(1 + \frac{k^2}{2}\right) \qquad (2.6.1)$$

where λ_u is the undulator period and k is the undulator strength linked to the undulator period and on axis magnetic field intensity by

$$k = \frac{e\,B\,\lambda_u}{2\pi\,m_o\,c^2} \qquad (2.6.2)$$

or in practical units

$$k = \frac{B_0\,[KG]\,\lambda_u\,[cm]}{10.7} \qquad (2.6.3)$$

as we will see in the following, in the case of pure permanent magnet of $SmCo_5$ type arranged in the Halbach configuration (see fig. 11), the on axis magnetic field is given by

$$B_0\,[KG] = 15.5 \cdot \exp\left(-\pi\,g/\lambda_u\right) \qquad (2.6.4)$$

where g is the undulator gap. The above formulae yields a first idea of what we should expect.

The relative bandwidth of the radiation emitted in undulators is one of the most important features and we will try to understand it at a qualitative level.

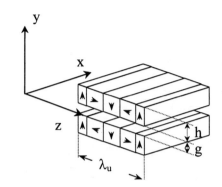

Figure 11   Arrays for PPM undulator in the Halbach configuration.

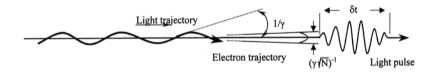

Figure 12   Characteristics of the emission light pulse in undulator magnet

The spectral broadness of the synchrotron radiation is related to the difference between electron and photons flight times. an undulator induces an orbit such that the angle between the trajectory slope and the observation direction is always less then $1/\gamma$ (see fig.12). The observer will therefore receive an e.m. pulse with time duration proportional to the difference of photon and electron flight times, namely

$$\Delta t = (1-\beta)\frac{N\lambda_u}{c} \sim \frac{N\lambda_u}{2\gamma^2 c} \qquad (2.6.5)$$

where N is the number of undulator periods. The electron orbit is almost periodic, with period λ_u, the signal received by the observer will be periodic too, with wavelength (see equation 2.1.6)

$$\lambda = \frac{\lambda_u}{2\gamma^2}\left(1 + \gamma^2\langle\theta^2\rangle\right) \qquad (2.6.6)$$

where $\langle\rangle$ means average over one undulator period. Since $\langle\theta^2\rangle \cong \frac{1}{\sqrt{2}}\frac{k}{\gamma}$ we obtain equation 2.6.1 and since the bandwidth is inversely proportional to the time duration of the e.m. pulse received by the observer, we get

$$\frac{\Delta\omega}{\omega} \cong \frac{1}{2N} \qquad (2.6.7)$$

The parameter k controls the deviation angle from the undulator axis, but also the harmonic content of the emitted radiation, as we will see in the following. (For further comments on the properties of undulator radiation from a heuristic point of view see problems 28-29).

Let us now discuss the problem from a more quantitative point of view. We consider therefore a transverse (or linearly polarized) undulator, in which the magnetic field on the midplane in the vertical direction and to a good approximation varies along z with the following dependence

$$\mathbf{B} \equiv B_0\left(0, \sin(k_u\, z), 0\right) \quad , \quad k_u = \frac{2\pi}{\lambda_u} \qquad (2.6.8)$$

The equations of motion of an electron moving in the above field are calculated from the Lorentz equation and read (see fig. 11 for the axis orientation)

$$\ddot{z} = -\frac{e\,\dot{x}}{m_0\,\gamma\,c}\, B_0 \sin\left(k_u\, z\right) \qquad (2.6.9)$$

$$\ddot{x} = \frac{e\,\dot{z}}{m_0\,\gamma\,c}\, B_0 \sin\left(k_u\, z\right)$$

The second equation can be immediately integrated, with the initial condition $\dot{x}_{t=0} = -\frac{ck}{\gamma}$, as

$$\dot{x} = -\frac{c\,k}{\gamma} \cos\left(k_u\, z\right) \qquad (2.6.10)$$

the total velocity is unaffected by the interaction so that

$$\dot{z}^2 = \beta^2 c^2 \left[1 - \frac{k^2}{\beta^2 \gamma^2} \cos^2\left(k_u\, z\right)\right] \qquad (2.6.11)$$

By expanding up to the order $\frac{k^2}{\gamma^2}$ and by integrating the velocities to get the trajectory, we end up with

$$\mathbf{r}(t) \equiv \left(-\frac{c}{\omega_u}\frac{k}{\gamma}\sin\left(\omega_u\, t\right)\, ,\, 0\, ,\, \beta^* c\, t - \frac{k^2}{8\gamma^2 \omega_u}\sin\left(2\,\omega_u\, t\right)\right) \qquad (2.6.12)$$

$$\beta^* = \beta \left[1 - \frac{1}{4}\frac{k^2}{\gamma^2}\right]$$

By exploiting equation 2.3.5c and using for the observation vector $\mathbf{n}$ its expansion at $1/\gamma^2$[‡]

[‡]Note that $\psi_v = \psi \sin\varphi$, $\vartheta = \psi \cos\varphi$, $\psi^2 = \psi_v^2 + \vartheta^2$.

$$\mathbf{n} \cong \left(\psi\cos\varphi, \psi\sin\varphi, 1 - \frac{1}{2}\psi^2\right)$$

we finally get (see problem II.30)

$$\frac{d^2\mathcal{I}}{d\omega d\Omega} = \frac{4\,e^2 N^2 \gamma^4}{c} \sum_{m=1}^{\infty} \frac{m^2}{\left(1 + \frac{k^2}{2} + \gamma^2\psi^2\right)} \cdot \left[\left|S_m^{(1)}\right|^2 + \left|S_m^{(2)}\right|^2\right] \cdot [\sin c\,(\nu_m/2)]^2 \quad (2.6.13a)$$

where

$$\nu_m = 2\pi N\left(m - \frac{\omega}{\omega_1}\right) \quad (2.6.13b)$$

$$\omega_1 = \frac{2\gamma^2 \omega_u}{\left(1 + \frac{k^2}{2} + \gamma^2\psi^2\right)}$$

and

$$S_m^{(1)} = \psi\cos\varphi\, J_m(\zeta_\omega, \xi_\omega) + \frac{k}{2\gamma}\{J_{m-1}(\zeta_\omega, \xi_\omega) + J_{m+1}(\zeta_\omega, \xi_\omega)\} \quad (2.6.13c)$$

$$S_m^{(2)} = \psi\sin\varphi\, J_m(\zeta_\omega, \xi_\omega)$$

with

$$\zeta_\omega = -m\zeta_\psi \quad,\quad \xi_\omega = -m\xi_\psi \quad,\quad \zeta_\psi = \frac{2k\gamma\psi\cos\varphi}{1 + \frac{k^2}{2} + \gamma^2\psi^2} \quad,\quad \xi_\psi = \frac{k^2}{4\left(1 + \frac{k^2}{2} + \gamma^2\psi^2\right)}$$
(2.6.13d)

The functions $J_n(x, y)$ are two variable Bessel function provided by the generating functions

$$\sum_{n=-\infty}^{+\infty} t^n J_n(x, y) = \exp\left[\frac{x}{2}\left(t - \frac{1}{t}\right) + \frac{y}{2}\left(t^2 - \frac{1}{t^2}\right)\right] \quad (2.6.14)$$

and the relevant properties are summarized in problems II.31-32.

It is also worth noting that on axis, equation (2.6.13a) reduces to ($\xi = \xi_{\psi=0}$)

$$\frac{d^2\mathcal{I}}{d\omega d\Omega} = \frac{e^2 N^2 \gamma^2}{c} \sum_{m=1}^{\infty} F_m(\xi) \cdot \sin c^2(\nu_m/2) \quad (2.6.15a)$$

where

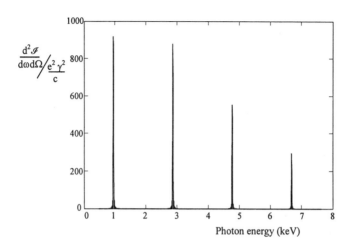

Figure 13   The undulator on axis spectral intensity, normalized to $\frac{e^2\gamma^2}{c}$, vs photon energy. $N = 50$, $k = \sqrt{2}$, e-beam energy $100\,MeV$.

$$F_m(\xi) = \left(\frac{4\,m\,\xi}{k}\right)^2 \left[J_{\frac{m-1}{2}}(m\,\xi) - J_{\frac{m+1}{2}}(m\,\xi)\right]^2 \quad (2.6.16)$$

The on axis frequency dependence of the undulator spectral intensity is shown in fig. 13.

For further comments see also problems II.33-34.

The quantity related to (2.6.15a) we will more frequently use in the following is the Flux ($\#\,of\,photons/s/0.1\%\,bandwidth$) defined as

$$\mathcal{F}_m = 1.431 \cdot 10^{14}\, N\, Q_m\, I\,[A] \quad (2.6.17)$$
$$Q_m = \frac{m\,k^2}{1+k^2/2}\left[J_{\frac{m-1}{2}}(m\,\xi) - J_{\frac{m+1}{2}}(m\,\xi)\right]^2$$

and an example is provided in fig.14

The problems (II.28-36) should be carefully considered by the reader, they are intended as a complement to the topics treated in this section and yield a deeper understanding of the various arguments we have just touched on.

2.7   Concluding remarks

The structure of the undulator spectrum is essentially due to the fact that radiation from different periods interferes coherently, thus producing sharp peaks at harmonics

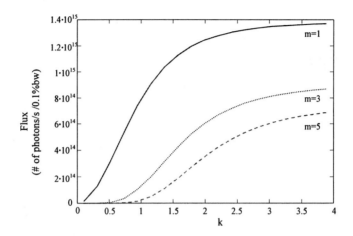

Figure 14 Photon flux as a function of the undulator parameter k for the $1^{\underline{st}}$, $3^{\underline{rd}}$ and $5^{\underline{th}}$ harmonics. $N = 50$, $I = 0.2\,A$.

of the fundamental. When k increases, interference effects become less important and the structure of the spectrum tends to that of wiggler (see fig.15 where we have reported the on axis spectrum for increasing k values). When $k \gtrsim 10$ radiation from different parts of the electron trajectory adds incoherently. The flux distribution is therefore given by $2N$ times the appropriate formula for bending magnet[§].

Regarding the bending magnet radiation, by inspecting fig.7, we realize that the variation intensity, with vertical angle ψ_v of the parallel component of the bending magnet spectrum is well described by a Gaussian if $\lambda/\lambda_c \leq 30$, with r.m.s. width (see problem II.37)

$$\sigma_{\psi_v}\,[mrad] \sim \frac{0.29}{E\,[GeV]}\left(\frac{\lambda}{\lambda_c}\right)^{0.45} \qquad (2.7.1)$$

This simple result offers the possibility of a further practical consideration. We have seen that electron beams have finite transverse dimensions and divergences, we expect therefore that this fact may produce a reduction of the radiation observed on axis. As to the bending magnet (as well as wiggler) we find that (see problem II.36) the reduction due to the vertical e-beam divergence $\sigma_{y'}$ is

$$\frac{d^2 \mathcal{F}}{d\Omega\,(d\omega/\omega)}\bigg|_{\psi=0} = \frac{d\mathcal{F}}{d\vartheta}\frac{1}{\sqrt{2\pi\left(\sigma_{\psi_v}^2 + \sigma_{y'}^2\right)}} \qquad (2.7.2)$$

[§]N is the number of undulator periods, for further comments see chapter VI.

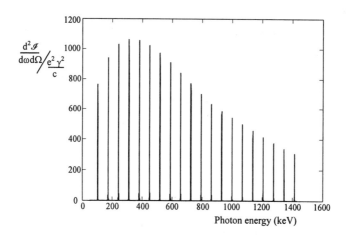

Figure 15 The undulator on axis spectral intensity, normalized to $\frac{e^2\gamma^2}{c}$, vs photon energy up to $39^{\underline{th}}$ harmonic. $N = 50$, $k = 3$, e-beam energy $1\,GeV$.

where $\frac{dF}{d\vartheta}$, expressed in $[photons \cdot s^{-1} \cdot mrad^{-1} \cdot (0.1\%\ b.w.)]$, is

$$\frac{dF}{d\vartheta} = 2.457 \cdot 10^{13}\ E\,[GeV]\ I\,[A]\ S\,(\varepsilon/\varepsilon_c) \qquad (2.7.3)$$

(for further comments see also chapter VI).

In the case of the undulator the angular distribution of the nth is concentrated in a narrow cone, whose half width is given by

$$\sigma_\psi \simeq \sqrt{\frac{\lambda_u}{L_u}} = \frac{1}{\gamma}\sqrt{\frac{1+k^2/2}{2\,N_u}} \qquad (2.7.4)$$

very roughly speaking we can also conclude (see problem II.37) the on axis, flux reduction is provided by

$$\overline{\mathcal{F}}_u = \frac{\mathcal{F}_u}{2\pi\sqrt{\left(\sigma_\psi^2 + \sigma_{y'}^2\right)\left(\sigma_\psi^2 + \sigma_{x'}^2\right)}} \qquad (2.7.5)$$

The extra contribution depending on $\sigma_{y'}$ is due to the fact that, unlike bendings or wigglers, the electron beam divergence is not negligible in the horizontal plane.

In this chapter we have presented the essential features of the spectral properties of the bending, undulator and wiggler radiation, the forthcoming next two chapters are devoted to the relevant coherence properties.

2.8 Problems

1) Find a physical argument to show that the synchrotron radiated power is a relativistic invariant.
2) Show that equation (2.1.12a) holds for transverse acceleration.
3) Discuss the case of power lost by longitudinal acceleration and show that

$$P = \frac{2}{3} \frac{e^2 a^2}{c^3} \gamma^6$$

(hint: $a' = \gamma^3 a$).

4) Recast the result of problem 3 in the form

$$P = \frac{2}{3} \frac{r_0}{mc} \dot{p}^2$$

where p is the particle momentum and r_0 its classical radius.

5) Find an argument to show that transverse acceleration is more efficient than longitudinal to produce synchrotron radiation.

(hint: write $\dot{p}^2 \cong \frac{\dot{E}F}{c\beta}$, where $\dot{E}$ is the rate of energy change and F is the force acting on the particle, then note that $\frac{P}{\dot{E}} \simeq 2/3 \frac{r_0 F}{mc\beta}$).

6) Derive electric and magnetic fields (2.2.3) from (2.2.2) potentials.
7) Derive (2.1.6) from (2.1.5) in the case of circular motion and the relativistic limit.
8) Show that in the non relativistic limit

$$\frac{dP}{d\Omega} = \frac{e^2}{\pi c} \dot{\beta}^2 \sin^2 \Theta$$

where Θ is the angle between the acceleration axis and the radiation direction.

(hint: show first that in the non relativistic limit $\mathbf{E}(t) = \frac{e}{c} \frac{\mathbf{n} \times (\mathbf{n} \times \dot{\boldsymbol{\beta}})}{R}$).

9) Derive (2.3.5c) from (2.3.5b).

(hint: use integration by part).

10a) Derive (2.3.5c) from (2.3.5b) in the case of circular motion

(hint: with reference to fig.6 note that in the radiation integral $\frac{d^2 I}{d\Omega d\omega} = \frac{e^2 \omega^2}{4\pi^2 c} \left| \int_{-\infty}^{+\infty} \frac{\mathbf{n} \times (\mathbf{n} \times \boldsymbol{\beta})}{(1-\mathbf{n}\cdot\boldsymbol{\beta})^2} \exp\left[i\omega\left(t' - \mathbf{n} \cdot \frac{\mathbf{r}(t')}{c}\right)\right] dt \right|$ the $\mathbf{r}$ $\boldsymbol{\beta}$ and $\mathbf{n}$ vectors are provided by

$$\mathbf{r}(t') = \rho[1 - \cos(\omega_\rho t'), \, 0, \, \sin(\omega_\rho t')]$$
$$\boldsymbol{\beta}(t') = \beta[\sin(\omega_\rho t'), \, 0, \, \cos(\omega_\rho t')] \quad , \quad \omega_\rho = \frac{\beta c}{\rho}$$
$$\mathbf{n} \equiv \left(0, \, \psi_v, \, 1 - \psi_v^2/2\right)$$

expand at the third order in t' and use the integral representation of the Airy functions).

10b) With reference to problem (10a) show that an alternative solution may be found by using the Bessel function expansion

$$e^{ix\sin\vartheta} = \sum_{n=-\infty}^{+\infty} e^{in\vartheta} J_n(x)$$

within this framework show that the bending magnet spectrum consists of a quasi-continuum of harmonics of the fundamental $\omega_0 \cong \frac{c}{\rho}$ show that the critical frequency corresponds to the harmonic $n_c \sim 1.124 \cdot 10^{10} E^3 [GeV]$.

11) Discuss the role of ω_c and its physical meaning.

12) Prove equation 2.3.12b. Integrate 2.3.6 over the frequencies and show that

$$\frac{dI}{d\Omega} = \frac{7}{16}\frac{\gamma^5}{\rho}\frac{1}{(1+\gamma^2\psi^2)^{5/2}}\left[1+\frac{5\gamma^2\psi^2}{7(1+\gamma^2\psi^2)}\right] .$$

13) Show that on axis ($\psi_v = 0$) equation 2.3.15b reduces to

$$\frac{d^2 I}{d\Omega d\omega/\omega} = 1.325 \cdot 10^{13} \cdot E^2 [GeV] \cdot I_b[A] \cdot H_2\left(\frac{\varepsilon}{\varepsilon_c}\right)$$

$$H_2\left(\frac{\varepsilon}{\varepsilon_c}\right) = \left(\frac{\varepsilon}{\varepsilon_c}\right)^2 K_{2/3}^2\left(\frac{\omega}{2\omega_c}\right) .$$

14) show that by integrating 2.3.14 on the vertical angle one gets the spectral distribution per unit horizontal angle (units: $W/(mrad\ horizontal)$)

$$\frac{d^2 P}{d\vartheta d\varepsilon} = 6.347 \cdot e^{-3} \cdot E[GeV] \cdot I_b[A] \cdot S\left(\frac{\varepsilon}{\varepsilon_c}\right) .$$

15) Show that, in practical units, the angular distribution of the total power becomes

$$\frac{dP}{d\Omega}\left[W/mrad^2\right] = 5.42\, E^4[GeV]\, B[T]\, I_b[A]\, \frac{1}{(1+\gamma^2\psi_v^2)^{5/2}}\left[1+\frac{5\gamma^2\psi_v^2}{7(1+\gamma^2\psi_v^2)}\right]$$

Note also that on axis

$$\frac{dP}{d\Omega}\left[W/mrad^2\right] = 5.42\, E^4[GeV]\, B[T]\, I_b[A]$$

and that integration on the vertical angle yields

$$\frac{dP}{d\vartheta} = 4.22\, E^4[GeV]\, B[T]\, I_b[A]$$

(units: $W/(mrad\ horizontal)$).

16)¶ Show that to extend the above formulae to protons the results must be scaled by an appropriate factor χ^n where $\chi = \frac{m_e}{m_p}$. Therefore

---
¶The authors consider this exercise important

$$\varepsilon_{c,p} = \chi^3 \varepsilon_c \, , \, \lambda_{c,p} = \chi^3 \lambda_c \, , \, \left(\frac{d^2P}{d\Omega d\omega}\right)_p = \chi^2 \frac{d^2P}{d\Omega d\omega} \, ,$$

$$\left(\frac{d^2F}{d\Omega d\omega/\omega}\right)_p = \chi^2 \frac{d^2F}{d\Omega d\omega/\omega} \, , \, \left(\frac{d^2P}{d\vartheta d\omega}\right)_p = \chi \frac{d^2P}{d\vartheta d\omega} \, ,$$

$$\left(\frac{d^2F}{d\vartheta d\omega/\omega}\right)_p = \chi \frac{d^2F}{d\vartheta d\omega/\omega} \, , \, \left(\frac{dP}{d\Omega}\right)_p = \chi^5 \frac{dP}{d\Omega} \, , \, \left(\frac{dP}{d\vartheta}\right)_p = \chi^4 \frac{dP}{d\vartheta} \, .$$

17) Derive equation 2.4.10a.
18) Derive equation 2.4.10c.
19) Discuss the physical meaning of equation 2.4.12.
20) Derive the solution of equation 2.4.11 in the form 2.4.13.
21 derive equation 2.4.14 from equation 2.4.13 by using integration by parts and the condition $\alpha_\varepsilon \ll \Omega_s$.
22) Discuss the concept of isomagnetic machine and derive equation 2.4.26 from equation 2.4.25.
23) Discuss the validity of equation 2.4.28.
24) Derive equation 2.4.32 and discuss its physical meaning.
25) Show that the vertical emittance

$$\varepsilon_y = C_q \frac{\langle \beta_y \rangle}{\rho}$$

where $\langle \beta_y \rangle$ is the average n-value of the β-Twiss coefficient on the machine turn.

(hint: note that if a photon of energy u is emitted at angle ϑ_y with respect to the median plane, the corresponding change in angle of the electron is $\delta y' = \frac{u}{E_0} \vartheta_y$ and hence the change in the vertical oscillation amplitude is $\delta A^2 = \frac{u^2}{E_0^2} \vartheta_y^2 \beta_y(s)$).
26) Use the radiation integrals to show that

$$\tau_x = \frac{3 \, T_0}{r_0 \, \gamma^3} \frac{1}{I_2 - I_4} \, , \, \tau_y = \frac{3 \, T_0}{r_0 \, \gamma^3} \frac{1}{I_2} \, , \, \tau_\varepsilon = \frac{3 \, T_0}{r_0 \, \gamma^3} \frac{1}{2I_2 + I_4} \, ,$$

$$\left(\frac{\sigma_\varepsilon}{E_0}\right)^2 = C_q \gamma^2 \frac{I_3}{2I_2 + I_4} = \frac{C_q \gamma^2}{J_3} \frac{I_3}{I_2} \, , \, \varepsilon_{x,0} = C_q \gamma^2 \frac{I_5}{I_2 - I_4} = \frac{C_q \gamma^2}{J_x} \frac{I_5}{I_2} \, .$$

27) Consider the Fokker-Planck type equation

$$\frac{\partial}{\partial t} f(\varepsilon, t) = \left(\frac{\partial}{\partial \varepsilon} \varepsilon + \frac{\partial^2}{\partial \varepsilon^2}\right) f(\varepsilon, t)$$

$$f(\varepsilon, 0) = \frac{1}{\sqrt{2\pi \sigma_\varepsilon}} \exp\left(-\frac{\varepsilon^2}{2\sigma_\varepsilon^2}\right)$$

show that its solution reads

$$f(\varepsilon, t) = \frac{1}{\sqrt{2\pi}\sigma_\varepsilon(t)} \exp\left(-\frac{\varepsilon^2}{2\sigma_v^2(t)}\right)$$

$$\sigma_\varepsilon(t) = \sqrt{(\sigma_\varepsilon^2 - 1)e^{-2t} + 1} \ .$$

Show that 2.5.12 is the stationary solution for 2.5.11 if $\sigma_x = \sigma_v = 1$. Obtain a time dependent solution of 2.5.11 by using 2.5.12 as initial distribution and show that it is a direct generalization of the previous solution.

28) Use the Weizsäcker-Williams approximation to treat the undulator as an e.m. field of wavelength $2\lambda_u$ and use the usual Compton scattering formulae to show that the electron backscatters this field at a wavelength provided by equation 2.6.1. Within this framework discuss the role and the meaning of the term $(1 + k^2/2)$.

29) Show that the number of photons emitted by one electron is approximately

$$W_{ph} \simeq \frac{e^2}{\hbar c} N k^2 \quad , \quad \left(\frac{e^2}{\hbar c} = \frac{1}{137}\right)$$

(hint: use the analogy with the emission in bending magnets to show that the totally radiated power is $P \propto \frac{e^2}{c} k^2 \gamma^2 \omega_u$).

30) derive equation 2.6.13a from the radiation integral.

31) Show that

$$J_n(x,y) = \sum_{l=-\infty}^{+\infty} J_{n-2l}(x) J_l(y)$$

where $J_n(\cdot)$ denote ordinary Bessel Functions.

Derive the following recurrence properties

$$\frac{\partial}{\partial x} J_n(x,y) = \frac{1}{2}[J_{n-1}(x,y) - J_{n+1}(x,y)]$$

$$\frac{\partial}{\partial y} J_n(x,y) = \frac{1}{2}[J_{n-2}(x,y) - J_{n+2}(x,y)]$$

$$2n J_n(x,y) = x[J_{n-1}(x,y) + J_{n+1}(x,y)] + 2y[J_{n-2}(x,y) + J_{n+2}(x,y)] \ .$$

32) Show also that

$$J_n(x,0) = J_n(x)$$

and

$$J_n(0,y) = \begin{array}{ll} J_{n/2}(y) & \text{for } n \text{ even} \\ 0 & \text{for } n \text{ odd} \end{array}$$

33) Use the properties of two variable Bessel functions to derive equation 2.6.15a.
34) Show that for $\varphi = \pi/2$ odd harmonics are radiated in the σ mode (parallel to the horizontal plane of the orbit), even harmonics in the π mode (polarization vertical to the horizontal plane of the orbit), in particular show that

$$\left[\frac{d^2\mathcal{I}}{d\omega d\Omega}\right]^\sigma_{\varphi=\pi/2} = \frac{e^2 N^2 \gamma^2}{c} \sum_{m=1}^{\infty} \left\{\frac{4m\xi}{k}\left[J_{\frac{m-1}{2}}(m\xi) - J_{\frac{m+1}{2}}(m\xi)\right] \cdot \operatorname{sinc}(\nu_m/2)\right\}^2$$

$$\left[\frac{d^2\mathcal{I}}{d\omega d\Omega}\right]^\pi_{\varphi=\pi/2} = \frac{4e^2 N^2 \gamma^2}{c} \sum_{m=1}^{\infty} \left[\frac{m\gamma\psi}{\left(1+\frac{k^2}{2}+\gamma^2\psi^2\right)} J_{\frac{m}{2}}(m\xi) \cdot \operatorname{sinc}(\nu_m/2)\right]^2.$$

35) Show that to satisfy the Maxwell's equation at second order in the spatial coordinates the undulator field components should read

$$B_x(x,y,z) \cong \frac{B_0}{2} \delta k_u^2 x y \sin(k_u z)$$

$$B_y(x,y,z) \cong B_0 \left\{1 + \frac{k_u^2}{4}\left[\delta x^2 + (2-\delta) y^2\right]\right\} x y \sin(k_u z)$$

$$B_z(x,y,z) \cong B_0 k_u y \cos(k_u z)$$

Derive the equations of motion by including the effect of the spatial coordinates and show that betatron part reads

$$x_1'' = -k_\beta^2 \delta x_1 \qquad (\cdot)' = \frac{d}{dz}(\cdot)$$

$$y_1'' = -k_\beta^2 (2-\delta) y_1 \quad , \quad k_\beta = \frac{\pi k}{\lambda_u \gamma}$$

Discuss the meaning of δ and the role of k_β.

Show that the betatron motion induces a shift in the emission frequency, proportional to the betatron motion Hamiltonian, use this conclusion to evaluate the line inhomogeneous broadening due to the emission from a beam having a transverse distribution provided by equation 1.3.5.

36) Show that for a helical undulator i.e. for an undulator with on-axis field

$$B = B_0 \left[\cos(k_u z) \, , \, \sin(k_u z) \, , \, 0\right]$$

one gets

$$\frac{d^2\mathcal{I}}{d\omega d\Omega} = \frac{2e^2 N^2 \gamma^2}{c} \left(\frac{k}{1+k^2+\gamma^2\psi^2}\right) \cdot$$

$$\sum_{m=1}^{\infty} m \left\{J_{m+1}^2(m\xi_H) + J_{m-1}^2(m\xi_H) - 2\frac{(1+k^2)}{k^2} J_m^2(m\xi_H)\right\} \cdot \operatorname{sinc}^2(\nu_m/2)$$

where $\xi_H = \frac{2k\gamma\psi}{1+k^2+\gamma^2\psi^2}$, and $J_m(\cdot)$ are ordinary Bessel functions.

37) Discuss the limits of validity of 2.7.1, and propose an approximant form for $\lambda/\lambda_c > 30$.

38) Prove the relation 2.7.2.

(hint: recall that the e-beam angular distribution in the vertical plane is a Gaussian with r.m.s. $\sigma_{y'}$, then use the convolution theorem between two Gaussians).

39) Derive equation 2.7.5 and discuss the difference with 2.7.2.

Suggested bibliography

For the seminal papers on synchrotron radiation see:

1. D. Iwanenko and J. Pomeranckuk, Phys. Rev. , 65, 343 (1944).
2. J.P. Blewett, Phys. Rev. , 69, 87 (1946).
3. J. Schwinger, Phys. Rev. , 70, 798 (1946) and ibidem 75, 1912 (1949).
4. F.R. Elder, K.V. Langmuir and H.C. Pollock, Phys. Rev. , 74, 52 (1948).

   For a more systematic treatment:
5. J.D. Jackson, *Classical Electrodynamic*, J.Wiley & sons ed.
6. S. Krinsky, M.L. Perlman and R.E. Watson *Characteristics of Synchrotron Radiation and Its Sources* "Handbook on Synchrotron Radiation", Vol 1 North-Holland, Amsterdam (1983).
7. K.J. Kim, *Characteristic of Synchrotron Radiation*, AIP Conference Proceedings, 184, Vol. 1 AIP, 565, New York (1989).
8. A. Hofmann, Proc. of CERN-Accelerator School (CAS) on *Synchrotron Radiation and free Electron Lasers*, 90-03, 115 (1990).
9. G. Dattoli, A. Renieri and A. Torre *Lectures on Free Electron Laser and Related Topics* Chapt. II World Scientific, Singapore (1993).

   For the treatment of circular accelerator dynamics:
10. H. Bruk, *Accélérateurs Circulaires de Particules,* Presser Universitaire de France, Paris (1966).
11. M. Sands SLAC-121 (1970).
12. K. Steffen, Proc. of CERN-Accelerator School (CAS) on *General accelerator Physics*, 85-19, 409 (1994).
13. A. Wrülich, Proc. of CERN-Accelerator School (CAS) on *General accelerator Physics*, 94-01, 409 (1994).
14. R.P. Wolker, Proc. of CERN-Accelerator School (CAS) on *General accelerator Physics*, 94-01, 461 (1994).

    For the emission in magnetic undulators:
    see the already quoted books: K.J. Kim[7], G. Dattoli, A. Renieri and A. Torre[9] and:
15. D.F. Alfenov, Yu A. Bashmakov and E.G. Bessanov, Sov. J. Tech. Phys., 18, 1336 (1974).
16. B.M. Kincaid, J. Appl. Phys., 48, 2684 (1977).
17. W.B. Colson, Ph. D. Dissertation, Stanford University (1977).
18. R. Barbini, F, Ciocci, G. Dattoli and L. Giannessi, "La Rivista del Nuovo Cimento" 6 (1990).
19. P. Elleaume, "Laser Handbook" Vol VI ed. by W.B. Colson, C. Pellegrini and A. Renieri, North-Holland, Amsterdam (1990).
20. G. Dattoli and A. Torre, *Theory and Application of Generalized Bessel Functions* ARACNE ROME (1996).

Chapter 3
GENERALITIES ON FREE ELECTRON LASERS

We present a simplified point of view on the theory of Free Electron Lasers (FELs). We discuss low and high gain regimes, linac and Storage-Ring based devices. We exploit a treatment which privileges a pragmatic description. The physical aspects of the process are carefully analyzed and the mathematical picture is held at the level of practical formulae useful for the understanding and the design of different Free Electron Laser devices.

Albeit we develop a formalism general enough to cover all the aspects of Compton Free Electron Lasers, we will be mainly concerned with short vawelength devices, i.e. high energy linacs and Storage-Ring based Free Electron Lasers.

3.1 Introduction and low gain equations.

The theory of Free Electron Lasers (FEL) has been discussed in a large number of review articles and books, some of them are reported at the end of this chapter. We consider therefore more convenient to present the subject in a fairly heuristic way, by privileging pragmatic aspects. The discussion will be kept at a level of mathematical language employing simple formulae, useful for the understanding and the design of a FEL device.

In this chapter we will limit ourselves to the case of devices operating in the short wave-length region of the e.m. spectrum. We will therefore discuss the case of FELs based on Storage Rings or high energy Linacs.

The introductory part of the chapter will be devoted to the understanding of the crucial FEL mechanism, i.e. the gain.

In the previous chapter we have discussed the emission by relativistic charges propagating in magnetic undulators. We have seen that the central emission frequency in linked to the undulator parameters by

$$\lambda_R = \frac{\lambda_u}{2\gamma^2}\left(1 + \frac{k^2}{2}\right) \quad (3.1.1)$$

The line-shape of the emitted spectrum has the characteristic form of short-time emission process, namely

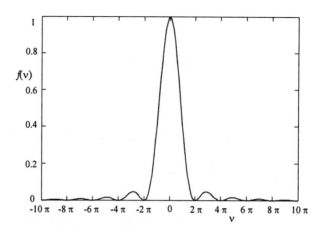

Figure 1 Line shape of the emitted spectrum.

$$f(\nu) = \left[\frac{\sin(\nu/2)}{\nu/2}\right]^2 \qquad (3.1.2)$$

where ν is the frequency detuning parameter linked to the central emission frequency by

$$\nu = 2\pi N \frac{\omega_R - \omega}{\omega_R} \quad , \quad \omega_R = \frac{2\pi c}{\lambda_R} = 2\gamma^2 \omega_u \left(1 + \frac{k^2}{2}\right)^{-1} \quad , \quad \omega_u = \frac{2\pi c}{\lambda_u} \qquad (3.1.3)$$

with N being the number of undulator periods. It is evident from fig.1 that the full-width half maximum of $f(\nu)$ is

$$\Delta\nu_{1/2} \sim 2\pi \qquad (3.1.4)$$

which allows to define a kind of "energy acceptance of the undulator emission linewidth", namely

$$\frac{\Delta\gamma}{\gamma} \sim \frac{1}{2N} \qquad (3.1.5)$$

The above relation is of paramount importance, as it will become more clear in the following. To start to appreciate this aspect of the problem, we will consider the line-shape deformation induced by the emission from a non-monoenergetic e-beam, but having a relative energy distribution provided by

Introduction and low gain equations.

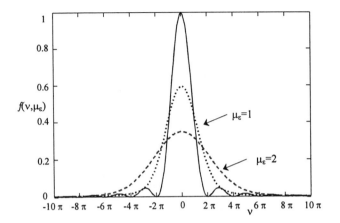

Figure 2 $f(\nu, \mu_\varepsilon)$ vs. ν for different μ_ε values in comparison with the case of monoenergetic e-beam. We have assumed $N = 50$.

$$f(\varepsilon) = \frac{1}{\sqrt{2\pi}\,\sigma_\varepsilon} e^{-\frac{\varepsilon^2}{2\sigma_\varepsilon^2}} \qquad (3.1.6)$$

Proceeding as indicated in problem III.1, we find that to account for the effect of a non negligible energy spread, (3.1.2) should be replaced by

$$f(\nu, \mu_\varepsilon) = 2 \int_0^1 (1-t) \cos(\nu t)\, e^{-\frac{\pi^2 \mu_\varepsilon^2}{2} t^2}\, dt \qquad (3.1.7)$$

where

$$\mu_\varepsilon = 4N\,\sigma_\varepsilon \qquad (3.1.8)$$

is the energy inhomogeneous broadening parameter, which controls the importance of the non-uniformity of the energy distribution on the emission process.

A more quantitative idea is provided by fig.2 where $f(\nu, \mu_\varepsilon)$ has been plotted vs. ν for different μ_ε values. For $\mu_\varepsilon \neq 0$ the curve broadens and the peak, centered at $\nu = 0$, decreases.

The behaviour of $f(0, \mu_\varepsilon)$ vs. μ_ε is provided in fig.3 and it is worth noting that it can be reproduced by the simple relation for μ_ε in the interval $(0, 2)$

$$f(0, \mu_\varepsilon) \cong \frac{1}{1 + 0.772\,\mu_\varepsilon^2 - 0.09\,\mu_\varepsilon^4}, \qquad (3.1.9)$$

the role of the magnetic undulator is more subtle than it may appear. It can, indeed, be viewed as a radiator, and thus as the element inducing into the initial

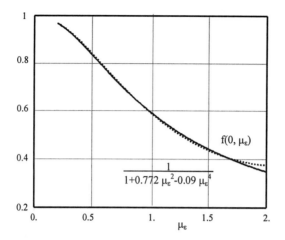

Figure 3 Behaviour of the spectrum amplitude at the central emission frequency as a function of μ_ε.

non-radiating longitudinal electron motion a transverse radiating component, which may also be responsible for the coupling with a TE field, copropagating with the electrons, and eventually for an amplification of the wave.

In more classical terms we can say that a TE field, quasi-resonating with λ_R, couples with the electrons by modulating the beam energy. The energy modulation transforms into a density modulation and coherent emission takes place when the bunching occurs at the wave-length of the TE field. All the FEL dynamics, including saturation, can be understood in terms of bunching, emission, higher order bunching, emission of higher harmonics, etc. We will come to this aspect of the problem in the following, for the moment we will try to derive the FEL gain formulae by following a simplified procedure.

We first note that $f(\nu)$ does not provide any information on the phase of the emitted radiation and on its time dependence, while the electrons are radiating inside the undulator. Denoting by $\tau = \frac{ct}{L_u}$ the interaction time, we introduce the quantity

$$\mathcal{J}(\nu, \tau) = \int_0^\tau e^{-i\nu\tau'} d\tau' \qquad (3.1.10)$$

so that

$$f(\nu) = |\mathcal{J}(\nu, 1)|^2 \qquad (3.1.11)$$

The function $e^{-i\nu\tau}$ defines the "frequency" and time dependence of the FEL "active-medium". The gain mechanism can be viewed as a balance between a photon

Introduction and low gain equations. 85

emission-absorption process. The two processes imply a down or up shift of the electron energy and thus a variation in ν, which may be decreased or increased. The balance implies that the FEL gain should be proportional to the derivative vs. ν of the active medium function. Furthermore the process occurs along all the undulator and must be extended to all the electrons of the e-beam. Denoting by $a(\tau)$ the dimensionless optical field amplitude* we may conclude that

$$\frac{da}{d\tau} = i\pi g_0 a(\tau) \int_0^\tau \tau' e^{-i\nu\tau'} d\tau' \qquad (3.1.12)$$

which is a consequence of the previous analysis, in fact

$$\tau' e^{-i\nu\tau'} = i\frac{\partial}{\partial \nu} e^{-i\nu\tau'} \qquad . \qquad (3.1.13)$$

The integral in eq.3.1.12 is due to the fact that all the contributions from all the elementary processes, in the interval $(0, \tau)$, are taken into account. Finally the coefficient g_0, proportional to the peak current, ensures that all the electrons in the beam are participating to the process. In the following we will precise the form of g_0, usually called small signal FEL gain coefficient, by providing its dependance on the other parameters, here we note that by assuming it is a small quantity (say less than 0.2) we can write the solution of equation 3.1.12 as (see problem III.2)

$$\frac{a(\tau) - a(0)}{a(0)} = g_0 \cdot g_1(\nu, \tau) \qquad (3.1.14)$$

where $g_1(\nu, \tau)$, usually called the first order complex gain function, is given by

$$g_1(\nu, \tau) = \frac{\pi}{\nu^3} \left\{ 2\left(1 - e^{-i\nu\tau}\right) - i\nu\tau \left(1 + e^{-i\nu\tau}\right) \right\} \qquad (3.1.15)$$

By defining the gain as

$$G_1(\nu, g_0, \tau) = \frac{|a(\tau)|^2 - |a(0)|^2}{|a(0)|^2} \qquad (3.1.16a)$$

we end up with (see problem III.3)

$$G_1(\nu, g_0, \tau) = 2 \operatorname{Re}[g_0 \cdot g_1(\nu, \tau)] = \frac{2\pi}{\nu^3} g_0 \left\{ 2(1 - \cos(\nu\tau)) - \nu\tau \sin(\nu\tau) \right\} \qquad (3.1.16b)$$

for $\tau = 1$, namely at the end of the interaction, equation 3.1.16b can be written in the form

$$G_1(\nu, g_0) = -\pi g_0 \frac{\partial}{\partial \nu} \left[\frac{\sin(\nu/2)}{\nu/2}\right]^2 \qquad (3.1.16c)$$

---
*Which will be more quantitatively discussed in the forthcoming sections.

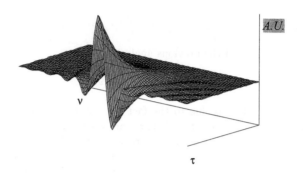

Figure 4 Field intensity $|a|^2$ vs. ν and τ. $\nu \equiv (-10\pi, 10\pi)$, $\tau \equiv (0, 1)$.

An idea of how the optical field evolves with the time is provided by fig.4, where we have reported the field intensity $|a|^2$ vs. ν and τ.

The evolution of the field intensity $|a|^2$ is otherwise shown in fig.5.

Furthermore the real and imaginary parts of the complex gain function, are plotted in fig.6.

In this treatment we have not included the effect of the e-beam energy distribution, which according to problem III.4 modifies the gain function as

$$G_1(\nu, \mu_\varepsilon) = 2\pi g_0 \int_0^1 t\,(1-t)\,\sin(\nu\,t)\,e^{-\frac{\pi^2 \mu_\varepsilon^2}{2} t^2}\,dt \qquad (3.1.17)$$

We must note that either for $\mu_\varepsilon = 0$ or $\mu_\varepsilon \neq 0$ the gain function is proportional to the derivative of $f(\nu)$. In the homogeneous case ($\mu_\varepsilon = 0$) the peak gain is located at $\nu_0 \simeq 2.6$ where

$$G_1(2.6) = 0.85\,g_0 \qquad (3.1.18)$$

in the case of a non negligible energy spread, the position of the peak gain shifts towards larger ν values, according to the relation

$$\nu^* \cong 2.6 + \mu_\varepsilon^{1.9} \qquad (3.1.19)$$

and the maximum gain behaves as (see problem III.5)

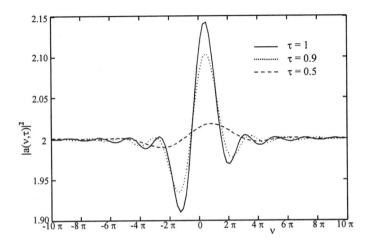

Figure 5 Field intensity $|a|^2$ vs. ν at different τ values.

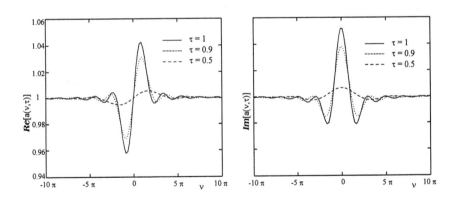

Figure 6 Real and imaginary parts of the dimensionless field amplitude vs. ν at different τ values.

$$G_1\left(\nu^*, \mu_\varepsilon\right) \cong \frac{0.85\, g_0}{1 + 1.7\, \mu_\varepsilon^2} \tag{3.1.20}$$

Equation 3.1.17 provides the effect of the energy spread on the gain function, the integral cannot be obtained in terms of elementary functions. Within this respect approximant functions may be very useful, as discussed in problem III.6. The role played by the approximant functions in FEL physics will be more carefully discussed in the following.

In this introductory part of the chapter, we have provided a simplified point of view on the theory of the FEL gain a deeper analysis will be provided in the forthcoming sections.

3.2 High-gain regimes

Even though not explicitly stated, the quantity controlling the FEL gain is the small signal gain coefficient g_0, which, in practical units, and for a linearly polarized undulator reads

$$g_0 = \frac{16\,\pi}{\gamma}\, \lambda_R\,[m]\, L_u\,[m]\, \frac{|J\,[A/m^2]|\, N^2}{1.7 \cdot 10^4}\, \xi\, f_b^2\,(\xi) \tag{3.2.1}$$

and

$$f_b\,(\xi) = J_0\,(\xi) - J_1\,(\xi) \quad , \quad \xi = \frac{1}{4}\frac{k^2}{1 + k^2/2} \tag{3.2.2}$$

is the Bessel factor discussed in the previous chapter and accounting for the modulation of the longitudinal motion and J is the e-beam current density. The derivation of the gain equation in the previous section was based on the assumption that g_0 is a small quantity. When g_0 is larger, and we will be more quantitatively precise below, equation (3.1.12) itself is not correct to describe the evolution of the complex optical field amplitude. Equation (3.1.12) as it stands is more appropriate for a device in which the evolution of the optical field does not grow self-consistently, namely without contributing to the active medium itself.

Within this respect an obvious modification of equation (3.1.12) is

$$\frac{da}{d\tau} = i\,\pi\, g_0 \int_0^\tau \tau'\, e^{-i\nu\tau'}\, a(\tau - \tau')\, d\tau' \tag{3.2.3}$$

The convolution form (3.2.3) ensures that the field grows preserving memory of the interaction at the previous times.

Equation 3.2.3 contains more informations than its analog 3.1.12. It can be solved by following many different procedures and we note that, from the mathematical point of view, it is a Volterra type integro-differential equation which can be solved by using e.g. a naive series of the type

High-gain regimes

$$\frac{da}{d\tau} = \sum_n g_0^n a_n(\nu,\tau) \quad , \quad a_n(\nu,0) = a(0) \quad , \quad a_n(\nu,\tau) = a(0)\, g_n(\nu,\tau) \quad (3.2.4)$$

which are always converging whatever large g_0. Accordingly from 3.2.4 and 3.2.3 we obtain the solution from the recursive relation

$$\frac{d}{d\tau} a_n = i\pi \int_0^\tau \tau' e^{-i\nu\tau'} a_{n-1}(\tau-\tau')\, d\tau' \quad (3.2.5)$$

the first order solution $n=1$ coincides with that derived in the previous section and for $n=2,3$ we obtain (see problem III.7)

$$g_2(\nu,\tau) = \frac{\pi^2}{6\nu^6}\left\{60 - 24\,i\nu\tau - 3\nu^2\tau^2 - e^{-i\nu\tau}\left[60 + 36\,i\nu\tau - 9\nu^2\tau^2 - i\nu^3\tau^3\right]\right\}$$

$$g_3(\nu,\tau) = \frac{\pi^3}{120\nu^7}\{6720 - 2520\,i\nu\tau - 360\nu^2\tau^2 + 20\,i\nu^3\tau^3 - \quad (3.2.6)$$
$$- e^{-i\nu\tau}[6720 + 4200\,i\nu\tau - 1200\nu^2\tau^2 - 200\,i\nu^3\tau^3 + 20\nu^4\tau^4 + i\nu^5\tau^5]\}$$

By defining the gain as the relative field intensity variation at $\tau=1$ (see equation 3.1.16a) we obtain (see problem III.8)

$$G(\nu,g_0) = g_0\,G_1(\nu) + g_0^2\,G_2(\nu) + g_0^3\,G_3(\nu) \quad (3.2.7a)$$

where $G_1(\nu)$ is provided by equation 3.1.16c and $G_{2,3}(\nu)$ are given below

$$G_2(\nu) = \frac{\pi^2}{3\nu^6}\left\{84(1-\cos\nu) - 60\nu\sin\nu + 3\nu^2 + 15\nu^2\cos\nu + \nu^3\sin\nu\right\} \quad (3.2.7b)$$

$$G_3(\nu) = \frac{\pi^3}{60\nu^9}\{11520(1-\cos\nu) - 9000\nu\sin\nu + 360\nu^2 + 2880\nu^2\cos\nu$$
$$+480\nu^3\sin\nu - 20\nu^4(1+2\cos\nu) - \nu^5\sin\nu\}$$

A comparison between the three gain functions is given in fig.7.

The so far obtained results needs to be clarified from the physical point of view. The function $G_1(\nu)$ is the low gain function, which describes the optical field evolution in the so called low gain regime i.e. when the optical field may be assumed constant during one interaction time. The higher order corrections take into account the fact that the field may grow during the interaction time and are a measure of the deviation from the low gain regime. An idea of such deviation is provided by fig.8 where we have reported the gain function vs. ν for different g_0 values. It is evident that when g_0 increases the gain looses its characteristic antisymmetric shape and the maximum gain provided by

$$G^*(g_0) = 0.85\,g_0 + 0.19\,g_0^2 + 4.12\cdot 10^{-3}\,g_0^3 \quad (3.2.8)$$

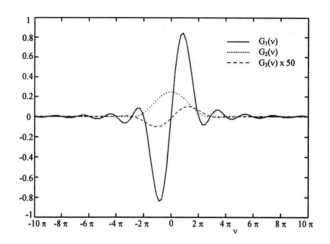

Figure 7 Comparison between the three gain functions $G_1(\nu)$, $G_2(\nu)$, $G_3(\nu)$ vs. ν.

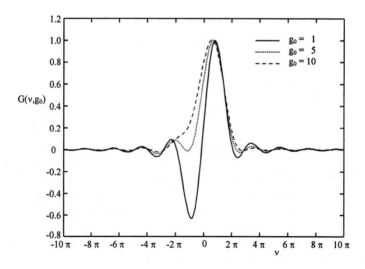

Figure 8 Gain curve vs ν. Comparison for three different values of g_0. The curves are normalized to the peak value.

High-gain regimes

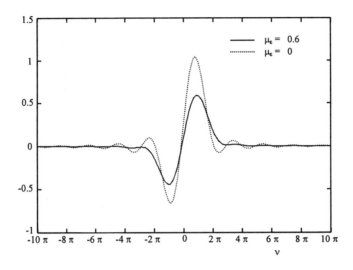

Figure 9 Effect of μ_ε on the gain curve. The curves are normalized to the peak value.

shifts towards smaller ν-values.

From a more quantitative point of view we can conclude that if $g_0 \lesssim 0.2$ the higher order terms will provide a correction of the maximum gain not larger than 5%.

The above results hold for the homogeneously broadened regime, the inclusion of effects like the relative energy spread is quite straightforward and in fact one gets (see problem III.10)

$$\frac{da}{d\tau} = i\pi g_0 \int_0^\tau \tau' e^{-i\nu\tau' - 1/2(\pi\mu_\varepsilon)^2 \tau'^2} a(\tau - \tau') d\tau' \qquad (3.2.9)$$

The effect of a non negligible μ_ε on the gain is provided in fig.9. It appears evident that the high gain corrections are more severely affected by inhomogeneous broadening effects than the corresponding low gain counterpart.

Even though high gain corrections up to third order, may accurately describe FEL devices with a large interval of g_0 values ($g_0 \leq 20$), the case of FELs operating in the single pass high gain regime require a non perturbative approach to equation 3.2.3.

By noting that, by introducing the formalism of negative derivatives, equation 3.2.3 can be rewritten as

$$e^{i\nu\tau} \mathcal{D}_\tau a(\tau) = i\pi g_0 \mathcal{D}_\tau^{-2} \left[e^{i\nu\tau} a(\tau) \right] \qquad (3.2.10)$$

where the notation $\mathcal{D}_\tau = \frac{d}{d\tau}$ has been adopted for brevity's sake. Its negative counterpart follows from the notion of Cauchy repeated integrals, namely

$$\mathcal{D}_\tau^{-n} g(\tau) = \frac{1}{(n-1)!} \int_0^\tau (\tau - \xi)^{n-1} g(\xi) d\xi \qquad (3.2.11)$$

It is therefore evident (see problem III.11) that, the dimensionless field amplitude $a(\tau)$ satisfies the following third order ordinary differential equation

$$\left[\mathcal{D}_\tau^3 + 2i\nu \mathcal{D}_\tau^2 - \nu^2 \mathcal{D}_\tau\right] a(\tau) = i\pi g_0 a(\tau) \qquad (3.2.12a)$$

with initial conditions

$$a(0) = a_0 \quad , \quad \left[\frac{da}{d\tau}\right]_{\tau=0} = 0 \quad , \quad \left[\frac{d^2 a}{d\tau^2}\right]_{\tau=0} = 0 \qquad (3.2.12b)$$

The solution of the above equation can be obtained by following different procedures.

The final result is that the dimensionless FEL amplitude can be cast in the form

$$a(\tau) = \sum_{j=1}^{3} a_j e^{-i(\nu + \delta v_j)\tau} \qquad (3.2.13a)$$

where the δv_j are the solutions of the algebraic equation

$$\delta v^2 (\nu + \delta v) = \pi g_0 \qquad (3.2.13b)$$

and a_j satisfies the coupled linear equations

$$\sum_{j=1}^{3} a_j = a_0 \quad , \quad \sum_{j=1}^{3} a_j \delta v_j = -a_0 \nu \quad , \quad \sum_{j=1}^{3} a_j \delta v_j^2 = a_0 \nu^2 \qquad (3.2.13c)$$

By solving the cubic equation 3.2.13b and the system 3.2.13c we end up with the general solution of the optical field evolution in the form (see problem III.12)

$$a(\tau) = a_0 \frac{e^{-\frac{2}{3}i\nu\tau}}{3(\nu + p + q)} \cdot \qquad (3.2.14a)$$
$$\cdot \left\{ (-\nu + p + q) e^{-\frac{i}{3}(p+q)\tau} + (2\nu + p + q) e^{+\frac{i}{6}(p+q)\tau} \cdot \right.$$
$$\left. \left[\left(1 - \frac{i\sqrt{3}\nu}{p-q}\right) e^{-\frac{\sqrt{3}}{6}(p-q)\tau} + \left(1 + \frac{i\sqrt{3}\nu}{p-q}\right) e^{\frac{\sqrt{3}}{6}(p-q)\tau}\right] \right\}$$

where

$$p = \left[\frac{1}{2}\left(r + \sqrt{d}\right)\right]^{\frac{1}{3}}, \quad q = \left[\frac{1}{2}\left(r - \sqrt{d}\right)\right]^{\frac{1}{3}}, \qquad (3.2.14b)$$

$$r = 27\pi g_0 - 2\nu^3, \quad d = 27\pi g_0 \left(27\pi g_0 - 4\nu^3\right)$$

by using the same gain definition 3.1.16a we can get a further idea of how the gain curve modifies with increasing g_0 as already shown in fig.8. It is evident that, while g_0 increases, the gain curve gets a symmetric form, which is not dissimilar from a Gaussian. According to fig.10 two useful approximation of the high gain curve are provided by

$$G_H(\nu, g_0) \cong \frac{1}{9} e^{\sqrt{3}(\pi g_0)^{1/3}} \left[1 + \frac{1}{9}\left(6 - \sqrt{3}\nu\right) \frac{\nu}{(\pi g_0)^{1/3}}\right] \qquad (3.2.15a)$$

and

$$G_h(\nu, g_0) \cong \frac{1}{9} e^{\sqrt{3}(\pi g_0)^{1/3}} \cdot e^{-\left[\frac{(\nu-\sqrt{3})^2 + 8 \cdot 10^{-2}\frac{\nu^2}{g_0^{1/3}}}{3\sqrt{3}(\pi g_0)^{1/3}}\right]} \qquad (3.2.15b)$$

The first reproduces the gain function in a small interval around the maximum, located at $\nu = \sqrt{3}$. The second reproduces the curve all over the ν interval with a noticeable accuracy, the small term $8 \cdot 10^{-2}\frac{\nu^2}{g_0^{1/3}}$ has been introduced to account for a slight intrinsic residual asymmetry, which tends to disappear with increasing g_0.

In general on resonance the gain can be cast in the form

$$G(\tau) = \frac{2}{9}\left\{\cosh\left[(\pi g_0)^{1/3}\sqrt{3}\tau\right] + 2\cos\left[(\pi g_0)^{1/3}\frac{3}{2}\tau\right] \cdot \cosh\left[(\pi g_0)^{1/3}\frac{\sqrt{3}}{2}\tau\right] - \tau\right\}$$

$$\cong \frac{1}{9} e^{\sqrt{3}\tau(\pi g_0)^{1/3}} \qquad (3.2.16)$$

The last equality holds for very large g_0 values. Albeit trivial this result indicates that within the context of high gain devices the concept of small signal gain coefficient is not particularly significant. We note indeed that for relativistic particles, the dimensionless time τ can be written in terms of the longitudinal coordinate as $\tau = z/(N\lambda_u)$ we can therefore recast 3.2.16 in the form

$$G(\tau) \cong \frac{1}{9} e^{4\pi\sqrt{3}\rho\frac{z}{\lambda_u}} \qquad (3.2.17a)$$

where

$$\rho = \frac{1}{4\pi}\left[\frac{\pi g_0}{N^3}\right]^{1/3} \qquad (3.2.17b)$$

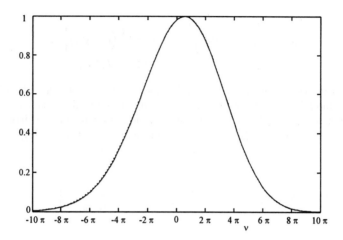

Figure 10 High gain curve vs ν, normalized to the maximum value, for $g_0 = 10^3$. Comparison between the exact gain formula and G_h.

is the high-gain growth-rate parameter and is independent of the number of undulator periods (see equation 3.2.1). Equation 3.2.17a may be therefore exploited to study the amplification of a seed signal while it is propagating along the z-direction.. The importance of the inhomogeneous broadening effects is no more specified by parameters of the type μ_ε, which involve the number of undulator periods, but by an analogous quantity involving ρ, this can be easily understood going back to equation 3.2.15b, which indicates that the "energy acceptance" of the high gain curve is (see problem III.13)

$$\frac{\delta\gamma}{\gamma} \sim \rho \qquad (3.2.18)$$

The inhomogeneous broadening parameters should be defined accordingly, therefore the parameter accounting for the energy spread high gain detrimental effect is

$$\tilde{\mu}_\varepsilon = \frac{1}{\rho} 2\,\sigma_\varepsilon \qquad (3.2.19)$$

and the high gain growth scale, including inhomogeneous broadening conditions, should be written as

$$\rho(\tilde{\mu}_\varepsilon) = \frac{\rho}{1 + 0.16\,\tilde{\mu}_\varepsilon^2} \cdot e^{-0.034\,\tilde{\mu}_\varepsilon^2} \qquad (3.2.20)$$

and equation 3.2.17a should be modified by replacing ρ with $\rho(\tilde{\mu}_\varepsilon)$.
For a more extended and rigorous treatment the reader is addressed to problem III.14 and to the bibliography at the end of the chapter.

3.3 FEL strong signal regime

The analysis of the previous section can be considered correct for a FEL device operating in the so called small signal regime, i.e. when non-linear effects due to saturation are negligible. The key parameter of this approximation is the field dimensionless amplitude and the condition regime, in terms of a writes $|a| \ll 1$.

To appreciate the above inequality more quantitatively we remind that the Colson's dimensionless amplitude can be written as

$$|a| = \sqrt{0.8}\, \pi^2 \left(\frac{I}{I_s}\right)^{1/2} \qquad (3.3.1)$$

where I is the optical field power density and I_s is the saturation power density (or more imprecisely the saturation intensity) which in practical units reads

$$I_s \left[\frac{MW}{cm^2}\right] \cong 6.9 \cdot 10^2 \left(\frac{\gamma}{N}\right)^4 \frac{1}{[\lambda_u\,[cm]\, k\, f_b(\xi)]^2} \qquad (3.3.2)$$

It is therefore evident that the small signal condition requires that the FEL optical power be much less than the saturation power. The role of I_s is that played in conventional lasers, it amounts indeed to the optical power halving the small signal gain.

The crucial importance of the FEL saturation intensity will be appreciated in the following. Before entering into this specific problem, we remind that the FEL dynamics is governed by the coupled pendulum and Maxwell equations. Limiting ourselves to the 1-D case these equations takes the very concise form

$$\frac{d^2\zeta}{d\tau^2} = |a|\cos(\zeta + \varphi) \qquad (3.3.3)$$
$$\frac{d}{d\tau}a = -2\pi g_0 \left\langle e^{-i\zeta}\right\rangle_{\zeta_0,\nu_0}$$

where $\zeta(\tau)$ is the electron relative phase specified by

$$\zeta(\tau) = -\omega\, t + (k + k_u)\, z\ ,\quad \omega = k\cdot c\ ,\quad k_u = \frac{2\pi}{\lambda_u} \qquad (3.3.4)$$

where z is the longitudinal coordinate, φ the phase of the field amplitude and $\left\langle e^{-i\zeta}\right\rangle_{\zeta_0,\nu_0}$ represents the average on the initial phase and energy distribution. It is particularly instructive to derive the small signal gain equation from equation 3.3.3.

Expanding indeed ζ up to the first order in the field amplitude strength, we get (see problem III.15)

$$\zeta \simeq \zeta_0 + \nu_0 \tau + \delta\zeta \qquad (3.3.5)$$

and

$$\delta\zeta \simeq \int_0^\tau d\tau' \, (\tau - \tau') \operatorname{Re}\left[a\left(\tau'\right) e^{i(\zeta_0 + \nu_0 \tau')}\right] \qquad (3.3.6)$$

which once inserted in the second of 3.3.3 yields (see problem III.16) equation 3.2.3.

It is obvious that the possibility of obtaining higher order corrections (see problem III.17,18) relies on the possibility of finding solutions of perturbative nature by exploiting the field strength as expansion parameter. Albeit conceptually simple, the problem is significantly complicated from the analytical point of view. We sketch therefore the concluding points addressing the reader to problems III.17,18 and to the bibliography at the end of the chapter

The strategy one can follow to deal with this type of problems is

a) to derive a perturbative expansion in the field strength up to a given order
b) to use a Padé-type approximant to accelerate the convergence of the series.
The field at $\tau = 1$ can be written in the form of the following expansion

$$a = a_0 \left\{ 1 + 2\pi i \, g_0 \sum_{k=0}^{2} g_{1,k} \, x^k \right\} \qquad (3.3.7)$$

where

$$g_{1,k} = h_{1,k+1} \cdot r^k \quad , \quad r = 0.8\,\pi^4 \quad , \quad x = \frac{I}{I_s} \qquad (3.3.8)$$

and

$$h_{1,s} = A_{1,s}\, e^{-i\alpha_{1,s}\nu\tau} \cdot e^{-\frac{\nu^2}{2(\eta_{1,s})^2}} - i^{\delta_{s,2}+\delta_{s,3}} \cdot B_{1,s}\, \nu^{3+\delta_{s,1}} e^{-i\beta_{1,s}\nu} e^{-\frac{\nu^2}{2(\eta_{1,s})^2}} \qquad (3.3.9)$$

with the coefficients $(A, B, \alpha, \beta, \gamma, \eta)$ given in Tab.(VIII.1a). The truncation of the series is just due to the fact that the evaluation of higher order terms is rather cumbersome. Equation 3.3.4 as it stands is not sufficiently accurate to reproduce the correct behaviour of the dimensionless field amplitude at saturation, i.e. when x is not much less than unity. The use of Padé approximants can be very helpful to accelerate the convergence of the series, so that (see problem III.19), we recast 3.3.4 in the form

$$a(\nu, x) = a_0 \left[1 + 2\pi i \, g_0 \, A(\nu, x)\right] \qquad (3.3.10)$$

$$A(\nu, x) = \frac{g_{1,0}(\nu)}{1 - \frac{g_{1,1}(\nu)}{g_{1,0}(\nu)} x - \left[\frac{g_{1,2}(\nu)}{g_{1,0}(\nu)} - \left(\frac{g_{1,1}(\nu)}{g_{1,0}(\nu)}\right)^2\right] x^2}$$

It is remarkable that 3.3.7 yields the correct description of the field dimensionless amplitude saturation for large x values ($x \leq 3$).

The use of the previous results yields a first important information, i.e. that the gain can be expressed as a simple function of x, namely

$$G(x) = G^* \frac{1 - e^{-\beta x}}{\beta x} \quad , \quad \beta = a\frac{\pi}{2} \tag{3.3.11}$$

where a is a number around 1, more precisely $0.9 \leq a \leq 1.045$, the upper limit seems to fit better the numerical results.

A formula analogous to that of conventional lasers can be derived from 3.3.10, by exploiting the Padé approximant method we get indeed (see problem III.10)

$$G(x) = G^* \frac{1}{1 + a_1 x + a_2 x^2} \tag{3.3.12}$$

where $a_1 + a_2 = 1$ and if $a = 1.0145$ the coefficient a_1 can be fixed around 0.797. equation 3.3.12, as well 3.3.11, implicitly contains the condition

$$G(1) = \frac{1}{2} G^* \tag{3.3.13}$$

i.e. the small signal gain is halved for optical power density equating the saturation power density. Other remarkable properties of the function $G(x)$ will be discussed below.

Equation 3.3.11 can be exploited in a number of interesting way and it is particularly useful to derive from 3.1.1 the intracavity power rate- equation growth, which can be written in the form

$$x_{n+1} = x_n + [(1 - \eta) G(x_n) - \eta] x_n \tag{3.3.14}$$

where n denotes the round trip number and η the cavity losses. An idea of the intracavity power growth is provided by fig.11, where we have reported the dimensionless intracavity power density vs. n.

The equilibrium intracavity power x_e can be obtained as a solution of the equation

$$G(x_e) = \frac{\eta}{1 - \eta} \tag{3.3.15}$$

The analysis of the numerical data and a naive perturbation theory has shown that x_e can be reproduced by a simple formula of the type (see problem III.21)

$$x_e \cong \frac{2}{\pi} \frac{1 - \eta}{\eta} G^* \left[1 - e^{-h(G^*, \eta)}\right] \tag{3.3.16}$$

$$h(G^*, \eta) = \frac{1.8}{1 + G^*} \frac{(1 - \eta) G^* - \eta}{\eta}$$

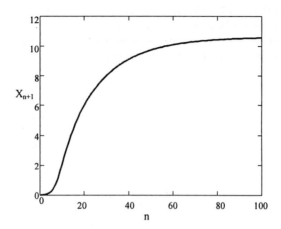

Figure 11 Dimensionless intracavity power density vs the round trip number n. $a = 1.0145$, $\eta = 0.05$.

If we limit ourselves to the low gain case and neglect the last contribution in the square bracket, by assuming $\eta \ll 1$ we obtain for the intracavity equilibrium intensity

$$I_e \simeq \frac{1.7}{\pi \eta} g_0 I_s \tag{3.3.17}$$

since

$$g_0 I_s = \frac{1}{2N} P_E \tag{3.3.18}$$

where P_E is the electron beam power density, we end up with the important conclusion that the intracavity equilibrium power density is

$$I_e \simeq \frac{1.7}{\pi \eta} \frac{1}{2N} P_E \tag{3.3.19}$$

From which it also follows that the output power density (roughly speaking the cavity losses time the equilibrium intracavity power) is proportional to the e-beam power and the proportionality factor is just provided by the FEL energy acceptance. It is now worth noting that the S-shaped curve yielding the intracavity power evolution vs. n can be reproduced by the discrete version of the logistic function, namely (see problem III.22)

$$x(n) = x_0 \frac{e^{[(1-\eta)G^* - \eta]n}}{1 + \frac{x_0}{x_e}\left[e^{[(1-\eta)G^* - \eta]n} - 1\right]} \tag{3.3.20}$$

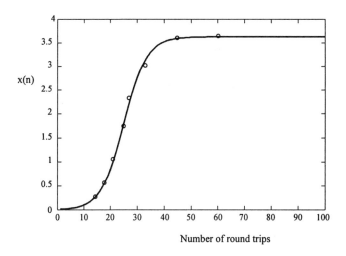

Figure 12 Dimensionless intracavity power $x = \frac{I}{I_s}$ vs. the number of round trips n. Comparison between analytical (continuous line) and numerical (circles) results. $a = 1.0145$, $\eta = 0.05$, $G^* = 0.3$.

where x_e can be evaluated by using either equation 3.3.15 or 3.3.16. A comparison between analytical and numerical results is given in fig.12 and it is evident that the analytical function 3.3.20 provides a reliable tool of analysis.

The so far discussed results are relevant to the low gain case, but they can be generalized to the high gain regime quite straightforwardly.

The high gain equilibrium power is linked to the e-beam power via the

$$I_E = \rho\, P_E \tag{3.3.21}$$

The logistic equation 3.3.20 will replaced by its continuous counterpart namely

$$I(z) = \frac{1}{9} I_0 \frac{e^{4\pi \rho \sqrt{3}\, z/\lambda_u}}{1 + \frac{I_0}{9 \rho P_E}\left[e^{4\pi \rho \sqrt{3}\, z/\lambda_u} - 1\right]} \tag{3.3.22}$$

where I_0 is the initial seed.

It is evident that in this case the evolution variable is no more the number of round trips but the longitudinal coordinate z.

Even though the FEL saturation process is a fairly complicated mechanism the scaling relations we have presented are accurate enough to provide a reasonably accurate description. Albeit we will deserve further comments on the topics discussed here in the concluding section of the chapter, we remind that, from the point of view of the electrons, saturation can be viewed as due to an induced energy spread, which reads (see problem III.24)

$$\sigma_i(x) \cong \frac{0.433}{N} e^{-\frac{\beta}{4}x} \left[\frac{\beta x}{1 - e^{-\beta x}} - 1 \right] \quad (3.3.23a)$$

where N is the number of undulator periods.

Equation 3.3.23a is correct for $x \leq 3$ and it is worth noting that for small x value it reduces to

$$\sigma_i(x) \cong \frac{0.387}{N} \sqrt{x} \quad (3.3.23b)$$

The results of this section provides the basis of the forthcoming discussion.

3.4 Storage Ring FEL dynamics

In this section we discuss a simplified picture of the dynamics of FELs operating with Storage-Rings ($\mathcal{SR}$).

According to the discussion of chapter II (section 4) the $\mathcal{SR}$ longitudinal dynamics is characterized by two competing processes: the damping due to the synchrotron radiation emission, the diffusion mechanism associated with the quantum nature of the emission itself. The inclusion of FEL can be qualitatively understood as a further beam heating due to the multiturn FEL interaction. The FEL produces an additional energy spread, which degrades the gain and counteracts the laser process itself which is inhibited when the gain is lowered at the level of the cavity losses. An idea of the power attainable with a $\mathcal{SR}$-FEL can be obtained in terms of a simple argument. According to the discussion of the previous sections the small signal gain is reduced significantly when the relative energy spread is of the order of the FEL energy acceptance, i.e. when

$$\delta E \sim \frac{1}{2N} E \quad (3.4.1)$$

the energy variation for $\mathcal{W}_e$ particle in the bunch will be

$$\xi_L \sim \frac{\mathcal{W}_e}{2N} E \quad (3.4.2)$$

when the FEL induced energy spread satisfies the condition 3.4.1 the energy transferred to the optical field is given by 3.4.2 and the laser process stops. When the FEL is no more active, the damping mechanism cools the beam and after a time comparable with the damping time a new laser pulse can start, the average laser power will be therefore provided by

$$P_L \sim \frac{1}{2N} \frac{\mathcal{W}_e E}{\tau_s} = \frac{1}{2N} P_s \quad (3.4.3)$$

where P_s is the power lost via synchrotron radiation in one machine turn and τ_s is the longitudinal damping time (see also chapter II).

The above simple example is perhaps illuminating on the type of problems one can deal with in the analysis of $\mathcal{SR}$-FELs, which can be treated by using a more complete model, which is a generalization of the rate equation 3.3.14. We must however keep in mind the following points

a) the FEL induces an e-beam energy spread and a consequent bunch lengthening, which cumulate turn by turn,

b) the $\mathcal{SR}$-FEL gain is therefore degraded by the induced inhomogeneous broadening and by the peak current reduction due to the bunch lengthening.

The $\mathcal{SR}$-FEL gain can be accordingly parametrized as

$$G(x) = 0.85\, g_0\, E(x)\, B(x)\, \frac{1 - e^{-\beta x}}{\beta x} \quad (3.4.4)$$

where

$$E(x) = \frac{1}{1 + 1.7\, \mu_\varepsilon^2(0) \left[1 + \frac{\sigma_i^2(x)}{\sigma_\varepsilon^2(0)}\right]} \quad , \quad B(x) = \frac{1}{\sqrt{1 + \frac{\sigma_i^2(x)}{\sigma_\varepsilon^2(0)}}} \quad (3.4.5)$$

where $\sigma_\varepsilon(0)$ is the natural energy spread, $\mu_\varepsilon(0) = 4\, N\, \sigma_\varepsilon(0)$ and $E(x), B(x)$ accounts for the induced energy spread and bunch lengthening respectively.

The rate equations yielding the $\mathcal{SR}$-FEL dynamics are therefore provided by

$$\begin{aligned} x_{n+1} &= (1 - \eta)\, [G(x_n) + 1]\, x_n \\ \sigma_{i,n+1}^2 &= \left(1 - 2\frac{T}{\tau_s}\right) \left[\sigma_{i,n}^2 + \left(\frac{0.433}{N}\right)^2 e^{-\frac{\beta x_n}{2}} \left(\frac{\beta x_n}{1 - e^{-\beta x_n}} - 1\right)\right] \end{aligned} \quad (3.4.6)$$

They are a clear generalization of equation 3.3.14 and the second equation accounts for the turn by turn damping of the induced energy spread, T is the machine revolution period, which is assumed to be equivalent to a cavity round trip. In a $\mathcal{SR}$-FEL the saturation intensity is a large quantity we can therefore safely assume $x \ll 1$ and expand the functions of x up to first order. Since the numerical stability of the finite difference system is not a priori ensured, it is convenient to transform it into a first order system of non linear differential equations (see problem III.24)

$$\begin{aligned} \frac{d\tilde{x}}{dt} &= \frac{\tilde{x}}{T_0} \left[\frac{1}{\sqrt{1 + \tilde{\sigma}^2}} \cdot \frac{1}{1 + 1.7\, \mu_\varepsilon^2(0)\, (1 + \tilde{\sigma}^2)} - r\right] \\ \frac{d\tilde{\sigma}^2}{dt} &= -\frac{2}{\tau_s} \left(\tilde{\sigma}^2 - \tilde{x}\right) \end{aligned} \quad (3.4.7a)$$

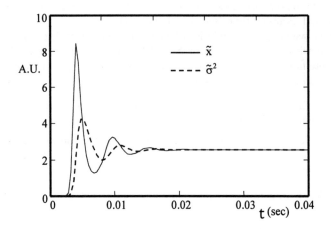

Figure 13 Evolution of the intracavity power $\tilde{x}$ and of the induced energy spread $\tilde{\sigma}^2$ vs time.

where

$$\tilde{x} = \mu x \quad, \quad \mu = \left(\frac{0.433}{N}\right)^2 \frac{\beta}{4} \frac{\tau_s}{T \sigma_\varepsilon^2(0)} \quad, \quad (3.4.7b)$$

$$T_0 = \frac{T}{0.85\, g_0} \quad, \quad \tilde{\sigma}^2 = \frac{\sigma_i^2(x)}{\sigma_\varepsilon^2(0)} \quad, \quad r = \frac{\eta}{0.85\, g_0}$$

Equations 3.4.7a can be used to model the system dynamics and an example is provided by figs. 13 which show the evolution of the intracavity power ($\tilde{x}$) and of the induced energy spread ($\tilde{\sigma}^2$).

According to the previous qualitative discussion, the system evolution is characterized by a fast blow up of the intracavity intensity and of the induced energy spread. The system successively relaxes and after a number of damped oscillations reaches eventually saturation.

The equilibrium output power is linked to $\tilde{x}_e$, i.e. the stationary value of $\tilde{x}$, by the relation (see problem III.25)

$$I_{out} = \frac{1}{4N} \chi P_s \qquad (3.4.8a)$$

where χ is a kind of efficiency factor defined as

$$\chi = 1.422\, r\, \tilde{x}_e\, \mu_\varepsilon^2(0) \qquad (3.4.8b)$$

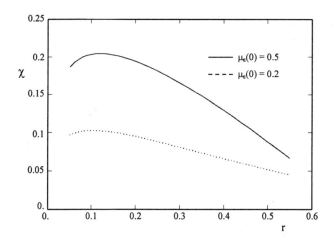

Figure 14 Efficiency function χ vs r for different $\mu_\varepsilon(0)$ values.

It is otherwise clear that $\tilde{x}_e$ is obtained from equation 3.4.7a by setting $\frac{d\tilde{x}}{dt} = \frac{d\tilde{\sigma}^2}{dt} = 0$ and is solution of the algebraic equation

$$(1+\tilde{x}_e)\left[1 + 1.7\,\mu_\varepsilon^2(0)\,(1+\tilde{x}_e)\right]^2 = \frac{1}{r^2} \tag{3.4.9}$$

(see problem III.26 for further comments), an idea of the values reached by χ is given by fig.14 where we have plotted χ vs. r for different $\mu_\varepsilon(0)$ values.

In this section we have just provided a feeling of how the problem should be treated and what we should expect as power output, further comments will be given in chapter VII.

3.5 FEL Optical Klystron devices

The small signal gain coefficient g_0 is proportional to N^3, undulators with large number of periods would therefore provide larger gain, if inhomogeneous broadening effects are negligible. Large N implies long undulators and some time the space for them is not available as e.g. on a straight section of a $\mathcal{SR}$.

To overcome such a problem the arrangement shown in fig.15 has been proposed. Two undulators with identical number of periods N are separated by a drift section, where the electrons follow a curved path of length $L_D \gg L_u$, where L_u is the length of one undulator.

A FEL device operating with the above arrangement instead of a single undulator is called Optical Klystron (O.K.).

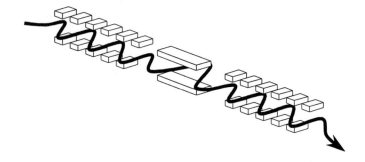

Figure 15   Scheme of the Optical Klystron.

We introduce the dispersive section equivalent number of periods $N_D = \frac{L_D}{L_u}$ and define a parameter of crucial importance for O.K.- FEL, namely

$$\delta = \frac{N_D}{N} \qquad (3.5.1)$$

The importance of an O.K. device is readily understood, in the first undulator an e-beam interacting with a transverse TE wave undergoes an energy modulation, which, in the drift section, transforms into a density modulation, and coherent emission occurs into the second undulator. The importance of the δ parameter can be understood from fig.16 representing the line shape of the spontaneously emitted power by an electron moving inside the O.K. magnet (see problem III.27). The emission curve has the typical structure of an interference pattern and the width of the individual fringe is inversely proportional to $1 + \delta$ since the gain is provided by the derivative of $S(\nu, \delta)$ with respect to ν, we can expect a significant enhancement with respect to the case of zero drift section length.

The O.K.- FEL small signal gain can indeed be written as

$$G = -2\pi g_0 \frac{\partial}{\partial \nu}\left\{\left(\frac{\sin(\nu/2)}{(\nu/2)^2}\right)^2 [1 + \cos(\nu(1+\delta))]\right\} \qquad (3.5.2)$$

where g_0 is the small signal gain referred to the single undulator. It is easy to show that if $\delta = 0$ equation 3.5.2 reduces to the small FEL gain equation (see problem III.28). An idea of the shape of the gain function vs. ν is provided in fig.17 and its maximum value reads

$$G^*_{O.K.} \sim 0.85\left[8\, g_0\left(1 + p\frac{\delta}{1 + q/\delta}\right)\right] \qquad (3.5.3a)$$

where $p \cong 0.913$ and $q \cong 0.057$.

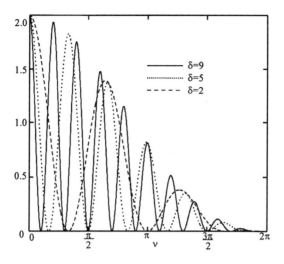

Figure 16 Spontaneous emission pattern for three different values of the drift section parameter δ.

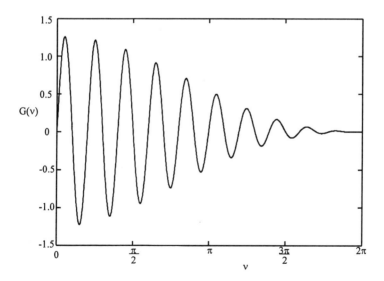

Figure 17 O.K. Gain vs ν for $g_0 = 0.0002$ and $\delta = 9$.

Very roughly speaking we can conclude that the O.K.-FEL is equivalent to a conventional FEL device having a small signal gain coefficient

$$g_{O.K.} = 8\, g_0 \left(1 + p\frac{\delta}{1 + q/\delta}\right) \tag{3.5.3b}$$

Even though trivial, this result can be conveniently exploited. We first note that the parameter $g_{O.K.}$ may control the high gain corrections, as for conventional FEL case, the extension of equation 3.2.8 to the O.K. case can be indeed written, up to $g_{O.K.}^3$, as (see problem III.29)

$$G_{O.K.}^*(g_0, \delta) \simeq 0.85\, g_{O.K.} + a\, g_{O.K.}^2 + b\, g_{O.K.}^3 \tag{3.5.4}$$

where $a = 0.185$ and $b \simeq 5.6 \cdot 10^{-5}$.

An other important point is the dependence of the O.K.-FEL gain on the e-beam energy spread. Limiting ourselves to the low gain case we get for the maximum gain

$$\begin{aligned} G_{O.K.}^*(\mu_\varepsilon, \delta) &= 0.85\, g_{O.K.} \cdot e^{-1.228(\mu_\varepsilon(\delta))^2} \\ \mu_\varepsilon(\delta) &= 2\, \mu_\varepsilon \cdot (1 + \delta) \end{aligned} \tag{3.5.5}$$

It is evident that an O.K.-FEL device is more sensitive to inhomogeneous broadening effects. This is due to the fact that the energy acceptance of an O.K.-FEL device is reduced by a factor $(1 + \delta)$ with respect to the ordinary FELs. The O.K.-FELs are therefore useful for accelerators with good e-beam qualities.

The effect of the e-beam energy spread on the O.K. gain is shown in fig.18.

The saturation mechanism of this kind of FEL device is not dissimilar from the ordinary case, the only difference being that the saturation intensity should be defined by including the δ parameter. It is not difficult (see problem III.30) to realize that the O.K. saturation intensity is

$$I_{S,\delta}\left[MW/cm^2\right] \cong \frac{6.9 \cdot 10^2}{f(\delta)^2} \left(\frac{\gamma}{N_T}\right)^4 \frac{1}{[\lambda_u\,[cm]\, k\, f_B(\xi)]^2} \tag{3.5.6}$$

where $N_T = 2N$ and $f(\delta)$ is provided by

$$f(\delta) = 1 + r\,\frac{\delta}{1 + s/\delta}\;,\quad r = 0.979\;,\quad s = 0.13 \tag{3.5.7}$$

A gain saturation formula, analogous to that discussed for conventional FEL holds for the O.K. case too and reads

$$G_{O.K.}^*(x_\delta, \delta, g_0) \simeq \frac{1}{2} G_{O.K.}^*(g_0, \delta) \left[1 + (1 - x_\delta)\, e^{-x_\delta\left(a + b\, g_0^{1/3}\right)}\right] \tag{3.5.8}$$

$$x_\delta = \frac{I}{I_{S,\delta}} \qquad a = 0.22\,,\; b = 0.204$$

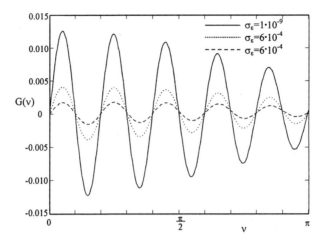

Figure 18 O.K. gain vs ν for three different values of σ_ε. $N = 20$, $\delta = 9$.

it is valid for $G^*_{O.K.}(g_0, \delta) \leq 16$, $\delta \leq 21$ and $x_\delta \leq 2.5$ with an accuracy better than 4% (see fig. III.20 for a comparison with the numerical results).

We must however underline that although 3.3.11 can be derived analytically, equation 3.5.8 is mainly empirical, obtained by a fitting procedure of the numerical data and its validity limited to the above quoted range of parameters, which however cover most of the practical cases.

Further comments on the O.K.- FEL devices will be given in the chapters relevant to the insertion devices and in problems III.31, III.32, III.33, III.34 at the end of the chapter.

3.6 Concluding remarks

In the previous sections we have just touched on the main aspects of the FEL theory. Many points have been, however, left open and this section is devoted at some complementary comments, helpful to provide a more complete treatment.

In the introductory section we have mentioned the problem of the inhomogeneous broadening, but we have dealt with the case relevant to the beam energy spread only. It is well known that a further source of inhomogeneous broadening is that associated with the e-beam emittance. The small signal FEL equation, which includes these effects, should be written as

$$\frac{d}{d\tau}a = i\pi g_0 \int_0^\tau d\tau'\, \tau'\, \frac{e^{-i\nu_0 \tau' - (\pi \mu_\varepsilon \tau')^2/2}}{R_x(\tau')\, R_y(\tau')}\, a(\tau - \tau') \tag{3.6.1}$$

where $(\eta = x, y)$

$$R_\eta(\tau) = \sqrt{\left(1 + \alpha_\eta^2\right)\left(1 - i\pi\,\mu_\eta\,\tau\right)\left(1 - i\pi\,\mu_\eta'\,\tau\right)} \cdot \alpha_\eta^2 \qquad (3.6.2)$$

$$\mu_\eta = \frac{4N}{1 + \frac{k^2}{2}} \gamma\, k_\eta^2 \frac{\varepsilon_{n,\eta}}{\gamma_\eta} \quad , \quad \mu_\eta' = \frac{4N^2}{1 + \frac{k^2}{2}} \gamma\, \frac{\varepsilon_{n,\eta}}{\beta_\eta}$$

The quantities $(\alpha_\eta, \beta_\eta, \gamma_\eta)$ are the Twiss parameters, k_η is the betatron wave number and $\varepsilon_{n,\eta}$ is the normalized emittance. The new terms R accounts for the gain degradation due to the beam emittance and to the beam matching conditions inside the undulator, for further comments see problems III.35-36. We have also discussed the high corrections to the small signal dynamics but we have not clarified how the low gain formula should be modified to include these effects. To this aim we note that a procedure based on Padé approximants and continued fraction method leads to a more general formula of the type (see problem III.37)

$$G^*(g_0, \mu_\varepsilon) = \frac{G^*(g_0)}{1 + 1.7\,\mu_\varepsilon^2\left(1 + 0.3\frac{g_0}{1+0.2\,g_0}\right) + 0.22\,\mu_\varepsilon^4\left(1 + 0.13\,g_0\right)} \qquad (3.6.3)$$

Furthermore the formulae we have given for the gain saturation are valid in the low gain homogeneously broadened regime. They can however be easily generalized. Equation 3.3.11 should indeed be modified by replacing G^* with $G^*(g_0, \mu_\varepsilon, \ldots)$ and x should be redefined as

$$x = \frac{I}{I_S(g_0, \mu_\varepsilon, \ldots)} \qquad (3.6.4)$$

where I is the optical power density and I_S the saturation intensity including the effect of inhomogeneous and high gain corrections (see problem III.38). An example of dependence of the saturation intensity on g_0 and μ_ε is provided by the following relation

$$I_S(g_0, \mu_\varepsilon) \simeq \frac{A(g_0, \mu_\varepsilon)}{B(g_0, \mu_\varepsilon)} \cdot I_S \qquad (3.6.5a)$$

where

$$A(g_0, \mu_\varepsilon) = 1 + 1.97\,\mu_\varepsilon^2 \frac{1 + 1.9 \cdot 10^{-2} g_0}{1 + 0.15 \frac{\mu_\varepsilon^4}{1+7\cdot 10^{-2} g_0}} \qquad (3.6.5b)$$

$$B(g_0, \mu_\varepsilon) = 1 + 0.182 \frac{\frac{9}{1+0.45\,\mu_\varepsilon^2}}{1 + 6.7 \cdot 10^{-2} \frac{9}{1+1.5\cdot 10^{-2}\,\mu_\varepsilon^2}}$$

Concluding remarks

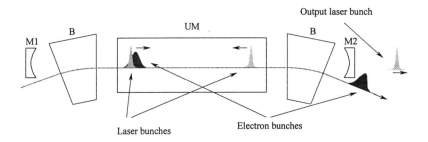

Figure 19 FEL oscillator scheme.

and I_S is the value provided by equation 3.3.2.

It is particularly important to stress that the product $I_S(g_0, \mu_\varepsilon) \cdot G^*(g_0, \mu_\varepsilon)$ is independent of g_0 and μ_ε for a large range of values. This ensures that the FEL output efficiency has only a weak dependence on the inhomogeneous and high gain corrections (see problem III.39).

A further important point, not touched at all, is the dependence of the FEL dynamics on the temporal structure of the e-beam. As it is well known when a FEL oscillator operates with a bunched e-beam, provided by a radio-frequency accelerating field, a natural mode-locking arises (see fig.19), each electron packet produces an optical packet which goes back and forth in the optical cavity, regarding this fact one must keep in mind a number of points:

a) to ensure overlapping between electrons and photons, the cavity length should be adjusted in such a way that it equals the distance between two adjacent electron bunches,

b) the FEL interaction tends to slow down the optical packets in the interaction region and thus a further correction, with respect to the prescription a), is necessary to ensure synchronism between electron and photons.

Within this respect, two parameters play a major role
1) the coupling parameter

$$\mu_c = \frac{\Delta}{\sigma_z} \quad , \quad \Delta = N \lambda_R \qquad (3.6.6)$$

where Δ is the slippage distance, i.e. the length cumulated between electrons and photons in one undulator passage (see problem III.40)

2) the cavity desynchronism or detuning parameter

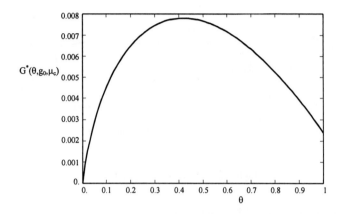

Figure 20   FEL gain vs. θ.

$$\vartheta = 4\frac{|\delta\mathcal{L}|}{g_0 \Delta} \qquad (3.6.7)$$

where $\delta\mathcal{L}$ is the cavity correction length to compensate interaction slow-down of the optical radiation (see problem III.41).

Limiting ourselves to the low gain case, a formula which includes the corrections due slippage and cavity adjustment effects is (see fig.20)

$$G^*(\vartheta, g_0, \mu_c) = -0.85\, g_0 \frac{\vartheta}{\vartheta_s}\left\{\ln\left[\frac{\vartheta}{\vartheta_s}\left(1+\frac{1}{3}\mu_c\right)\right]-1\right\} \qquad (3.6.8a)$$

$$\vartheta_s = 0.456$$

the maximum gain is located at

$$\vartheta^* = \frac{\vartheta_s}{1+\frac{1}{3}\mu_c} \qquad (3.6.8b)$$

where it takes the value

$$G^*(g_0, \mu_c) = \frac{0.85\, g_0}{1+\frac{1}{3}\mu_c} \qquad (3.6.8c)$$

(see problem III.42) for further comments.

Saturation effects can also be easily included, by modifying equation 3.6.8a as follows (see problem III.43)

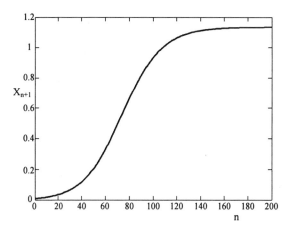

Figure 21   Dimensionless intracavity power density vs the round trip number n. $a = 1.0145$, $\eta = 0.01$.

$$G^*(\vartheta, g_0, \mu_c, x) = -0.85 \frac{\vartheta}{\vartheta_s} \left\{ \ln\left[\frac{\vartheta}{\vartheta_s}\left(1 + \frac{1}{3}\mu_c\right) \cdot f(x)\right] - 1 \right\} \quad (3.6.9)$$

$$f(x) = \frac{\beta x}{1 - e^{-\beta x}}$$

Equation 3.6.9 can be exploited to derive rate equations of the type 3.3.14 (see problem III.43 and fig.21 for further comments)

and it is worth notify that the maximum equilibrium intracavity power is given by

$$x_e^*(g_0, \mu_c, \eta) = \frac{2}{\pi} G^*(g_0, \mu_c) \cdot \left(1 - e^{-h(g_0, \mu_c)}\right) \quad (3.6.10a)$$

where

$$h(g_0, \mu_c) = \frac{1.8}{1 + G^*(g_0, \mu_c)} \cdot \frac{(1 - \eta)\, G^*(g_0, \mu_c) \cdot \eta}{\eta} \quad (3.6.10b)$$

We stress that the maximum output power is obtained at $\vartheta_e < \vartheta^*$ (see fig.22) and we have not a simple formula capable of reproducing the equilibrium intracavity power as a function of ϑ.

It is obvious that such a dependance can be derived from the equation (see problem III.44)

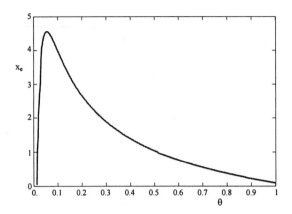

Figure 22 Dimensionless equilibrium intracavity power as a function of ϑ. $a = 1.0145$, $\eta = 0.01$.

$$G^* (\vartheta, g_0, \mu_c, x_e) = \frac{\eta}{1 - \eta} \qquad (3.6.10c)$$

For other practical aspects concerning the above points see problem III.45-49.

In this chapter we have given the basic elements of the FEL theory, the problems at the end of the chapter are a useful complement for a more complete understanding. Most of the points we have touched on in the previous sections will be reconsidered, and some times with more details, in chapter VII.

3.7 Problems

1) Derive equation 3.1.7.

Hint: use the convolution form

$$S(\nu, \mu_\varepsilon) = \int_{-\infty}^{+\infty} S(\nu + 4\pi N \varepsilon) \, f(\varepsilon) \, d\varepsilon$$

and use the integral representation

$$\left(\frac{\sin(\nu/2)}{\nu/2} \right)^2 = 2 \int_0^1 (1 - t) \cos(\nu t) \, dt$$

Extend 3.1.7 outside the interval $(0, 1)$.

2) Solve equation 3.1.12 with the assumption $g_0 \ll 1$ and $a(0) \neq 0$.

3) Derive the first order complex gain function (equation 3.1.15) and the first order gain equation 3.1.16b.

4) Derive equation 3.1.17 and discuss its physical meaning.
5) derive equation 3.1.20 by using a Padé-like approximation.
6) Prove that in the region of interest $(-2\pi, 2\pi)$ the FEL gain function is well reproduced by the approximant

$$\Gamma_1(\nu) = \frac{\pi}{3} e^{-\nu^2/40} \sin(\nu/2)$$

Use this expression to get an analytical approximation of the FEL gain inhomogeneous broadening (see e.g. G. Dattoli, L. Giannessi, P.L. Ottaviani and A. Segreto to J. Appl. Phys. 77, 6162 (1995)).
7) Use the naive series expansion 3.2.4 to solve equation 3.2.3 and specify the gain functions up to $n = 3$.
8) Use the naive series expansion to derive the gain functions $G_{1,2,3}(\nu)$.
9) Show that in the ν-space the relative energy distribution 3.1.6 reads

$$f(\nu, \mu_\varepsilon) = \frac{1}{\sqrt{2\pi}(\pi \mu_\varepsilon)} e^{-\frac{1}{2}\frac{\nu^2}{(\pi \mu_\varepsilon)^2}}$$

and discuss its physical meaning.
10) Derive equation 3.2.9.
11) Derive equation 3.2.12a and condition 3.2.12b.
 (Hint: multiply both sides of 3.2.10 by $\mathcal{D}_\tau^2$).
12) Derive equation 3.2.14a.
13) Use equations 3.2.15a, 3.2.15b to show that the relative energy width of high-gain FEL processes is provided by equation 3.2.17a, compare this result with the case relevant to the low gain regime and discuss its physical meaning.
14) Show that in the case of high gain regime the FEL growth rate parameter can be derived from the integral equation

$$\frac{1}{2}\left(1 + i\sqrt{3}\right)\chi = \int_0^{(\pi g_0)^{1/3}} ds\, s\, e^{-\frac{1+i\sqrt{3}}{2}\chi s} e^{-\frac{\mu_\varepsilon^2}{(\pi g_0)^{1/3}} s^2} e^{-\frac{i\nu_0}{(\pi g_0)^{1/3}} s}$$

derived from equation 3.2.9 after setting

$$a(\tau) = a_0\, e^{-\frac{(\pi g_0)^{1/3}}{2}\left(1+i\sqrt{3}\right)\chi\tau}$$

Use a numerical procedure to derive the dependance of χ on ν_0 and $\tilde{\mu}_\varepsilon$.
(see e.g. W.B. Colson, J.C. Gallardo, and P.M. Bosco to Phys. Rev. A34, 4875 (1986)).
15) show that

$$\frac{d\zeta}{d\tau} = \nu$$

(hint: rewrite ζ as

$$\zeta = \frac{L_u}{c}\left[(k+k_u)\frac{zc}{L_u} - \omega\tau\right] \quad , \quad k_u = \frac{2\pi}{\lambda_u}$$

and remind that in the ultrarelativistic limit

$$\frac{dz}{dt} \sim c\left[1 - \frac{1}{2\gamma^2}\left(1 + k^2/2\right)\right]).$$

16) Use the small signal approximation to derive equation 3.3.2 from the pendulum equation 3.3.3 and remind that

$$\langle f(\zeta_0) \rangle_{\zeta_0} = \frac{1}{2\pi}\int_0^{2\pi} d\zeta_0 \, f(\zeta_0)$$
$$\langle g(\nu) \rangle_{\nu_0} = \frac{1}{\sqrt{2\pi}\,\pi\mu_\varepsilon}\int_{-\infty}^{+\infty} d\nu \, g(\nu) \, e^{-\frac{(\nu-\nu_0)^2}{2(\pi\mu_\varepsilon)^2}}$$

17) Use the pendulum equation to show that the field evolution with the inclusion of saturation

$$\frac{da}{d\tau} = -2\pi g_0 \, e^{-i\nu_0\tau}\frac{J_1(\rho)}{\rho}\int_0^\tau (\tau-\tau')\, a(\tau') \, e^{-i\nu_0\tau'} \, d\tau'$$

where $J_1(\rho)$ is the cylindrical Bessel function of $1^{\underline{st}}$ order, and

$$|\rho|\,e^{i\psi} = \int_0^\tau (\tau-\tau')\,a(\tau')\,e^{-i\nu_0\tau'}\,d\tau'$$

Discuss the validity of the approximation.
(hint: use $\zeta = \zeta_0 + \nu_0\tau + \int_0^\tau(\tau-\tau')\,\mathrm{Re}\left[a(\tau')\,e^{i(\zeta_0+\nu_0\tau')}\right]\,d\tau'$ and the integral representation of Bessel functions).

18) Use a naive expansion in terms of the field strength $|a|$ to solve the pendulum equation.
(hint: $\zeta = \sum_n |a|^n \zeta_n$).

19) Derive equation 3.3.10 and propose alternative Padé approximants.
20) Derive equation 3.3.12 from 3.3.11.
21) Use a numerical or an analytical procedure to derive 3.3.16.
22) Find an argument to derive 3.3.20 from 3.3.14, or from any other point discussed in the chapter.
23) Justify equation 3.3.23a.
(hint: use the gain saturation formula 3.3.11 and equation 3.2.20).
24) Derive equation 3.4.7a from 3.4.6.
25) Derive equation 3.4.8a and discuss its physical meaning.
26) Discuss the physical meaning of equation 3.4.9, get the approximate solution

$$\tilde{x}_e = \left[\frac{1}{a\,r}\right]^{3/2} - \left(1 + \frac{2}{3\,a}\right) \quad , \quad a = 1.7\,\mu_\varepsilon^2\,(0)$$

valid for $\mu_\varepsilon(0) \gg 0.2\,r$ and compare it with the exact solution.
27) Use the argument of the introductory section to obtain $S(\nu, \delta)$ in fig.16
28) Show that equation 3.5.2 for $\delta = 0$ reduces to the ordinary FEL low-gain equation.
29) Compare equation 3.5.4 and 3.2.8 and find an argument to state the validity of the first.
30) Find a convincing physical argument to justify equation 3.5.6.
   (hint: look at the O.K.-FEL gain curve and note that its maximum is locate at $\nu^* \simeq 2.6/f(\delta)$).
31) Compare FEL and O.K.-FEL gain equations and find criteria to establish when O.K.-FEL device is convenient with respect to an ordinary FEL, discuss the concept of "convenience".
32) Find a rate equation for the intracavity evolution of an O.K.-FEL oscillator.
33) Derive the output equilibrium power.
34) Extend the results of problems III.32-33 to an O.K.-$\mathcal{SR}$ FEL.
35) Derive equation 3.6.1 by assuming for the e-beam a Gaussian distribution in the transverse phase-space.
36) Define the optimum e-beam matching, i.e. that minimize the effect of the inhomogeneous broadening due to the e-beam emittance, in particular show that it is achieved for

$$\alpha_\eta = 0 \quad , \quad \beta_\eta = \frac{\gamma}{\pi\,k}\lambda_u \quad , \quad \eta = x, y$$

37) Find a procedure to derive equation 3.6.3 and discuss its physical meaning.
38) Find an argument to explain why the saturation intensity depends on g_0, μ_ε ....
39) Explain why in general the product $I_S \cdot G_{MAX}$ should not be dependent on the small signal gain and inhomogeneous broadening parameters.
40) Show that the slippage distance is just $\Delta = N\,\lambda_R$.
41) Show that the average packet velocity inside the interaction region is linked to $\delta\mathcal{L}$ by

$$v_L = \frac{c}{1 + 2\frac{\delta\mathcal{L}}{L_u}}$$

where L_u is the length of the undulator.
42) Derive equation 3.6.8b and 3.6.8c from equation 3.6.8a.
43) Justify equation 3.6.9, derive a rate equation of the type 3.3.14 and study the evolution of the optical intracavity signal for different ϑ values.
44) Find the equilibrium intracavity power from equation 3.6.10c and discuss the limits of validity of equation 3.6.10a.

45) Discuss the dependence of the equilibrium intracavity power on ϑ and μ_c.
46) Include the cavity detuning effects in the analysis of the $\mathcal{SR}$ FEL evolution.
47) Find the equilibrium intracavity power of a $\mathcal{SR}$ FEL as a function of ϑ.
48) Write the logistic equation including the effect of the ϑ parameter.
49) Discuss the physical meaning of the parameter ϑ_S.

Suggested bibliography

The literature on Free Electron Lasers is wide and disseminated in specialistic papers and review articles. The interested reader is however addressed to the following review articles and book:

1. G. Dattoli and A. Renieri, *Experimental and theoretical aspects of Free Electron Lasers*, Laser Handbook vol. 4 ed. by M. L. Stitch and M. S. Bass, North Holland Amsterdam (1984).
2. J. C. Marshall, *Free Electron Lasers*, Mac Millan publishing company, New York (1985).
3. P. Luchini and H. Motz, *Undulators and free electron lasers*, Claredon Press Oxford (1990).
4. C. A. Brau, *Free electron lasers*, Academic press Oxford (1990).
5. Laser Handbook vol. 6 ed. by W. B. Colson, C. Pellegrini and A. Renieri, North Holland Amsterdam (1990).
6. G. Dattoli, A. Renieri and A. Torre, *Lectures on the free electron laser theory and related topics*, World Scientific Singapore (1995).
   In particular the analysis of the strong signal regime discussed in section 3 has benefitted from:
7. G. Dattoli, L. Giannessi, P. L. Ottaviani and A. Segreto, Free Electron Laser saturation: an analytical description, Phys. Plasmas, vol.2, 4325 (1995).
8. G. Dattoli, Logistic functions and evolution of free electron laser oscillators, J. Appl. Phys. 84, 2393 (1998).
   For the analysis of Storage Ring FELs the reader is addressed to the contribution by:
9. J.M. Ortega and D. A. G. Deacon in Laser Handbook vol. 6 as referenced previously.
   Most of the analysis developed in sec. 4 is based on:
10. P. Elleaume, J. Phys. (Paris) 45, 997 (1984).
11. G. Dattoli, L. Giannessi, P. L. Ottaviani and A. Renieri Nucl. Instrum. Methods in Phys. Res. A365, 559 (1995).
    As to the Optical- Klystron the reader is addressed to the contribution of :
12. W. B. Colson in Laser Handbook vol. 6 as referenced previously[5],
    while for the specific treatment of section 4 to:
13. G. Dattoli and P. L. Ottaviani, Gain parametrization formulas in Optical- Klystron configurations, IEEE J-QE 35, 27 (1999).

Chapter 4
OPTICAL SYSTEMS IN THE GEOMETRICAL AND WAVE OPTICS FRAMEWORK

Summary

The basic approach of geometrical and wave optics to the propagation, and corresponding transformations, of an optical beam through optical systems is briefly described. The conceptual and formal differences are stressed.

4.1 Introduction

The optical properties of a radiation source manifest into specific attributes of the emitted radiation, as, for instance, the frequency bandwidth in which radiation is emitted, the degree of spatial coherence, i.e. the capability of producing high-visibility interference pattern, as well as the accuracy with which radiation can be collimated into a nicely small angular divergence beam, or focussed into a small spot or further monochromatized without severe loss of available power. These attributes assure the possibility of a precise control of the photon flux from the source for specific applications and studies.

Visible radiation sources have evolved from little controllable lightbulbs, having a very large spectral content and angular extent, to the highly controllable laser sources, which in some cases produce radiation beams with narrow spectral and angular width, limited primarily by the finite wavelength.

An overall characterization of these properties is provided by the source *brightness*, i.e. the radiant power per unit source area and per unit solid angle, into which the radiation is emitted. More precisely, we should talk of the *spectral* brightness, that is the brightness per unit spectral bandwidth.

High brightness means that the radiation can propagate through a long distance without significant spread, can be focussed to a small spot and can produce sharp interference fringes.

It clearly emerges from the discussion of the previous sections that synchrotron radiation sources like visible light sources are steadily evolving towards the development of an advanced technology to produce VUV and X-ray wavelength photon flux with the laser-like properties of narrow bandwidth and emission cone. Fig. 1 shows the time averaged spectral brilliance ($photons/(s \cdot mrad^2 \cdot mm^2 \cdot 0.1\% \, bandwidth)$),

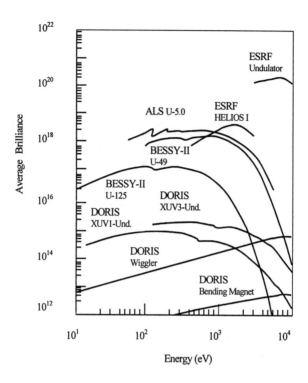

Figure 1 Average brilliance for some second and third generation synchrotron radiation sources.

as a function of the photon energy, achieved at a variety of second and third generation synchrotron radiation sources. It is evident that the synchrotron radiation sources provide soft and hard X-rays of several orders of magnitude brighter than those presently attainable from conventional laboratory sources.

Synchrotron radiation sources - present and future - are extremely bright not only because of the increasingly higher available photon flux values - attainable by using long undulators and low-emittance high-current electron beams - but also because of the smaller product of the source area times the solid angle into which photons are radiated. Such product is tried to approach the limit set by the wavelength of the emitted radiation. It is indeed a well established result of wave optics that the smallest possible diameter × divergence angle product is fixed by the radiation wavelength according to

Introduction

$$d \cdot \theta \simeq \lambda \qquad (4.1.1)$$

the *optical* Heisenberg uncertainty principle.

Radiation fields emitted by sources satisfying the relation (4.1.1) are said *diffraction limited* and would have a perfectly defined phase front, suitable therefore for interferometry and holography, i.e. phase sensitive experiments.

The product (4.1.1) directly involves the optical phase space variables. It stands out indeed as the phase space area associated with a linear source of size d emitting radiation within the maximum angle θ. Thereby, it is naturally suggested to frame the analysis of the optical properties of the synchrotron radiation sources as well as of the relevant transformations when propagating through any optical medium in the optical phase space.

In this connection, the concept of brightness, which is shown to be related to the density distribution of light rays in phase space, turns to be basic for the description of phenomena related to the propagation of wavefields.

Apart from its geometrical (in phase space) significance, a formal definition of the brightness function can be given as a function of the phase space variables, involving the mutual coherence function of the optical field -which is in some sense a measure of the statistical similarity between the signal at two different source points.

However, the identification of the brightness function with the source brightness should be done with some care as the brightness function cannot be directly related to physically measurable quantities. It must be regarded therefore as a useful mathematical construction in terms of which the analysis of optical propagation greatly simplifies. In fact, although its definition involves the optical wave function, whose propagation requires the quite intrigued methods of wave optics, the brightness function reveals the very attractive property of transforming through optical systems according to the simple matrix formalism of ray optics.

In this Chapter, we will exploit the brightness function and the relevant formalism to study the optical characteristics of the synchrotron radiation sources. Thus providing a rigorous method to analyze the propagation and imaging characteristics of synchrotron radiation, which are usually studied by means of geometrical optics, as well as to account for the effects of the electron beam distribution in position and angle.

Then, we deserve some space to discussing the basic concept of optical phase space (section 2), introduced within the context of Hamiltonian optics. Limiting ourselves to consider paraxial propagation and rotationally symmetric optical systems, we illustrate the matrix formalism of ray optics, as arising just from the linear approximation and the basic methods of Hamiltonian optics (section 3). Section 4 is devoted to discuss the approach to optical propagation within the framework of wave optics; the analysis of section 3 is indeed restated in the language proper to wave optics.

4.2 The optical phase space

The phase space concept is the natural by-product of the Hamiltonian formulation of classical mechanics, in which the dynamics of a mechanical system with n degrees of freedom is described in terms of n generalized independent coordinates $(q_1, q_2, ..., q_n)$ and the same number of canonically conjugate variables $(p_1, p_2, ..., p_n)$. The Cartesian space of these $2n$ coordinates is the *phase space*. The state of a single particle at a certain time, for instance, is represented by a point in the 6-dimensional phase space Γ_6, given by the three Cartesian coordinates $\mathbf{q} = (q_x, q_y, q_z)$ and the relevant momenta $\mathbf{p} = (p_x, p_y, p_z)$. The motion of the particle in real space corresponds to a trajectory in phase space. Then, the state of an ensemble of identical and noninteracting particles at a certain time corresponds to a set of points in Γ_6 space, with which is associated a real density in phase space and a corresponding distribution function of density $\rho(\mathbf{q}, \mathbf{p}, t)$, defined so that $\rho(\mathbf{q}, \mathbf{p}, t)dV$ provides the number of particles -or representative points - in the element of volume $dV = dq_x dq_y dq_z dp_x dp_y dp_z$ in the vicinity of the point $(\mathbf{q}, \mathbf{p})$. The domain occupied by the representative points of the N particles moves through phase space as the particles move in real space. However, as the total number of points remain constant (if the particles move through a lossless medium), the Liouville invariance of the density of representative point along the trajectory of any given point is assured as well as the invariance of the volume of the phase space domain, even though its shape may deform considerably.

Moreover, since at a given time the phase space trajectories corresponding to different particles nowhere intersect, representative points on the domain boundary remain on the boundary at every instant of time. Thus, the evolution of the whole assembly of particles may be accounted for by the evolution of the boundary of the associated phase space domain.

Within the framework of charged particle beam dynamics, the volume (or, to be more precise, the hypervolume) enclosed by the boundary of the phase space domain is usually referred to as the (normalized) beam *emittance*, apart from a multiplicative constant factor. The normalized emittance remains unchanged if the beam propagates through a non dissipative medium and hence it is an intrinsic parameter of the beam.

Phase space concepts and methods provide the natural tool to the physics and engineering of the transport of charged particle beams, whilst in contrast they seemed to be alien to light optics. Only recently, the phase space concepts with the relevant mathematical machinery has been applied to light optics, thus providing an overall context where both wave and geometrical optics can be framed.

Phase space representation of light optics is the natural by-product of the Hamiltonian formulation of geometrical optics, which then we will briefly review.

The Hamiltonian of ray optics is given by the well known expression

$$H(x, y, p_x, p_y, z) = -\sqrt{n^2 - p_x^2 - p_y^2} \qquad (4.2.1)$$

The optical phase space

In the above, (x, y, z) denote Cartesian coordinates, $n(x, y, z)$ is the refractive index of the medium accounting for all its optical properties, p_x and p_y are the *optical direction cosines*, given by the product of the refractive index times the direction cosines of the light ray* with respect to the x, y axis; explicitly

$$p_x = n\frac{dx}{ds}, \qquad p_y = n\frac{dy}{ds}, \qquad (4.2.2)$$

ds being the infinitesimal arclength measured along the ray path.

The optical Hamilton's equations readily follow as

$$\frac{dq_x}{dz} = \frac{\partial H}{\partial p_x} \qquad \frac{dq_y}{dz} = \frac{\partial H}{\partial p_y} \qquad (4.2.3)$$
$$\frac{dp_x}{dz} = -\frac{\partial H}{\partial q_x} \qquad \frac{dp_y}{dz} = -\frac{\partial H}{\partial q_y}$$

where for future convenience we switched on the notation $x \to q_x$, $y \to q_y$.

Given the initial values for the position (q_x, q_y) and the optical direction cosines (p_x, p_y) of a ray at some z_i, it is possible in general to solve the Hamilton's eqs. (4.2.3) for the position and the optical momenta at any other z.

Since (q_x, q_y, p_x, p_y) are canonically conjugate variables with respect to the Hamiltonian (4.2.1), we can construct a four-dimensional phase-space Γ_4, where a ray is represented by a point with coordinates (q_x, q_y, p_x, p_y). A collection of rays distributed over a range of possible positions and directions fills a certain region in the optical phase-space, with which we can associate a density distribution function $\rho(q_x, q_y, p_x, p_y, z)$. The change in the ray bundle as it propagates through some optical medium corresponds to the motion of the representative points as they move through phase-space in accordance with the equations of motion (4.2.3).

While the exact motion of each representative point in the inherent phase space domain (i.e. each ray in the bundle) is uniquely determined by the initial conditions, it is rather impractical to calculate an exact solution for the whole bundle of rays. It is convenient to exploit the phase-space representation and relevant methods to provide a statistical description of the behaviour of the assembly of rays, regarded as an ensemble of identical systems differing over a range of initial conditions.

Geometrical optics is basically concerned with the behaviour of a single ray when it passes through optical media. Thus, we firstly describe the methods of geometrical optics, in particular of linear and Gaussian optics, to turn then to the substantially differing tools of wave-optics (section 4). Finally in the next chapter we illustrate the phase space picture to the propagation of optical radiation, that allows

---

*Within the context of geometrical optics, the notion of light ray as thin beam is quite intuitive. However, it is a geometrical abstraction and is not physically observable due to diffraction. One may think of rays as optical beams, i.e. optical radiation having small angular divergence, with narrow transverse size, but large compared with wavelength. In wave optics rays are defined as the orthogonal trajectories to phase fronts of a light wave.

124 Optical systems in the geometrical and wave optics framework

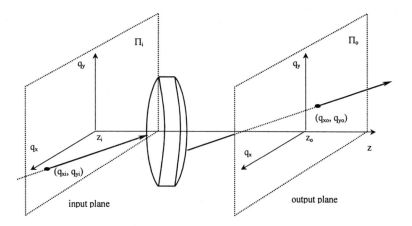

Figure 2 Optical system and relevant input and output planes, with respect to which the coordinates of the incoming and emerging rays are determined.

to recover the formalism of ray-optics and to take a full account of the diffraction effects of wave-optics.

Let us consider therefore a light ray propagating through an optical system, composed of some optical imaging devices. It is a common situation that the optical elements are aligned along some axis, the so-called *optical axis*, which is normally taken as z-axis and oriented as the direction of ray propagation assumed to go from the left to the right. The light rays therefore enter the system at the left and emerge from it at the right. Let us fix two planes $\prod_i$ and $\prod_o$, transversal to the optical axis, respectively located at the left - the input or object plane - and at the right -the output or image plane - with respect to the optical system; for instance at z_i and z_o (fig. 2). It is evident that the incoming ray is completely determined by the coordinates (q_{x_i}, q_{y_i}) of the point of intersection of the ray with the input plane $\prod_i$, and by its optical direction cosines (p_{x_i}, p_{y_i})

$$p_{x_i} = n_i \sin \alpha_{x_i} \,, \quad p_{y_i} = n_i \sin \alpha_{y_i} \qquad (4.2.4)$$

where $n_i \equiv n(q_{x_i}, q_{y_i}, z_i)$ and $(\alpha_{x_i}, \alpha_{y_i})$ are the angles that the incoming ray makes with the $q_x - z$ and $q_y - z$ planes (fig. 3). Similarly, the outgoing ray is specified by the intersection (q_{x_o}, q_{y_o}) with the output plane $\prod_o$ and its optical direction cosines (p_{x_o}, p_{y_o}) at this plane.

The problem of ray optics is then to determine the input-output relations of the optical system, relating the coordinates of the outgoing ray to those of the incoming ray

The optical phase space

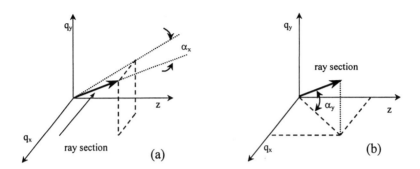

Figure 3 Geometry inherent to the definition of the optical direction cosines (a) p_x and (b) p_y.

$$q_o = q_o(q_i, p_i)$$
$$p_o = p_o(q_i, p_i) \qquad (4.2.5)$$

where shortly $q_i(q_o)$ denotes both $q_{x_i}(q_{x_o})$ and $q_{y_i}(q_{y_o})$, and similarly $p_i(p_o)$.

The above corresponds to a transformation in the optical phase-space, which is symplectic:

$$\begin{array}{ccc} z_i & & z_o \\ (q_{x_i}, q_{y_i}, p_{x_i}, p_{y_i}) & \longrightarrow & (q_{x_o}, q_{y_o}, p_{x_o}, p_{y_o}) \\ & symplectic\ transformation & \end{array} \qquad (4.2.6)$$

The symplecticity of the transformation (4.2.6) is a straightforward consequence of the structure of eqs. (4.2.1), ruling the ray propagation.

To put the concept of symplecticity into more formal terms, we consider the 4×4 unit symplectic matrix $\mathbf{J}$ given as

$$\mathbf{J} = \begin{pmatrix} 0 & \mathbf{I} \\ -\mathbf{I} & 0 \end{pmatrix} \qquad (4.2.7)$$

where $\mathbf{I}$ is the 2×2 unit matrix. The above writing is a quite general convention of representing $2n \times 2n$ matrices, which are indeed partitioned into four $n \times n$ submatrices.

Furthermore, denoting the components of the phase-space vectors by indices ranging from 1 to 4, according to

$$\mathbf{u} = \begin{pmatrix} q_x \\ q_y \\ p_x \\ p_y \end{pmatrix} = \begin{pmatrix} u_1 \\ u_2 \\ u_3 \\ u_4 \end{pmatrix}, \qquad (4.2.8)$$

we can write eqs. (4.2.1) in the form

$$\frac{du_j}{dz} = J_{jl}\frac{\partial H}{\partial u_l}, \qquad j,l = 1,..,4 \qquad (4.2.9)$$

where the summation over the repeated indices is understood. Transforming $\mathbf{u}$ into $\mathbf{v}$ with

$$M_{ij} \equiv \frac{\partial v_i}{\partial u_j} \qquad (4.2.10)$$

we easily get the equations for $\mathbf{v}$

$$\frac{dv_j}{dz} = M_{jl}J_{lk}M_{hk}\frac{\partial H}{\partial v_h} \qquad (4.2.11)$$

Requiring that the transformation (4.2.11) leave the form of the Hamilton's equations invariant, we end up with the matrix relation

$$\mathbf{MJM^T = J} \qquad (4.2.12)$$

which expresses the condition for $\mathbf{M}$ to be symplectic. It is easy to prove that if $\mathbf{M}$ satisfies the relation (4.2.12) also $\mathbf{M}^{-1}$ does, thus allowing to rewrite the symplecticity condition in the equivalent form

$$\mathbf{M^T J M = J} \qquad (4.2.13)$$

Let us note that $\mathbf{M}$ as introduced in (4.2.9) is the Jacobian matrix relevant to the transformation $\mathbf{u} \to \mathbf{v}$.

It is an immediate consequence of (4.2.12-4.2.13) that

$$\det \mathbf{M} = \pm 1 \qquad (4.2.14)$$

being $\det \mathbf{J} = 1$ and $\det \mathbf{M} = \det \mathbf{M^T}$.

The ambiguity on the sign in (4.2.14) can be solved by noting that in a transformation elements of volume are linked through the Jacobian of the transformation (i.e. $\det \mathbf{M}$) as $dV_i = \det \mathbf{M}\, dV_o$. The Liouville theorem, according to which $dV_i = dV_o$, assures that $\det \mathbf{M} = +1$.

For a more detailed discussion on symplectic groups as $Sp(4,R)$ and $Sp(2,R)$, which have relevance in linear and Gaussian optics respectively, the reader should consult the suggested references.

The optical phase space 127

After these formal comments, we turn now to the problem of geometrical optics to determine the transformation (4.2.5) of light-ray coordinates generated by the Hamiltonian (4.2.1) through eqs. (4.2.3).
For future convenience we rewrite eqs. (4.2.3) as

$$\frac{d}{dz}\mathbf{u}(z) = -\{H(q_x, q_y, p_x, p_y, z), \mathbf{u}(z)\} \qquad (4.2.15)$$

in terms of Poisson brackets, whose expression for any arbitrary functions f and g is known to be

$$\{f, g\} = \sum_\alpha \left(\frac{\partial f}{\partial q_\alpha}\frac{\partial g}{\partial p_\alpha} - \frac{\partial f}{\partial p_\alpha}\frac{\partial g}{\partial q_\alpha}\right) \qquad (4.2.16)$$

with the index α running over the degrees of freedom ($\alpha = x, y$ in the case we are considering).
In particular,

$$\{q_\alpha, q_\beta\} = \{p_\alpha, p_\beta\} = 0; \quad \{q_\alpha, p_\beta\} = \delta_{\alpha\beta}, \quad \alpha, \beta = x, y \qquad (4.2.17)$$

Let us recall the properties of Poisson brackets, which can be synthesized into the following

$$\begin{aligned}
\textit{antisymmetry}: & \quad \{f, g\} = -\{g, f\} \\
\textit{linearity}: & \quad \{f, ag + bh\} = a\{f, g\} + b\{f, h\} \\
\textit{derivation property}: & \quad \{f, gh\} = \{f, g\}h + g\{f, h\} \\
\textit{Jacobi identity}: & \quad \{f, \{g, h\}\} = \{\{f, g\}, h\} + \{g, \{f, h\}\}
\end{aligned} \qquad (4.2.18)$$

for any complex number a and b.
For any space of smooth functions $\Upsilon(V)$ on a finite-dimensional vector space V, we consider the operators $:f:$ defined as the left Poisson bracket by f in $\Upsilon(V)$; in symbols

$$:f: \equiv \{f, \cdot\} = \frac{\partial f}{\partial q}\frac{\partial}{\partial p} - \frac{\partial f}{\partial p}\frac{\partial}{\partial q} \qquad (4.2.19)$$

It can be easily proved that the operators (4.2.19), with $f \in \Upsilon(V)$, span a Lie algebra, say $T(V)$, with respect to the commutation between operators. In fact, as a straightforward consequence of the properties of Poisson brackets, the Lie product

$$[: f :, : g :] \equiv : f :: g : - : g :: f := \{f, g\}:, \qquad (4.2.20)$$

is antisymmetric, bilinear and obeys the Jacobi identity, that are the basic requisites for the linear space $T(V) = \{: f :, f \in \Upsilon(V)\}$ being a Lie algebra.

Accordingly, the Hamilton's eqs. (4.2.15) can be further recast in the operatorial form

$$\frac{d}{dz}\mathbf{u}(z) = -:H:\mathbf{u}(z) \qquad (4.2.21)$$

where $:H:$ is the operator associated with the Hamiltonian (4.2.1) under Poisson bracket, as prescribed by (4.2.19).

As is well known, due to the paramount relevance of equations of the type (4.2.3) or (4.2.21) in several and differing fields of physics, a large variety of both analytical and numerical methods have been devised to integrate them, which methods provide formal expressions to the exact or approximate solutions and define general procedures for numerical algorithms with defined degree of accuracy.

We will try to avoid the formal complexity of a fully general treatment, then we confine ourselves to linear optics, although the validity of the procedure we describe is not limited to linear optics. The methods we illustrate remain indeed conceptually valid within the wider context of geometrical optics, representing the general approach to the integration of eqs.(4.2.3) and (4.2.21), but the application of them to linear (and gaussian) optics is greatly simplified by the inherent paraxial approximation (and rotational symmetry), thus allowing to obtain nice results with nearly a *null* cost in terms of the formal machinery involved in.

As already noted, the paraxial propagation is the basic approximation of linear (or first-order) optics, which amounts to assuming that rays are confined close to the axis of the optical system. Thereby, only first-order expressions in the ray variables (q, p) are retained, second or higher powers being ignored. The circular functions of angles, for instance, are approximated as $\cos\vartheta \simeq 1$ and $\sin\vartheta \simeq \tan\vartheta \simeq \vartheta$. In explicit terms, it is assumed that

$$|x|, |x'| \ll 1 \quad \text{and} \quad |y|, |y'| \ll 1 \qquad (4.2.22)$$

Then, since the ray propagates nearly parallel to the optical axis, $ds \simeq dz$ and hence the optical direction cosines specify as

$$\begin{array}{l} p_x = n\frac{dx}{ds} \simeq \frac{dx}{dz} \equiv nx' \\ p_y = n\frac{dy}{ds} \simeq \frac{dy}{dz} \equiv ny' \end{array} \qquad (4.2.23)$$

and the slopes $x' = \frac{dx}{dz}$, $y' = \frac{dy}{dz}$ of the ray trajectory relative to the optical axis can be further approximated by the angles the ray makes with the $q_y - z$ and $q_x - z$ planes (fig. 4 and eqs.(4.2.4)).

The phase space appropriate to linear optics comprises therefore the 4-component vectors, specifying the intersections and the angles of the ray relative to some reference plane.

In accordance with the paraxial approximation the refractive index is taken to vary slightly along the transverse directions. Then, we expand it with respect to q_x, q_y and retain only terms up to second order as

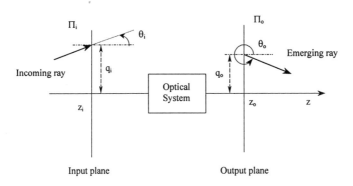

Figure 4   Ray-coordinates in the gaussian approximation.

$$n(q_x, q_y, z) \simeq n_0(z) - \frac{1}{2}n_{2x}(z)q_x^2 - \frac{1}{2}n_{2y}(z)q_y^2 - n_{2xy}(z)q_x q_y \quad (4.2.24)$$

where $n_0(z)$ is the refractive index along the z-axis, i.e.

$$n_0(z) = n(0, 0, z) \quad (4.2.25)$$

and the linear terms vanish according to the assumed axisimmetry.

The optical Hamiltonian becomes quadratic in the ray-coordinates according to

$$H(q_x, q_y, p_x, p_y, z) = \frac{1}{2n_0}(p_x^2 + p_y^2) + \frac{1}{2}n_{2x}q_x^2 + \frac{1}{2}n_{2y}q_y^2 + n_{2xy}q_x q_y - n_0 \quad (4.2.26)$$

showing a close correspondence — apart the unessential additive term n_0— with the Hamiltonian of two coupled harmonic oscillators.

A further simplification is allowed by the additional assumption of rotational symmetry as it is in most optical systems.

We consider rays which are coplanar with the optical axis, the so called *meridional rays*. Hence, by rotational symmetry, it is clearly sufficient to restrict our attention to rays lying in one fixed plane containing the z-axis. The optical phase space appropriate to gaussian optics in then the 2-dimensional linear space formed by vectors $\mathbf{u} = \begin{pmatrix} q \\ p \end{pmatrix}$, specifying at the chosen reference plane the ray height q

above the z-axis and the ray *reduced slope* $p = n_0 \vartheta$, the on-axis value of the refractive index times the angle ϑ of the ray relative to the optical axis (fig. 4). According to the usual practice, the angle ϑ is positive if a counterclockwise rotation carries the positive z-direction into the direction of the ray.

The optical Hamiltonian appropriate to Gaussian optics is therefore

$$H(q,p,z) = \frac{1}{2n_0(z)}p^2 + \frac{1}{2}n_2(z)q^2 \qquad (4.2.27)$$

resembling the classical Hamiltonian of the one-degree of freedom harmonic oscillator.

A closer look at the ray equation (4.2.21) and the Hamiltonian (4.2.27), reveals the possibility of rewriting (4.2.21) into the nicely compact matrix form

$$\frac{d}{dz}\begin{pmatrix} q \\ p \end{pmatrix}(z) = \begin{pmatrix} 0 & \frac{1}{n_0(z)} \\ -n_2(z) & 0 \end{pmatrix}\begin{pmatrix} q \\ p \end{pmatrix}(z) \qquad (4.2.28)$$

where the 2×2 matrix can be regarded as the matrix representation of the operator $-:H:$

$$-:H:=-\mathbf{H} = \begin{pmatrix} 0 & \frac{1}{n_0(z)} \\ -n_2(z) & 0 \end{pmatrix} \qquad (4.2.29)$$

In accordance with the paraxial approximation, the input-output relations implicit in eqs. (4.2.5) can be written in the explicit linear form

$$\begin{pmatrix} q_o \\ p_o \end{pmatrix} = \begin{pmatrix} A & B \\ C & D \end{pmatrix}\begin{pmatrix} q_i \\ p_i \end{pmatrix} \qquad (4.2.30)$$

which can be understood as the first-order expansion of the general relations (4.2.5) with respect to the ray-variables.

The 2×2 matrix

$$\mathbf{M}(z_o, z_i) = \begin{pmatrix} A & B \\ C & D \end{pmatrix} \qquad (4.2.31)$$

whose entries are in general functions of the reference coordinates z_o and z_i, is know as the *ray transfer matrix*.

A direct calculation shows that as the Jacobian of the transformation $\begin{pmatrix} q_i \\ p_i \end{pmatrix} \rightarrow \begin{pmatrix} q_o \\ p_o \end{pmatrix}$ is unit, the determinant of $\mathbf{M}$ equals 1: $\det \mathbf{M} = \mathbf{I}$. For a 2×2 matrix this represents the necessary and sufficient condition that the matrix be symplectic.

Gaussian optics deals with the determination of the ray matrix relative to arbitrary optical systems, which according to (4.2.30) fully characterizes the paraxial propagation of light rays through the system.

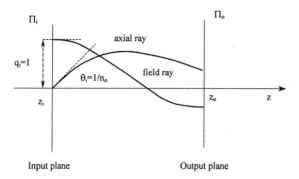

Figure 5 Cos-like and sin-like solutions to the paraxial ray-equation.

Then, inserting the input-output relation (4.2.30) into the ray equation (4.2.28), one gets the equation for $\mathbf{M}(z, z_i)$, regarded as function of z:

$$\frac{d}{dz}\mathbf{M}(z, z_i) = -\mathbf{H}(z)\mathbf{M}(z, z_i) \qquad (4.2.32)$$

with the obvious initial condition $\mathbf{M}(z_i, z_i) = \mathbf{1}$, the identity matrix.

Equation (4.2.32) can be approached in different ways. According to the explicit form (4.2.31) for $\mathbf{M}$ and (4.2.27) for $\mathbf{H}$, eq. (4.2.32) provides the set of first order differential equations obeyed by the entries A, B, C and D, which separate into two sets for each column; explicitly

$$\begin{array}{ll} \frac{dA}{dz} = \frac{C}{n_0} & \\ \frac{dC}{dz} - n_2 A & A(z_i) = 1, \ C(z_i) = 0 \\ \frac{dB}{dz} = \frac{D}{n_0} & \\ \frac{dD}{dz} - n_2 B & B(z_i) = 0, \ D(z_i) = 1 \end{array} \qquad (4.2.33)$$

It is evident that the entries A, C as B, D separately obey the ray eq. (4.2.28). They represent two particular solutions to the ray-equation corresponding to different initial conditions. Thus, the ray $\begin{pmatrix} A \\ C \end{pmatrix}$ is a paraxial ray which leaves the input plane z_i parallel to the axis; it is called *field ray*. On the other hand, the solution $\begin{pmatrix} B \\ D \end{pmatrix}$ represents a paraxial ray that has null height on the input plane z_i (see fig. 5); it is called *axial ray*.

In this connection, the input-output relations (4.2.30) state that the general solution to the ray equation is obtained as linear combination of axial and field rays with coefficients fixed by the initial values (q_i, p_i).

An alternative approach to eq. (4.2.32) takes advantages of the group theoretical methods, which allow to represent **M** as the exponential function of **H** in a small neighbor of z_i. Taking as valid the exponential representation of **M** over all the involved z-interval, we write

$$\mathbf{M}(z_o, z_i) = \exp\left\{-\mathbf{H}(z_o - z_i)\right\} \tag{4.2.34}$$

which translates into the language of group theory the circumstance that the Hamiltonian (4.2.29) generates the transformation (4.2.30) of the ray coordinates from z_i to z_o.

Strictly speaking, the expression (4.2.34) is correct only if **H** does not depend on z. In general more involved expressions should be assigned to **M**, accounting for both the *chronological* (with respect to z) ordering, arising when $\mathbf{H}(z)$ does not commute with itself at any z_1 and z_2, and the symplecticity of **M**. For a detailed analysis of the topic, which is out of the purposes of the present book, the reader is referred to the suggested references. However, let us stress that as the exponential (4.2.34) is still a 2×2 matrix as in (4.2.31), the procedure above leading to eqs. (4.2.33) provides a practicable technique to solve the matrix eq. (4.2.32) on account of both the z-ordering and the symplecticity of **M**.

Gaussian optics is practically all contained into eq. (4.2.30), which defines any arbitrary optical system as a black box fully characterized by the transfer matrix **M**, and into eqs.(4.2.32) and (4.2.34), which trace the direction to determine the optical matrix **M** once specified the geometry and optical properties of the system, i.e. n.

In the next section we consider in detail the basic optical systems within the framework of gaussian optics. We derive the relevant ray-matrices and analyze their effect on single light rays in real space, deserving the analysis of the behaviour of a collection of rays in section 4, where optical systems act on wave distributions.

4.3 Quadratic Hamiltonian and optical ray matrices

We apply the technique envisaged in the previous section to construct the optical matrices, describing the transformations acted in real space by Gaussian optical systems on light rays. The basic ingredient of the strategy is

- the optical Hamiltonian $H(q, p, z)$ corresponding to the system under study, which within the context of gaussian optics is quadratic in the ray-coordinates;

whereas the main tools are

- the Poisson brackets, which allow to work out the Lie operator $: H :$ and the associated matrix **H**

and

- the operation of exponentiating : H : or $\mathbf{H}$, that yields the optical propagator, or more properly within the formalism of Gaussian optics, the optical transfer matrix form z_i to z_o.

In most cases we consider z-independent Hamiltonians, so that the writing (4.2.34) is meaningful. When considering z-depending Hamiltonian, we will apply in a quite naïve fashion ordering and factorization procedures, which are rigorously discussed in the quoted references.

We consider therefore as basic Hamiltonians the quadratic polynomials $\frac{1}{2}q^2$, $\frac{1}{2}p^2$ and $\frac{1}{2}qp$; any bilinear function of q and p can be expressed as linear combination of the above polynomials. We work out the relevant Lie operators and Lie matrices via Poisson bracket and then analyze the corresponding transformations $\exp\{-:H:\Delta z\}$ or $\exp\{-\mathbf{H}\Delta z\}$ in order to verify if they may describe constructible optical systems.

Denoting by $\hat{K}_+, \hat{K}_-$ and $\hat{K}_3$ the Lie operators associated with the above quadratic polynomials under Poisson bracket, we can write

$$\hat{K}_+ \equiv\; :\tfrac{1}{2}q^2: = q\tfrac{\partial}{\partial p}$$
$$\hat{K}_- \equiv\; :\tfrac{1}{2}p^2: = -p\tfrac{\partial}{\partial q} \qquad (4.3.1)$$
$$\hat{K}_3 \equiv\; :\tfrac{1}{2}qp: = \tfrac{1}{2}\left(p\tfrac{\partial}{\partial p} - q\tfrac{\partial}{\partial q}\right)$$

which correspond to the matrices

$$\mathbf{K}_+ = \begin{pmatrix} 0 & 0 \\ 1 & 0 \end{pmatrix}, \quad \mathbf{K}_- = \begin{pmatrix} 0 & -1 \\ 0 & 0 \end{pmatrix}, \quad \mathbf{K}_3 = \frac{1}{2}\begin{pmatrix} -1 & 0 \\ 0 & 1 \end{pmatrix} \qquad (4.3.2)$$

It is worth noting that the above matrices obey the relations of commutation

$$[\mathbf{K}_+, \mathbf{K}_-] = -2\mathbf{K}_3, \quad [\mathbf{K}_3, \mathbf{K}_\pm] = \pm\mathbf{K}_\pm \qquad (4.3.3)$$

that characterize the algebra $sp(2, R)$ of the generators of the symplectic group $Sp(2, R)$ [†].

We consider now the optical transformations generated by the operators (4.3.1). Since $\frac{1}{2}p^2$ is just the Hamiltonian of free evolution in free space we expect the system $e^{a\mathbf{K}_-}$ for any a real to act on the ray, changing its position but leaving the optical momentum unchanged; in other words $e^{a\mathbf{K}_-}$ describes the translation in real space. Similarly, $\frac{1}{2}q^2$ accounts for free propagation in p-space. Then the system $e^{b\mathbf{K}_+}$ for any b real leaves unchanged the q-coordinate but translates the momentum: $e^{b\mathbf{K}_+}$ describes translations in the spatial frequency space.

[†]It is worth stressing that as the 2×2 Hamiltonian matrix $\mathbf{H}$ has null trace only three of the four entries are linearly independent, in perfect accordance with the dimensionality of the algebra $sp(2, R)$, associated with the three quadratic polynomials.

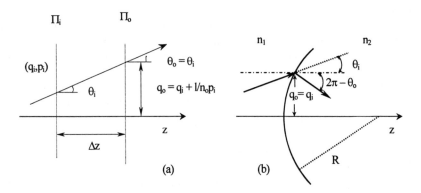

Figure 6 The basic elementary optical systems: (a) free-space section with uniform refractive index n_0 and (b) refracting surface between regions of differing refractive index with curvature radius R.

We specify the two transformations in more formal terms. By direct exponentiation of the matrix $\mathbf{K}_-$ in (4.3.2) or the operator $\hat{K}_-$ in (4.3.1) — noting that $\frac{\partial}{\partial q}$ generates shifts with respect to q — we easily get

$$e^{-\frac{\Delta z}{n_0}\mathbf{K}_-}\begin{pmatrix} q \\ p \end{pmatrix} = \begin{pmatrix} q + \frac{\Delta z}{n_0} \\ p \end{pmatrix} = \begin{pmatrix} 1 & \frac{\Delta z}{n_0} \\ 0 & 1 \end{pmatrix}\begin{pmatrix} q \\ p \end{pmatrix} \qquad (4.3.4)$$

which describes a translation, where the ray continues to travel in a straight line between two reference planes lying in the same medium with refractive index n_0 and separated by the distance Δz (fig. 6a).

Similarly, by direct exponentiation of $\mathbf{K}_+$ or $\hat{K}_+$-noting now that $\frac{\partial}{\partial p}$ generates shifts in the p-coordinate- we have

$$e^{-n_2\Delta z \mathbf{K}_+}\begin{pmatrix} q \\ p \end{pmatrix} = \begin{pmatrix} q \\ p - n_2\Delta z q \end{pmatrix} = \begin{pmatrix} 1 & 0 \\ -n_2\Delta z & 1 \end{pmatrix}\begin{pmatrix} q \\ p \end{pmatrix} \qquad (4.3.5)$$

which describes the refraction at the boundary surface between two regions of differing refractive index, the two reference planes being taken immediately to the left and immediately to the right of the surface, respectively (fig. 6b). The nonzero off-diagonal entry in the matrix (4.3.5) is just minus the surface *refracting power*: $\mathsf{P} = \frac{\Delta n}{R}$, if Δn represents the difference between the refractive indices at the output and input planes and R specifies the curvature of the surface.

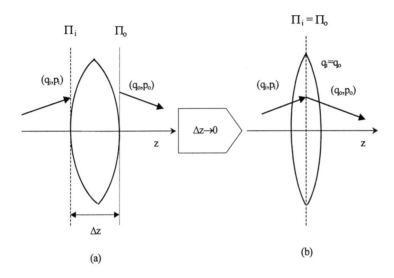

Figure 7 Ray geometry relative to (a) a thick lens and (b) a thin lens.

Equivalently, the ray matrix in (4.3.5) may represent a thin lens with focal length $f = \frac{1}{n_2 \Delta z}$. We recall that the so called *thin lens* consists of two refracting surfaces with negligible separation in between, thus a ray incident at a given point on one surface emerges at the same height from the other surface. In this case, the reference planes z_i and z_o can be both conveniently taken to coincide with the plane of the lens, although the plane z_i relates to incident rays whilst the plane z_o relates to rays emerging from the lens (fig.7b).

It is needless to say that the optical systems described by (4.3.4) and (4.3.5) represent the basic elementary systems of Gaussian optics: free-space sections and refracting surfaces or thin lenses. Any other optical system can be built combining into appropriate products the operators and then the matrices relative to free sections and refracting surfaces or thin lenses. The ray propagation through a thick lens, for instance, may be understood as the following sequence: 1) refraction at the front surface, 2) free propagation through the lens and 3) another refraction at the back surface (see fig. 7a) and Problem 2).

To prove the generality of this assertion, we rewrite the optical matrices, that according to (4.3.4) and (4.3.5) account for the transformation acted on the ray-coordinates by free-space section of length d and thin lens with focal length f [‡]

[‡]We prefer to consider thin lenses rather than refracting surfaces as elementary optical systems;

$$\mathbf{T}(d) = \begin{pmatrix} 1 & d \\ 0 & 1 \end{pmatrix}, \quad \mathbf{L}(f) = \begin{pmatrix} 1 & 0 \\ -\frac{1}{f} & 1 \end{pmatrix} \tag{4.3.6}$$

where d should be more properly understood as the reduced length $d = \frac{\Delta z}{n_0}$, if Δz is the length of the section and n_0 the refractive index. In the following, unless otherwise specified, we will implicitly refer to reduced distances and thin lenses, when talking of diffraction sections and lenses.

Both matrices (4.3.6) have determinant equal to unity: $\det \mathbf{T} = 1$ and $\det \mathbf{L} = 1$, thus confirming that both are symplectic and hence according to (4.2.13): $\mathbf{T}^T \mathbf{JT} = \mathbf{J}$ and $\mathbf{L}^T \mathbf{JL} = \mathbf{J}$.

To prove that any optical system can be realized as an appropriate sequence of elementary optical transformations, we prove that any 2×2 symplectic matrix $\mathbf{M} = \begin{pmatrix} A & B \\ C & D \end{pmatrix}$ can be obtained as product of $\mathbf{T}$ and $\mathbf{L}$-like matrices. If $C \neq 0$ we can factorize $\mathbf{M}$ as

$$\begin{pmatrix} A & B \\ C & D \end{pmatrix} = \mathbf{T}(d_2) \mathbf{L}(\frac{1}{C}) \mathbf{T}(d_1) \tag{4.3.7}$$

which on account of (4.3.6) fixes the lengths d_1 and d_2 as

$$1 + d_1 C = A, \quad 1 + d_2 C = D \tag{4.3.8}$$

On the other hand, if $C = 0$ we can arrange the basic systems (4.3.6) in the sequence

$$\begin{pmatrix} A & B \\ 0 & D \end{pmatrix} = \mathbf{L}(f_2) \mathbf{T}(d) \mathbf{L}(f_1) \tag{4.3.9}$$

and choose $d = f_1 + f_2 = B$ and $A = -\frac{f_2}{f_1}$ or $f_1 = \frac{B}{1-A}$ and $f_2 = -\frac{AB}{1-A}$.

Thus any 2×2 symplectic matrix can arise as an optical matrix. In addition, any optical system can be realized by a suitable sequence of straight sections and thin lenses, which thereby stand out as basic optical systems.

In formal terms, we proved the isomorphism between the symplectic group $Sp(2, R)$ and the group of gaussian optical systems.

The above is a general assertion holding for linear optical systems, which are described by 4×4 symplectic matrices; the inherent group is indeed $Sp(4, R)$. In that case, the $\mathbf{T}$ and $\mathbf{L}$-like matrices are given as

$$\mathbf{T}(d) = \begin{pmatrix} \mathbf{I} & d\mathbf{I} \\ 0 & \mathbf{I} \end{pmatrix}, \quad \mathbf{L}(\mathbf{F}) = \begin{pmatrix} \mathbf{I} & 0 \\ -\mathbf{F} & \mathbf{I} \end{pmatrix} \tag{4.3.10}$$

where $\mathbf{I}$ is the 2×2 unit matrix and $\mathbf{F} = \mathbf{F}^T$ is a symmetric matrix.

---

in practice the difference is only verbal as the corresponding ray-matrices have the same form.

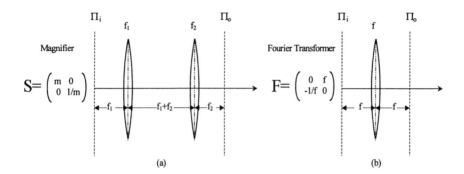

Figure 8   Optical configurations of (a) a magnifier and (b) a Fourier transformer.

We consider now the optical system e^{-cK_3}. As K_3 is composed by two commuting operators, we can decouple the exponentials, thus obtaining the scaling matrix

$$e^{-cK_3}\begin{pmatrix} q \\ p \end{pmatrix} = \begin{pmatrix} m & 0 \\ 0 & \frac{1}{m} \end{pmatrix}\begin{pmatrix} q \\ p \end{pmatrix} \qquad (4.3.11)$$

where the scale factor $m \equiv e^{c/2}$ is called the *magnification*. The system in (4.3.11) works as a magnifier, producing a magnification of the ray-height: $q_o = mq_i$ and conversely a demagnification by the same factor of the ray reduced slope: $p_o = \frac{1}{m}p_i$.

According to the previous discussion, it is readily verified that the magnifier (4.3.11) can be designed as the following sequence (fig.8a))

$$S(m) = T(f_2)L(f_2)T(f_2)T(f_1)L(f_1)T(f_1) \qquad (4.3.12)$$

where f_1 and f_2 are linked through the relation $f_1 m = -f_2$.

As a final comment, we note that the sequence (fig.8b))

$$F(f) \equiv T(f)L(f)T(f) = \begin{pmatrix} 0 & f \\ -\frac{1}{f} & 0 \end{pmatrix} \qquad (4.3.13)$$

entering the above arrangement realizes a Fourier transforming optical system (i.e. the spatial frequencies of the input signal are displayed on the output plane).

With the language of geometrical optics, the system (4.3.13) connects the focal planes of the lens f. Therefore, parallel rays emerge to pass through the same point in the back focal plane, as incident rays passing through a given point in the front focal plane are imaged as parallel rays.

We consider now the ray propagation along a lens-like medium, described by the square-law

138   Optical systems in the geometrical and wave optics framework

$$n(q,z) = n_0 - \frac{1}{2}n_2 q^2 \qquad (4.3.14)$$

with n_0 and n_2 constant.

The relevant optical matrix is obtained by means of the previously described strategy, which provides the set of equations (4.2.33) for the entries A, B, C, D. It is an easy matter to solve eqs. (4.2.33) in the case where n_0 and n_2 are z-independent. We obtain that the field ray $\begin{pmatrix} A \\ C \end{pmatrix}$ is represented by the cosine solution to the harmonic equation $\frac{d^2}{dz^2}y + \frac{n_2}{n_0}y = 0$, as the axial ray $\begin{pmatrix} B \\ D \end{pmatrix}$ is just the sine-solution. Thus, the optical matrix between the planes z_i and z_o is

$$\mathbf{M}(z_o, z_i) = \begin{pmatrix} \cos\zeta & \sqrt{n_0 n_2}\sin\zeta \\ -\frac{1}{\sqrt{n_0 n_2}}\sin\zeta & \cos\zeta \end{pmatrix}, \quad \zeta \equiv \sqrt{\frac{n_2}{n_0}}(z_o - z_i) \qquad (4.3.15)$$

It is evident that for a focusing index profile, i.e. $n_2 > 0$, the above matrix produces an oscillatory path of the ray around the z axis, while for a defocusing profile: $n_2 < 0$, the circular functions are replaced by hyperbolic sine and cosine and hence the matrix (4.3.15) gives rise to a progressive increase of the ray height above the optical axis. The property of the ray to remain close to the optical axis or to escape far away from it characterizes the *stability* of the optical system. It can be proved that a stability condition can be inferred, which requires $|Tr\mathbf{M}| = |A + D| \leq 2$.

It is worth considering the role of the magnifier (4.3.11) within the realm of gaussian optical systems. In this connection, we note that besides the realizations (4.3.7) and (4.3.9), appropriate embedding of $\mathbf{T}, \mathbf{L}$ and $\mathbf{S}$-like matrices can be arranged to be equivalent to any arbitrary optical system. In formal terms, the following decomposition can be realized

$$\begin{pmatrix} A & B \\ C & D \end{pmatrix} = \mathbf{S}(m)\mathbf{L}(f)\mathbf{T}(d) \qquad (4.3.16)$$

which gives for the parameters m, f and d the relations

$$m = A \quad f = -\frac{1}{AC} \quad d = \frac{B}{A} \qquad (4.3.17)$$

It is evident that the order of the matrices in the product (4.3.16) can be changed, providing optical systems with different components and geometries but equivalent as far as ray-tracing is concerned (Problem 3).

The relation (4.3.16) states in optical terms the property of the matrices $\mathbf{K}_+, \mathbf{K}_-, \mathbf{K}_3$ of being the generators of the group $Sp(2, R)$. Accordingly, any operator in $Sp(2, R)$ and hence any optical matrix (4.2.31) can be expressed in form of the product

$$\mathbf{M}(z) = e^{-c(z)\mathbf{K}_3} e^{-f(z)\mathbf{K}_+} e^{-d(z)\mathbf{K}_-} \qquad (4.3.18)$$

The magnification factor $m(z) = e^{c/2}$, the focal length $f(z)$ and the reduced distance $d(z)$ in the above factorization as well as in the others obtainable changing the order of the operators in the product can be obtained from relations of the type (4.3.16) or directly from the equations of motion (4.2.33) for the specific optical Hamiltonian (Problem 4). The interested reader may consult the relevant bibliography for a more detailed discussion.

4.4 Quadratic Hamiltonians and optical propagators

We discuss now some concepts and methods of wave optics. The treatment presented below has the character of a cursory discussion rather than that of a detailed formalistic analysis.

Geometrical and wave optics are intimately connected; the former can be obtained from the latter in the limit of very short wavelengths. Similarly, wave optics can be regarded as quantized geometrical optics. A *quantum* theory of light rays has been formulated, applying the same rules for transition from classical to quantum mechanics, according to which variables are replaced by linear operators.

As previously noted, geometrical optics is built on the concept of *ray of light*, whose trajectory is determined solving the ray-equation. Correspondingly, wave optics is based on the concept of optical wave function $\varphi(\vec{r}, z)$, whose behaviour is ruled by the scalar wave equation or its paraxial version. In wave optics the concept of a *ray* is not all elementary; its place is taken by the simplest solution to the wave equation: the plane monochromatic wave.

In addition, in linear ray optics an optical system is treated as a *black box* with an input plane and an output plane and an assigned 2×2 or 4×4 symplectic ray-matrix $\mathbf{M}$. Thus, the input-output relationships are formulated in matrix form (section 3)

$$\begin{pmatrix} q_o \\ p_o \end{pmatrix} = \begin{pmatrix} A & B \\ C & D \end{pmatrix} \begin{pmatrix} q_i \\ p_i \end{pmatrix} \qquad (4.4.1)$$

providing the ray variables at the output plane as linear functions of the ray coordinates at the input plane with coefficients determined by the geometry and optical features of the system (fig.9).

In wave optics, an optical system realizes still a mapping of a set of functions given on the input plane into a set of functions on the output plane. The tool of such a mapping is not the ray matrix as in (4.4.1) but the *optical transfer* (or *response*) *function* $\mathbf{g}(\vec{r}_o, \vec{r}_i; z_o, z_i)$, which can be calculated from the design data of the optical system. The input-output relations are indeed written in terms of an integral transform involving the input optical signal as well the response function of the system:

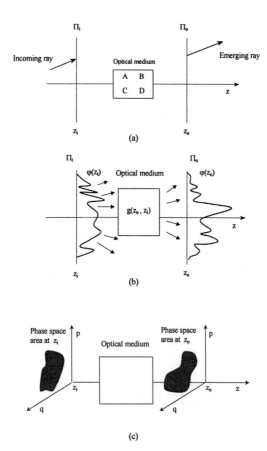

Figure 9 Optical systems: (a) ray optics, (b) wave optics, (c) phase space picture.

Quadratic Hamiltonians and optical propagators   141

$$\varphi(\vec{r}_o, z_o) = \int d\vec{r}_i \, g(\vec{r}_o, \vec{r}_i; z_o, z_i)\varphi(\vec{r}_i, z_i) \qquad (4.4.2)$$

We note that the vectors $\vec{r}_i$ and $\vec{r}_o$ in the above represent the space coordinates in the input and output planes z_i and z_o, i.e. $\vec{r} \equiv (q_x, q_y)$. Moreover, time t does not explicitly appear into eq. (4.4.2); this is because the functions φ are intended as the complex-valued amplitudes of a monochromatic component of the wave field at some temporal frequency ω.

It is worth noting that the transfer function $g(\vec{r}_o, \vec{r}_i; z_o, z_i)$ is also called *point-spread function* as it is the response of the system to the input signal $\varphi(\vec{r}, z_i) = \delta(\vec{r} - \vec{r}_i)$, which represents a point source.

It will be shown below that in the paraxial approximation the point-spread function can be expressed in terms of the entries A, B and D of the optical matrix of ray optics, thus providing a link between the two pictures (4.4.1) and (4.4.2) of an optical system.

We do not retrace the steps of the quantization procedure of geometrical optics, leading to the basic equation of wave optics, i.e. the scalar wave equation and then its paraxial version, but simply assume the results, which allow to identify a strategy to study the wave propagation through a nonhomogeneous medium, in other words to determine the response function of the medium. The discussion will rephrase that put forward for linear ray optics in section 3. A *recipe* can be suggested within the present context as well, whose basic *ingredient* is just as in ray optics

• the optical Hamiltonian H of ray optics, accounting for the optical properties of the medium or the optical system.

The main tools are

• the correspondence rule, that allows to obtain the optical Hamiltonian operator $\hat{H}$ associated to the classical function H. Then, the ray variables are replaced by the linear operators according to

$$q \to \hat{q} \qquad p \to \hat{p} = -\frac{i}{k}\frac{\partial}{\partial q} \qquad (4.4.3)$$

$$\{q, p\} = 1 \to \left[\hat{q}, \hat{p}\right] = \frac{i}{k} = i\,\lambda$$

where the inverse of the wavenumber $\lambda = \frac{1}{k} = \frac{\lambda}{2\pi}$ plays the role of the reduced Planck's constant $\hbar$.

• the operation of exponentiating the Hamiltonian operator $\hat{H}(\hat{q}, \hat{p}, z)$ to form the optical propagator $\hat{M}(z_o, z_i)$. Accordingly, we write eq. (4.2.34) in the operator form proper to wave optics as

$$\hat{M}(z_o, z_i) = \exp\left\{-\frac{i}{\lambda}\hat{H}(z_o - z_i)\right\} \qquad (4.4.4)$$

142   Optical systems in the geometrical and wave optics framework

The role of $\hat{M}$ in wave optics is that of propagate the optical signal from the input plane z_i to the output plane z_o; in symbols

$$\varphi(\vec{r}_o, z_o) = \hat{M}(z_o, z_i)\varphi(\vec{r}_i, z_i) \qquad (4.4.5)$$

which is the input-output relation (4.4.2) rewritten in terms of the optical propagator.

We consider the quadratic Hamiltonians of section 3, then according to (4.4.3) the corresponding operators -including the factor ik- are straightforwardly written as

$$H = \frac{1}{2}q^2 \quad \to \quad \hat{K}_+ = ik\frac{1}{2}\hat{q}^2 \qquad (4.4.6)$$

$$H = \frac{1}{2}p^2 \quad \to \quad \hat{K}_- = -i\frac{1}{2k}\frac{\partial^2}{\partial q^2}$$

$$H = \frac{1}{4}(qp + pq) \quad \to \quad \hat{K}_3 = \frac{1}{2}(q\frac{\partial}{\partial q} + \frac{1}{2})$$

where we used the same notation as in section 3 for the operator (4.3.1). The operators (4.4.6) obey the same rules of commutation as (4.3.3); in fact, they are the generators of the metaplectic group $Mp(2, R)$, which is homomorphic to the symplectic group $Sp(2, R)$.

It is worth stressing that to form the third operator in the table (4.4.6) we symmetrized the classical polynomial $\frac{1}{2}qp$ to take account of that $\hat{q}$ and $\hat{p}$ are non commuting operators. For $\hat{q}$ and $\hat{p}$ c-numbers, we recover the bilinear product considered in (4.3.1).

Inspecting the action of the optical propagators generated by $\hat{K}_\pm, \hat{K}_3$ through (4.4.4) on a given signal, we can infer the response function of the relevant optical system.

According to the results of section 3, the propagator

$$\hat{L}(f) = e^{-i\frac{k}{2f}\hat{q}^2} \qquad (4.4.7)$$

describes the effect of a thin lens with focal length f. The operator $\hat{L}(f)$ produces a simple multiplication by the phase factor $\exp\left(-i\frac{k}{2f}q^2\right)$, thereby the complex amplitudes $\varphi(q, z_i)$ and $\varphi(q, z_o)$ of the waves incident to and emerging from the lens are linked as

$$\varphi(q, z_o) = e^{-i\frac{k}{2f}q^2}\varphi(q, z_i) \qquad (4.4.8)$$

which represents the input-output relation for a thin lens.

As noted in section 3, for a thin lens the input and output planes are taken to coincide: $z_i = z_o$. However, the explicit specification by z_i and z_o is aimed at distinguishing the input signal from the outgoing, which will be also identified by means of appropriate subscripts as $\varphi(q, z_i) \equiv \varphi_i(q)$ and $\varphi(q, z_o) \equiv \varphi_o(q)$.

The operator $\hat{K}_-$ generates free propagation; then

$$\hat{T}(d) = e^{\frac{i}{2k} d \frac{\partial^2}{\partial q^2}} \qquad (4.4.9)$$

accounts for the diffraction of the optical field through the distance d.

To obtain the relevant input-output relations, we transform the complex amplitude from the space domain to the spatial frequency domain via Fourier transform. Free propagation in the frequency domain is accounted for by the simple plane-wave phase shift; transforming back to the spatial domain, we end up with the following relation

$$\varphi(q, z_o) = \frac{1}{\sqrt{i\lambda(z_o - z_i)}} \int dq' e^{i \frac{k}{2(z_o - z_i)}(q - q')^2} \varphi(q', z_i) \qquad (4.4.10)$$

governing the free propagation from z_i to z_o.

We turn now to consider the optical propagator

$$\hat{S}(m) = \exp\left\{-\frac{\alpha}{2}(q\frac{\partial}{\partial q} + \frac{1}{2})\right\}, \qquad m \equiv e^{-\frac{\alpha}{2}} \qquad (4.4.11)$$

which represents a scaling operator by the factor m; in fact, the relevant input-output relation writes as

$$\varphi(q, z_o) = \sqrt{m}\varphi(mq, z_i) \qquad (4.4.12)$$

It can be proved by direct calculation using the input-output relations (4.4.8), (4.4.10) and (4.4.12) (see Problem 7) that the expectation values of the operators $\hat{q}$ and $\hat{p} = -\frac{i}{k}\frac{\partial}{\partial q}$, namely $\langle \hat{q} \rangle$ and $\langle \hat{p} \rangle$, evolve in similar fashion as the ray coordinates q and p. That confirms the correspondence, intuitively assumed, between the optical propagators (4.4.7), (4.4.9) and (4.4.11) and the ray matrices of the optical systems they describe. In particular, it must be noted that the scaling operator $\hat{S}(m)$ corresponds to the ray matrix of a magnifier with magnification factor $\frac{1}{m}$.

The optical propagators are unitary operators (for which, namely, the Hermitian conjugate equals the inverse operator), whereas the ray-matrices are symplectic. Symplecticity and unitarity are indeed similar concepts, that apply to the classical context of c-number functions and to the quantum context of operator functions, respectively.

We have now all the tools to infer the input-output relation for a generic optical system described by the ray matrix $\mathbf{M} = \begin{pmatrix} A & B \\ C & D \end{pmatrix}$. To this end, we resort to the factorization (4.3.16), which within the present context yields

$$\hat{M} = \hat{S}(m)\,\hat{L}(f)\,\hat{T}(d) \tag{4.4.13}$$

giving $\hat{M}$ as the product of the diffraction, thin lens and scaling operators. The parameters m, f and d are inferred from (4.3.17) as

$$m = \frac{1}{A} \qquad f = -\frac{1}{AC} \qquad d = \frac{B}{A} \tag{4.4.14}$$

on account of that as previously noted the propagator $\hat{S}(m)$ with scale factor m corresponds in ray optics to a magnifier with magnification factor $\frac{1}{m}$.

With the rules assessed in (4.4.8), (4.4.10) and (4.4.12), one can easily get the required input-output relation for the optical system

$$\varphi(q, z_o) = \frac{1}{\sqrt{i\lambda B}} \int dq' \exp\left\{ i\frac{k}{2B}\left[Dq^2 + Aq'^2 - 2qq'\right] \right\} \varphi(q', z_i) \tag{4.4.15}$$

The response function $g(q, q'; z_o, z_i)$ is obviously given by the integral kernel

$$g(q, q'; z_o, z_i) = \frac{1}{\sqrt{i\lambda B}} \exp\left\{ i\frac{k}{2B}\left[Dq^2 + Aq'^2 - 2qq'\right] \right\} \tag{4.4.16}$$

where the matrix elements A, B, D connecting the reference planes are to be understood as functions of z_i and z_o.

The above expression deserves some comments as it seems to be meaningful only for $B \neq 0$. For optical systems with $B = 0$ as for instance thin lens and magnifier, the input-output relation (4.4.15) in the limit $B \to 0$ takes the simpler form [§]

$$\varphi(q, z_o) = \sqrt{D}\, e^{i\frac{k}{2}\frac{C}{D}q^2} \varphi(Dq, z_i) \tag{4.4.17}$$

which can also be established from the factorization of $B = 0$ matrices in terms of lens and magnifier matrices, as suggested in Problem 8.

Before going on, we introduce the Fourier transform $\tilde{\varphi}(\kappa, z)$ of the signal $\varphi(q, z)$ with respect to the transverse coordinate q

$$\tilde{\varphi}(\kappa, z) = \frac{1}{\sqrt{2\pi}} \int_{-\infty}^{+\infty} dq\, \varphi(q, z) e^{-i\kappa q} \tag{4.4.18}$$

whose inverse write in a similar form

$$\varphi(q, z) = \frac{1}{\sqrt{2\pi}} \int_{-\infty}^{+\infty} d\kappa\, \tilde{\varphi}(\kappa, z) e^{i\kappa q} \tag{4.4.19}$$

[§] We used the relation: $\lim\limits_{a\to\infty} \sqrt{\frac{a}{i\pi}} e^{iax^2} = \delta(x)$

More comments on the physical meaning of the integral (4.4.18) and on the spatial frequencies κ's will be given later. Now, we simply use the definition (4.4.18) and its inverse (4.4.19) to establish the link of the Fourier transform of a wave field to the far-field pattern. Fourier transforming both sides of (4.4.10) according to (4.4.18), one easily obtains the input-output relation for the Fourier transforms, that is

$$\tilde{\varphi}(\kappa, z_o) = e^{-i\frac{(z_o-z_i)}{2k}\kappa^2} \tilde{\varphi}(\kappa, z_i) \tag{4.4.20}$$

which, once inserted into the inverse relation (4.4.19), yields

$$\varphi(q, z_o) = \frac{1}{\sqrt{2\pi}} \int_{-\infty}^{+\infty} d\kappa \, e^{-i\frac{(z_o-z_i)}{2k}\kappa^2} \tilde{\varphi}(\kappa, z_i) e^{i\kappa q} \tag{4.4.21}$$

Taking $z_o = z_i + \ell$ and rehandling the argument of the exponential function, the above integral writes as

$$\varphi(q, z_i + \ell) = \frac{1}{\sqrt{2\pi}} e^{i\frac{\kappa}{2\ell}q^2} \int d\kappa \, e^{-i\frac{\ell}{2k}(\kappa - \frac{kq}{\ell})^2} \tilde{\varphi}(\kappa, z_i) \tag{4.4.22}$$

which in the limit $\ell \to \infty$ turns into

$$\varphi(q, z_i + \ell) \underset{\ell \to \infty}{\approx} e^{i\frac{\kappa}{2\ell}q^2} \sqrt{\frac{k}{i\ell}} \tilde{\varphi}(\frac{kq}{\ell}, z_i) \tag{4.4.23}$$

using the relation reported in the footnote.

The above relation expresses in formal terms the *optical reciprocity property*, according to which the far-field ($\ell \to \infty$) pattern in real space is equivalent, apart from some complex factor, to the spatial Fourier transform of the near-field distribution, evaluated at frequencies $\kappa = \frac{kq}{\ell}$.

4.5 Problems

1. Infer from the matrix relation (4.3.5) the thin lens law, i.e.

$$\frac{1}{z_1} + \frac{1}{z_2} = \frac{1}{f}$$

where z_1 and z_2 are the intersections of the ray with the z-axis respectively in the object and image spaces, and f denotes as usual the lens focal length.

2. Deduce the ray matrix relative to a thick lens, consisting of two refracting surfaces with curvature radius R_1 and R_2 separated by a distance $\Delta z = d$ on the optical axis. Assume furthermore the refractive index to be n_1 at input plane, n_2 at the output plane and n between the refracting surfaces.

3. Specify the parameters m, f and d for the other two possible arrangements of the matrices **S**, **L**, **T** to reproduce an arbitrary optical system $\begin{pmatrix} A & B \\ C & D \end{pmatrix}$ as in (4.3.16).

4. Write down the equations obeyed by the functions $m(z)$, $f(z)$ and $d(z)$ relevant to some of the possible factorizations of the type (4.3.16) for the lens-like medium (4.3.14).

5. Using the definitions of Fourier and inverse Fourier transforms, given in (4.4.18) and (4.4.19), deduce the input-output relation (4.4.10) relevant to free propagation.

6. Prove that the norm of the wave signal $\varphi(q,z)$ defined as

$$\|\varphi\|^2(z) = \int \varphi^*(q,z)\varphi(q,z)dq$$

is preserved when propagating through an arbitrary optical system characterized by the ray matrix $\mathbf{M} = \begin{pmatrix} A & B \\ C & D \end{pmatrix}$.

7. Evaluate the expectation values of the operators $\hat{q}$ and $\hat{p} = -\frac{i}{k}\frac{\partial}{\partial q}$, defined by the integrals

$$\langle \hat{q} \rangle = \frac{1}{\|\varphi\|^2} \int \varphi^*(q,z)\hat{q}\varphi(q,z)dq$$

$$\langle \hat{p} \rangle = \frac{1}{\|\varphi\|^2} \int \varphi^*(q,z)\hat{p}\varphi(q,z)dq$$

where $\|\varphi\|^2$ is defined in Problem 6, for the signal transformed by the operators (4.4.9), (4.4.7) and (4.4.11). Show that they evolve as the ray variables $\begin{pmatrix} q \\ p \end{pmatrix}$ when acted by the ray matrices corresponding to free-space section d, a thin lens f, and a magnifier $\frac{1}{m}$.

8. Factorize the ray matrix with $B = 0$ into the product of lens and magnifier-like matrices. Then, infer the optical input-output relation.

9. Deduce the input-output relation for a Fourier transformer, described by the ray-matrix (4.3.13). Show that the output signal $\varphi(q,z_o)$ is proportional to the Fourier transform of the input signal $\varphi(q,z_i)$, taken at spatial frequencies $\kappa = \frac{k}{f}q$.

10. Envisage one or more optical systems, which image the input signal $\varphi(q,z_i)$ as in the right-hand side of eq. (4.4.23), thus providing the far-field pattern of φ.

Suggested bibliography

As to the optics of synchrotron radiation, we suggest

1. C.K. Green "Spectra and Optics of Synchrotron Radiation", BNL 50522 (1976).
2. K.-J. Kim, Nucl. Instr.& Meth. A, **246**, 71 (1986).

   The Hamiltonian formulation of optics is presented in many textbooks, among which we quote
3. D. Marcuse "Light Transmission Optics" (Van Nostrand-Reinhold, Princeton, New Jersey 1972).
4. A.K. Ghatak and K. Thyagarajan "Contemporary Optics" (Plenum Press, New York 1978).
5. S. Solimeno, B. Crosignani and P. Di Porto "Guiding, Diffraction and Confinement of Optical Radiation" (Ac. Press, Orlando Fla, 1985).

   Geometrical optics and relevant methods are treated in
6. A.Yariv "Quantum Electronics" (J.Wiley & Sons, New York, 1975).
7. A.E. Siegman "Lasers" (Univ. Science Boos, Mill Valley, California 1986).
8. A. Nussbaum and R.A. Phillips "Contemporary Optics for Scientists and Engineers" (N. Holonyak Jr.).

   whilst wave optics and related techniques are illustrated in
9. M. Born and E. Wolf "Principles of Optics" (Pergamon Press, Oxford, 1980).
10. J.W. Goodman "Introduction to Fourier Optics" (Mc Graw Hill, New York, 1968).

    For a general view to the group theory and to the symplectic methods we suggest
11. R. Gilmore "Lie Groups, Lie Algebras and some of their Applications" (J. Wiley and Sons, New York, 1974).
12. V. Guillemin and S. Sternberg "Symplectic Techniques in Physics" (Univ. Press, New York, 1990).

# Chapter 5
WIGNER DISTRIBUTION AND SYNCHROTRON RADIATION SOURCES

Summary

The formalism relative to the optical Wigner distribution function is illustrated. The relevant main properties are discussed as well as the propagation law through optical systems. The method is then applied to the propagation of light beam from synchrotron radiation sources.

5.1 Introduction

Wigner method in quantum mechanics provides a means for associating a c-number function in phase-space with every operator which is a function of position and momentum operators. That allows to make phase space representation, and relevant methods, usable also to a discipline for which the classical concept of phase space is problematic due to the uncertainty principle. The immediate consequence of this process is the possibility of calculating expectation values of quantum mechanical observables in the classical manner rather than through the operator formalism of quantum mechanics.

As a specific example, let us consider a system with one degree of freedom. Classically, the system is represented in phase space by a distribution function $W_C(q,p)$, with q and p denoting as usual the classical conjugate variables. The ensemble average of any function $G(q,p)$ of the position and momentum can then be expressed as the integral

$$\langle G \rangle_C = \int\int dq\,dp\, G(q,p) W_C(q,p) \qquad (5.1.1)$$

In the usual formulation of quantum statistical mechanics, the system is represented by the density operator $\hat{\rho}\,(\hat{q},\hat{p})$, which is a function of the non commuting position and momentum operators. Then, the ensemble expectation value of any observable $\hat{G}\,(\hat{q},\hat{p})$ is obtained according to

$$\left\langle \hat{G} \right\rangle_Q = Tr(\hat{\rho}\hat{G}) \qquad (5.1.2)$$

where Tr denotes the trace. The subscripts C and Q in (5.1.1) and (5.1.2) mark the classical and quantum ambience.

The Wigner transform provides a strategy to recover the phase space representation within a quantum framework. Mapping indeed all the operators onto c-number functions of the conjugate variables q and p,

$$\hat{q} \to q, \quad \hat{p} \to p$$
$$\hat{\rho}(\hat{q},\hat{p}) \to W_Q(q,p) \qquad (5.1.3)$$
$$\hat{G}(\hat{q},\hat{p}) \to G(q,p)$$

according to a prescribed rule, one can express the quantum mechanical expectation value $\left\langle \hat{G} \right\rangle_Q$ in terms of phase space integration similar to the classical definition (5.1.1), namely

$$\left\langle \hat{G} \right\rangle_Q = \int\int dq\, dp\, G(q,p) W_Q(q,p) \qquad (5.1.4)$$

Then, the function $W_Q(q,p)$ can be regarded as the quantum analog of the classical phase space distribution function $W_C(q,p)$, providing an intermediate representation of the system between the purely quantum and purely classical pictures.

In some sense the mapping (5.1.3) allows to project quantum dynamics onto classical phase space.

The Wigner representation gives $W_Q(q,p)$ as the convolution integral

$$W_Q(q,p) = \frac{1}{2\pi\hbar} \int dq'\, \psi^*(q - \frac{q'}{2}) \psi(q + \frac{q'}{2}) e^{-i\frac{pq'}{\hbar}} \qquad (5.1.5)$$

where ψ is the wave function of the system, which is assumed in the above to have one degree of freedom and to be in the pure state $\psi(q)$.

The rule (5.1.5) is the inverse of the Weyl's rule, used to calculate quantum mechanical operators from classical quantities.

It is interesting to observe that many different mappings have been envisaged for constructing a quantum density in phase space, that correspond to different schemes to order the non commuting operators $\hat{q}$ and $\hat{p}$ when their products are involved. The Wigner representation is the most widely used as it uniquely possesses the maximum symmetry possible.

However, as the Wigner's theorem asserts, no one among the phase space distribution functions, arising from the possible mappings of the type (5.1.3), and hence also $W_Q(q,p)$ as given in (5.1.5), has all the properties of a true probability density. The Wigner distribution can take on negative values, for instance, although it gives the correct marginal distributions of position and momentum. Nevertheless, the Wigner phase space representation provides a very powerful technique for solving quantum mechanical problems, with its potentiality being enlarged by novel applications to other branches of physics, as for instance, to quantum optics in investigations concerning the coherence properties of light. We do not dwell here with the drawbacks inherent to the generalized phase-space distribution function formalism, inviting the interested reader to consult the quoted references for a deeper discussion of the topic. In section 2 we introduce the optical Wigner distribution function and illustrate the relevant propagation properties; we stress its potentiality as intermediate picture between ray and wave optics, yielding in addition an elegant description of optical signals and systems. We give as intuitive the identification of the Wigner distribution with brightness of the optical signal, the formal subtleties of such a correspondence being investigated in the suggested references. In section 3 the geometry inherent to the propagation through optical systems is investigated through the transformation of the ellipse phase space associated to the source brightness. The introduced concepts and methods are applied to the synchrotron radiation sources in section 4. As basic result of the formalism, we establish an important theorem concerning the effects of the electron beam transverse (in position and angle) and longitudinal distribution, which is accounted for in a very simple and intuitive fashion. Finally, in section 5 the undulator radiation brightness is calculated, using the results of section 4.

5.2 Optical Wigner distribution function

We consider a time-harmonic scalar optical signal, which can be described in the space domain by the complex amplitude $\varphi(\vec{q}, z)$, with $\vec{q} = \begin{pmatrix} q_x \\ q_y \end{pmatrix}$. We restrict ourselves to the one-dimensional case, in which the signals are functions only of one transverse coordinate: $\varphi(q, z)$. The extension to two dimensions is straightforward. Furthermore, for notational simplicity the z-dependence will be suppressed whenever possible and integration are intended to extend to $-\infty$ to $+\infty$, unless otherwise stated.

An equivalent description of the signal can be given in the transverse spatial frequency domain via the Fourier transform, introduced in Section 4.4.

We rewrite below the definition of $\tilde{\varphi}(\kappa, z)$

$$\tilde{\varphi}(\kappa, z) = \frac{1}{\sqrt{2\pi}} \int \varphi(q, z) e^{-i\kappa q} dq \qquad (5.2.1)$$

along with the inverse formula

$$\varphi(q, z) = \frac{1}{\sqrt{2\pi}} \int \tilde{\varphi}(\kappa, z) e^{i\kappa q} d\kappa \qquad (5.2.2)$$

In the above κ denotes wavenumber in the plane transverse to the z-axis. Inspecting fig. 1 with the graphic view of the relation of κ to the wavevector $\vec{k}$ in the medium, which is in the direction of propagation -i.e. perpendicular to the wave fronts- it is easy to establish the correspondence between κ and the ray momentum p of ray optics:

$$\kappa = kp \qquad (5.2.3)$$

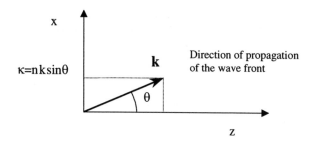

Figure 1 Spatial frequency κ and direction cosine of the ray in the direction of the wave front.

where k is the free-space wavenumber: $k = \frac{2\pi}{\lambda}$. A more rigorous proof of (5.2.3) is suggested in Problem 1.

Thus, (5.2.2) has the physical interpretation of an expansion of the optical beam in a set of planes waves propagating along slightly different directions. As the squared modulus of φ: $|\varphi(q,z)|^2$ gives the *positional power spectrum* of the signal at the plane z, so the squared modulus of $\tilde{\varphi}$: $|\tilde{\varphi}(\kappa,z)|^2$ represents the *directional power spectrum* at the same plane, that shows how the energy of the beam is distributed as a function of the direction.

It is a well known result of the Fourier transform theory that

$$\int |\varphi(q,z)|^2 \, dq = \int |\tilde{\varphi}(\kappa,z)|^2 \, d\kappa \tag{5.2.4}$$

The Wigner distribution function $\mathsf{B}(q,\kappa,z)$ of the optical signal $\varphi(q,z)$ is defined as a function of the Fourier conjugate variables (q,κ) through an integral transform similar to (5.1.5). We write

$$\mathsf{B}(q,\kappa,z) = \frac{nk}{2\pi} \int dq' \varphi^*(q - \frac{q'}{2}, z)\varphi(q + \frac{q'}{2}, z)e^{-i\kappa q'} \tag{5.2.5}$$

in the space domain, or equivalently in the spatial frequency domain

$$\mathsf{B}(q,\kappa,z) = \frac{nk}{2\pi} \int d\kappa' \, \tilde{\varphi}^*(\kappa - \frac{\kappa'}{2}, z) \, \tilde{\varphi}(\kappa + \frac{\kappa'}{2}, z)e^{i\kappa q'} \tag{5.2.6}$$

which follows from (5.2.5) on account of the inverse Fourier transform (5.2.2). For a more general writing the products in both (5.2.5) and (5.2.6) should be enclosed between sharp brackets $\langle ... \rangle$ to denote some ensemble average accounting for an arbitrary state of coherence of the optical signal. As they stand, the integral transform (5.2.5) and (5.2.6) are appropriate to completely coherent signals. For simplicity's sake, we omit this notation allowing for the implicit understanding that when fluctuations are present (partially coherent light) ensemble averages must be performed.

As the B-function is naturally defined as function of the Fourier conjugate variables q and κ, in the following we will refer to $q - \kappa$ as the phase plane, taking account of the relation (5.2.3) whenever needed.

It should be noted the close similarity of the expressions (5.2.5) and (5.2.6) as regards to the role played by the conjugate variables q and κ. As we will see in the following, it is just this symmetry that enables the B-function to describe the signal in space and spatial frequency, or position and direction, simultaneously. In fact, the q and κ variables play equivalent role in the Wigner distribution function. Interchanging the q and κ variables in any expression containing the B-function yields an expression that is the dual of the original one, in the sense that as the original expression describes a property in the space domain, the dual expression describes a similar property in the spatial frequency domain.

A straightforward example of this property is provided by the inverse relations of (5.2.5) and (5.2.6). Fourier transforming B in (5.2.5) and (5.2.6) with respect to κ and q, we obtain (for simplify notation we set $n = 1$ in the following)

$$\varphi(q)\varphi^*(q') = \frac{1}{k}\int \mathsf{B}\left(\frac{q+q'}{2}, \kappa\right) e^{i\kappa(q-q')} d\kappa \qquad (5.2.7)$$

$$\tilde{\varphi}(\kappa)\tilde{\varphi}^*(\kappa') = \frac{1}{k}\int \mathsf{B}\left(q, \frac{\kappa+\kappa'}{2}\right) e^{-i(\kappa-\kappa')q} dq$$

which appear as dual to each other according to the above stated meaning. A further interesting result follows from (5.2.7). For $q = q'$ the first relation yields

$$|\varphi(q)|^2 = \frac{1}{k}\int \mathsf{B}(q, \kappa) d\kappa \qquad (5.2.8)$$

The positional power spectrum is therefore obtained integrating the Wigner distribution function with respect to the spatial frequency.

Similarly, with $\kappa = \kappa'$ the second of (5.2.7) turns into

$$\left|\tilde{\varphi}(\kappa)\right|^2 = \frac{1}{k}\int \mathsf{B}(q, \kappa) dq \qquad (5.2.9)$$

thus yielding the directional power spectrum after integrating B with respect to the position variable.

When interpreting B as the brightness function, the above integrals have a direct link with some radiometric quantities. Specifically, the integral (5.2.9) is proportional, apart from the obliquity factor $\cos\theta$ - θ being the angle of observation- to the radiant intensity, which describes the angular distribution of the radiant power.

The relations (5.2.8) and (5.2.9) clearly indicate that the Wigner distribution function provides an intermediate description of the optical signal between the pure space distribution $\varphi(q,z)$ and the pure spatial frequency description $\tilde{\varphi}(\kappa,z)$, containing informations on both, that can easily be extracted. Then, it allows a description of the signal within the context of wave-optics formalism, but with some attributes which are peculiar of ray-optics picture. As we see, although defined in terms of the wave functions (or its Fourier transform) of wave optics, $B(q,\kappa,z)$ is propagated through an arbitrary linear optical system according to the matrix formalism of ray optics. In some way $B(q,\kappa,z)$ can be understood as the amplitude of a light ray passing through the point (q,z) in the direction $p=\frac{\kappa}{k}$.

The B-function has several properties, for a detailed account of which the reader is addressed to the specific references. Here, we mention its realness, which follows from (5.2.5) and (5.2.6), being

$$B(q,\kappa,z) = B^*(q,\kappa,z) \qquad (5.2.10)$$

In addition, the Wigner distribution function reproduces the properties of the optical signal in both space and frequency domain. It is evident that if the signal is limited to a certain q or κ interval and vanishes outside that interval, then the B-function is limited to the same interval as well.

Unfortunately, as already remarked, the Wigner distribution function can take negative values in some cases, thus prohibiting to be interpreted as a true power density distribution in phase-space, i.e. as a brightness. Then, the Wigner distribution function and relative phase-space formalism should be understood as mathematical constructions yielding a simpler description of propagation, which retains the physical content of wave optics and the formal methods of ray optics.

It is evident that the statistical valence of $B(q,\kappa,z)$ is appropriate to describe a collection of light rays, analogous to the phase space particle density distribution of the classical statistical mechanics. In a qualitative fashion we can state

$$B(q,\kappa,z)dqd\kappa \propto \frac{\text{number of rays in the phase space}}{\text{area } dqd\kappa \text{ at the plane } z} \qquad (5.2.11)$$

We turn now our attention to the input-output relationships for the Wigner distribution function of the signal when passing through an optical system. As remarked in Section 4.4, the input-output relation of an optical system linking the signal on the output plane z_o to that on the input plane z_i is formally expressed in terms of the response function $g(q,q')$ according to the integral transform (see (4.4.2))

$$\varphi(q,z_o) = \int g(q,q')\varphi(q',z_i)dq' \qquad (5.2.12)$$

where the z-dependence of g is implicitly understood, but omitted for notational simplicity: $g(q,q') = g(q,q';z_o,z_i)$.

As already stressed, the optical transfer function $g(q,q')$ represents the output signal for an input signal shaped as a δ-function: $\varphi(q',z_i) = \delta(q'-q)$.

A similar relation to (5.2.12) can be easily obtained in the dual κ-domain, involving the Fourier transforms of the signal at the input and output planes. Using the Fourier's reciprocal integral relations (5.2.1) and (5.2.2) we find

$$\tilde{\varphi}(\kappa, z_o) = \int \tilde{g}(\kappa, \kappa') \tilde{\varphi}(\kappa', z_i) d\kappa' \tag{5.2.13}$$

where $\tilde{g}(\kappa, \kappa')$ turns to be the symplectic Fourier transform of $g(q, q')$, namely

$$\tilde{g}(\kappa, \kappa') = \frac{1}{2\pi} \int \int dq dq' g(q, q') e^{-i(\kappa q - \kappa' q')} \tag{5.2.14}$$

It represents the response function of the system in the spatial frequency domain. It is evident that $\tilde{g}(\kappa, \kappa')$ is the response signal to a δ-signal in κ-space: $\tilde{\varphi}(\kappa', z_i) = \delta(\kappa' - \kappa)$, which in the q-domain represents a plane wave. Thereby, $\tilde{g}(\kappa, \kappa')$ is also called *wave response* of the system *.

According to the relation (5.2.12) and the first of the formulae (5.2.7), it is possible to express the Wigner distribution function $B(q, \kappa, z_o)$ at z_o in terms of the Wigner distribution function $B(q, \kappa, z_i)$ at z_i. Explicitly, we have

$$B(q, \kappa, z_o) = \int \int dq' d\kappa' G(q, q'; \kappa, \kappa') B(q', \kappa', z_i) \tag{5.2.15}$$

where $G(q, q'; \kappa, \kappa')$ has the rather intrigued expression

$$G(q, q'; \kappa, \kappa') = \frac{1}{2\pi} \int \int dq'' dq''' g^*(q - \frac{q''}{2}, q' - \frac{q'''}{2}) \cdot g(q + \frac{q''}{2}, q' + \frac{q'''}{2}) e^{-i(\kappa q'' - \kappa' q''')} \tag{5.2.16}$$

involving the impulse response function $g(q, q')$.

An equivalent expression for G can be worked out, using (5.2.13) and the second of (5.2.7), namely

$$G(q, q'; \kappa, \kappa') = \frac{1}{2\pi} \int \int d\kappa'' d\kappa''' \tilde{g}^*(\kappa - \frac{\kappa''}{2}, \kappa' - \frac{\kappa'''}{2}) \cdot \tilde{g}(\kappa + \frac{\kappa''}{2}, \kappa' + \frac{\kappa'''}{2}) e^{i(\kappa'' q - \kappa''' q')} \tag{5.2.17}$$

involving the wave response function $\tilde{g}(\kappa, \kappa')$.

The function $G(q, q'; \kappa, \kappa')$ represents the response of the system in the (q, κ)-space. In particular, it is just the response to an input signal $B(q', \kappa', z_i) = \delta(q' - q)\delta(\kappa' - \kappa)$, which in some sense can be regarded as a single ray with coordinates (q, κ). Hence, the function $G(q, q'; \kappa, \kappa')$ is called *ray response* of the system. However, it is worth stressing that there does not exist any signal $\varphi(q)$ leading to a Wigner

---

*Other equivalent input-output relations can be formulated in which the input and output signals are described in different domains: the input signal in the q-domain as the output in κ-domain, for instance. They will be not considered here.

distribution function $B(q', \kappa', z_i) = \delta(q' - q)\delta(\kappa' - \kappa)$, which thereby appears rather artificial.

Equations (5.2.16) and (5.2.17) can be inverted to give for instance

$$g^*(q - \frac{\rho}{2}, q' - \frac{\eta}{2}) g(q + \frac{\rho}{2}, q' + \frac{\eta}{2}) = \frac{1}{2\pi} \int\int d\kappa d\kappa' e^{i(\kappa\rho - \kappa'\eta)} G(q, q'; \kappa, \kappa') \quad (5.2.18)$$

According to the discussion in Section 4.4, the impulse response of an optical system is

$$g(q, q') = \sqrt{\frac{k}{2\pi i B}} \exp\left\{i\frac{k}{2B}\left[Dq^2 + Aq'^2 - 2qq'\right]\right\} \quad (5.2.19)$$

where A, B, C are the entries of the relevant ray matrix. Inserting the above expression into (5.2.16) provides the transfer function

$$G(q, q'; \kappa, \kappa') = \frac{k}{4\pi^2 B} \int dq'' \exp\left\{i\frac{k}{B}q''\left[Dq - q' - \frac{B}{k}\kappa\right]\right\} \cdot \int dq''' \exp\left\{i\frac{k}{B}q'''\left[Aq' - q - \frac{B}{k}\kappa'\right]\right\} \quad (5.2.20)$$

which turns into a δ-function, namely

$$G(q, q'; \kappa, \kappa') = \delta\left[q' - \left(Dq - \frac{B}{k}\kappa\right)\right] \delta\left[\kappa' - (A\kappa - kCq)\right] \quad (5.2.21)$$

Taking account of that the ray matrix inherent to the phase space vector $\mathbf{u} = \begin{pmatrix} q \\ \kappa \end{pmatrix}$ is †

$$\begin{pmatrix} q \\ \kappa \end{pmatrix} = \begin{pmatrix} A & \frac{B}{k} \\ kC & D \end{pmatrix} \begin{pmatrix} q' \\ \kappa' \end{pmatrix} \quad (5.2.22)$$

eq. (5.2.21) can be recast in the more compact form

$$G(q, q'; \kappa, \kappa') = \delta\left[\mathbf{u}' - \mathbf{M}^{-1}\mathbf{u}\right] \quad (5.2.23)$$

Then, according to (5.2.15) we can write the transformation law of the optical Wigner distribution function through an arbitrary optical medium

$$B(\mathbf{u}, z_o) = B(\mathbf{M}^{-1}\mathbf{u}, z_i) \quad (5.2.24)$$

---

†The ray matrix in 5.2.22 is obtained from the usual $q - p$ matrix through the product

$$\begin{pmatrix} 1 & 0 \\ 0 & k \end{pmatrix} \begin{pmatrix} A & B \\ C & D \end{pmatrix} \begin{pmatrix} 1 & 0 \\ 0 & \frac{1}{k} \end{pmatrix}$$

in order to take account of the relation 5.2.3.

which in formal terms states that the B-function at a given phase point on a transverse plane z is the same as the B-function at the corresponding phase point on another transverse plane, the correspondence being determined by the optical transformation, i.e. by the ray matrix $\mathbf{M}$. In particular, as the origin of the phase space transforms into itself for any optical matrix, the Wigner distribution function at the origin of phase space in invariant:

$$B(0, 0, z_o) = B(0, 0, z_i) \tag{5.2.25}$$

In eqs. (5.2.23) and (5.2.24) we used the symbol $\mathbf{M}$ to signify the matrix (5.2.22), i.e. the ray matrix relevant to the Fourier conjugate variables (q, κ), whilst in the previous Chapter the same symbol was used to denote the ray matrix relevant to the Hamilton conjugate variables (q, p). In the following we will continue to use the symbol $\mathbf{M}$ to signify both the matrices, as the specific context will make clear which matrix is involved in.

We invite the reader to deduce the G-function for some simple optical systems (Problem 2).

The basic result of this section is represented by the transformation law (5.2.24). It states the attractive property of the optical Wigner function of being, as the phase space density distribution function $\rho(\mathbf{u}, z)$, invariant along the trajectories of the phase space representative points.

5.3 Moments of Wigner distribution function and phase space ellipse

The results of the previous section specify the mathematics inherent to the transformations of the Wigner distribution function $B(\mathbf{q}, \boldsymbol{\kappa}, z)$ through any arbitrary optical medium, which occur according to the laws of linear ray-optics. Here, we approach the problem to specify the geometry inherent to such transformations, providing B with a graphic visualization in phase space.

To this end, we discuss the properties of the first and second-order momenta of the brightness function $B(\mathbf{u})$.

The zeroth-order moment defined as

$$P = \int\int B(\mathbf{u})d\mathbf{u} \tag{5.3.1}$$

represents the total radiant power at the considered frequency ω. As before, integrals are intended to go from $+\infty$ to $-\infty$ unless otherwise specified. To simplify notation, we will assume throughout the following normalization of $B(\mathbf{u})$:

$$\frac{1}{k}\int\int B(\mathbf{u})d\mathbf{u} = 1 \tag{5.3.2}$$

which according to (5.2.8) and (5.2.9) turns into

$$\int dq\, |\varphi(q)|^2 = 1, \qquad \int d\kappa\, \left|\widetilde{\varphi}(\kappa)\right|^2 = 1 \tag{5.3.3}$$

The first order moments of the Wigner distribution function are defined through the integrals

$$\langle \mathbf{u} \rangle = \frac{1}{P} \int \int \mathbf{u} B(\mathbf{u}) d\mathbf{u} = \begin{pmatrix} \frac{1}{k} \int q B(q,\kappa) dq d\kappa \\ \frac{1}{k} \int \kappa B(q,\kappa) dq d\kappa \end{pmatrix} \quad (5.3.4)$$

which once interpreting B as a probability density in phase space can be regarded as the expectation value of the vector $\mathbf{u} = \begin{pmatrix} q \\ \kappa \end{pmatrix}$. It is evident that being B a real-valued function the first order moment $\langle \mathbf{u} \rangle$ is real as well. Furthermore, exploiting (5.2.8), (5.2.9) and (5.3.1) we can write $\langle \mathbf{u} \rangle$ in the form

$$\langle \mathbf{u} \rangle = \begin{pmatrix} \langle q \rangle \\ \langle \kappa \rangle \end{pmatrix} = \begin{pmatrix} \int q |\varphi(q)|^2 dq \\ \int \kappa |\tilde{\varphi}(\kappa)|^2 d\kappa \end{pmatrix} \quad (5.3.5)$$

thus allowing to interpret $\langle q \rangle$ and $\langle \kappa \rangle$ as the *means* (or *centers of gravity*) of the signal $\varphi(q)$ and its angular spectrum $\tilde{\varphi}(\kappa)$, respectively.

It can be easily seen how the first order moment $\langle \mathbf{u} \rangle$ propagates through some optical system. According to the transformation law (5.2.24), we obtain

$$\begin{aligned} \langle \mathbf{u} \rangle = \int \int \mathbf{u} B(\mathbf{u}, z_o) d\mathbf{u} = \int \int \mathbf{u} B(\mathbf{M}^{-1}\mathbf{u}, z_i) d\mathbf{u} = \\ = \mathbf{M} \int \int \mathbf{u} B(\mathbf{u}, z_i) d\mathbf{u} = \mathbf{M} \langle \mathbf{u}(z_i) \rangle \end{aligned} \quad (5.3.6)$$

where we used that in phase space $d\mathbf{u} = d(\mathbf{M}^{-1}\mathbf{u}) = d(\mathbf{M}\mathbf{u})$.

The first order moments transform as the ray coordinates (q, κ), which then as already remarked can be interpreted as the centers of gravity of the positional $|\varphi(q)|^2$ and directional $|\tilde{\varphi}(\kappa)|^2$ spectrum.

We assume that $\langle \mathbf{u} \rangle$ is the null vector; this does not limit the validity of the forthcoming analysis, representing a mere shift of the coordinate axes in phase plane.

The second order moments have some relevance within the present context. We introduce the dyadic products $\mathbf{u} \cdot \mathbf{u}^T$ and $\langle \mathbf{u} \rangle \cdot \langle \mathbf{u} \rangle^T$, which provide the 2×2 symmetric matrices

$$\mathbf{u} \cdot \mathbf{u}^T = \begin{pmatrix} q^2 & q\kappa \\ \kappa q & \kappa^2 \end{pmatrix} \quad (5.3.7)$$

and

$$\langle \mathbf{u} \rangle \cdot \langle \mathbf{u} \rangle^T = \begin{pmatrix} \langle q \rangle^2 & \langle q \rangle \langle \kappa \rangle \\ \langle \kappa \rangle \langle q \rangle & \langle \kappa \rangle^2 \end{pmatrix} \quad (5.3.8)$$

The second order moments of the Wigner distribution function are defined as the entries of the matrix

$$\Sigma = \frac{1}{P} \left(\langle \mathbf{u} \cdot \mathbf{u}^T \rangle - \langle \mathbf{u} \rangle \cdot \langle \mathbf{u} \rangle^T \right) \quad (5.3.9)$$

which according to the assumption that $\langle \mathbf{u} \rangle$ equals the null vector and (5.3.1), turns to be the expectation value of the matrix (5.3.9), namely

$$\Sigma = \int\int \mathbf{u} \cdot \mathbf{u}^T B(\mathbf{u}) d\mathbf{u} = \begin{pmatrix} \langle q \rangle^2 & \langle q\kappa \rangle \\ \langle \kappa q \rangle & \langle \kappa \rangle^2 \end{pmatrix} \quad (5.3.10)$$

It is easy to identify the diagonal entries $\langle q \rangle^2$ and $\langle \kappa \rangle^2$ of Σ as the variances of the positional and angular spectrum, i.e.

$$\langle q \rangle^2 = \int dq q^2 |\varphi(q)|^2, \qquad \langle \kappa \rangle^2 = \int d\kappa \kappa^2 \left|\widetilde{\varphi}(\kappa)\right|^2 \quad (5.3.11)$$

which usually define the widths of the optical signal $\varphi(q)$ and its spatial Fourier transform $\widetilde{\varphi}(\kappa)$.

As known the uncertainty relation

$$\langle q \rangle^2 \langle \kappa \rangle^2 \geq \frac{1}{4} \quad (5.3.12)$$

holds.

Within the present context, the means (5.3.11) define the widths of the radiance function $B(q, \kappa)$ along the q and κ axis in phase space.

Similarly, the off-diagonal entry $\langle q\kappa \rangle$ in (5.3.10), which can be explicitly written as

$$\langle q\kappa \rangle = -\frac{i}{2} \int dq q \left[\varphi'(q)\varphi^*(q) - \varphi(q)\varphi^{*'}(q)\right] \quad (5.3.13)$$

where primes mean derivative with respect to the argument, yields informations about the $q - \kappa$ coupling of the B-dependence.

The inequality (5.3.12) deserves some comments. Rewriting (5.3.12) in the alternative form

$$\Delta q \Delta \kappa \geq \frac{1}{2} \quad (5.3.14)$$

where $\Delta q \equiv \sqrt{\langle q \rangle^2}$ and similarly $\Delta \kappa = \sqrt{\langle \kappa \rangle^2}$, it is immediate to recognize that the left-hand side represents an area in phase space. Accordingly, the optical uncertainty principle (5.3.14) states that the optical phase space is *granular*: there is a minimum possible phase space area associated with a source or an optical system, which is just $\frac{1}{2}$.

Let us say that in the (q, p) phase plane (5.3.14) writes in the more familiar form

$$\Delta q \Delta p \geq \frac{\lambdabar}{2} \quad (5.3.15)$$

resembling the Heisenberg uncertainty principle of quantum mechanics, with the correspondence $\lambdabar \to \hbar$.

The matrix Σ is an Hermitian operator, being real and symmetric; furthermore, it is non-negative definite. We recall that a matrix $\mathbf{A}$ is called nonnegative defined if for any vector a $\mathbf{u} \neq 0$ the bilinear form $\mathbf{u} \cdot \mathbf{A}\mathbf{u}$ is non negative, namely

$$\mathbf{u} \cdot \mathbf{A}\mathbf{u} \geq 0 \quad \forall \, \mathbf{u}$$

It is a well-established result of matrix theory that a necessary and sufficient condition that $\mathbf{A}$ is non negative defined is that all leading principal minors are positive, as it is indeed the case of Σ, being (5.3.13) non negative. As a consequence, the eigenvalues of Σ are non negative and hence det Σ is non negative as well: det $\Sigma \geq 0$.

We establish now how the second order moments propagate through an arbitrary optical system. Exploiting the relation (5.2.22) in (5.2.24), the transformation law for $\Sigma(z)$ straightforwardly follows in the form

$$\Sigma(z_o) = \mathbf{M}\Sigma(z_i)\mathbf{M}^\mathsf{T} \tag{5.3.16}$$

which after carrying out the calculations inherent to the product of the involved matrices yields

$$\begin{pmatrix} \Sigma_{qq} \\ \Sigma_{q\kappa} \\ \Sigma_{\kappa\kappa} \end{pmatrix}(z_o) = \begin{pmatrix} A^2 & 2\frac{AB}{k} & \frac{B^2}{k^2} \\ kAC & 2AD-1 & \frac{DB}{k} \\ k^2C^2 & 2kDC & D^2 \end{pmatrix} \begin{pmatrix} \Sigma_{qq} \\ \Sigma_{q\kappa} \\ \Sigma_{\kappa\kappa} \end{pmatrix}(z_i) \tag{5.3.17}$$

where we introduced the frequently encountered symbology to denote the entries of Σ, i.e. $\langle q \rangle^2 \equiv \Sigma_{qq}$, $\langle q\kappa \rangle = \langle \kappa q \rangle \equiv \Sigma_{q\kappa}$ and $\langle \kappa \rangle^2 \equiv \Sigma_{\kappa\kappa}$.

It is an easy matter to verify that according to the transformation law (5.3.16) the determinant of Σ is a constant of motion

$$\det \Sigma(z_o) = \det \Sigma(z_i) \quad \forall \, z_i, z_o \tag{5.3.18}$$

In many cases the brightness function, as the ray density distribution function in phase space, depends on q and κ through some quadratic form. This is the case for instance for Gaussian beams, we will discuss in more details later. In general, however, a quadratic dependence of B on q and κ can be regarded as a valid approximation in connection with paraxial optics. In most cases, the brightness function can be adequately modelled in the form of a Gaussian, which turns to be just the exact brightness function for Gaussian light. It is needless to say that the use of Gaussians to represent the brightness of a source greatly simplifies the discussion of the optical properties of the emitted light as well as the account of the propagation.

We consider thereby a B-function in the form as

$$\mathsf{B}(q, \kappa, z) = \frac{k}{2\pi\varepsilon} \exp\left\{-\frac{1}{2}\mathbf{u}^\mathsf{T}\Sigma\mathbf{u}\right\} \tag{5.3.19}$$

where the normalization (5.3.1) has been applied. The emittance ε, defined by

$$\varepsilon \equiv \sqrt{\det \mathbf{\Sigma}} = \sqrt{\Sigma_{qq}\Sigma_{\kappa\kappa} - \Sigma_{q\kappa}^2} \qquad (5.3.20)$$

corresponds to $\frac{1}{\pi}$ times the area occupied by the elliptical contour [‡]

$$\mathbf{u}^T\mathbf{\Sigma}\mathbf{u} = 1 \qquad (5.3.21)$$

which contains 39.3% of the optical power (at the frequency ω) according to (5.3.19).

It can be easily checked that the second order moments of (5.3.19) are just the entries of $\mathbf{\Sigma}$.

According to the transformation laws (??-5.3.16) one can see that ellipses in phase space transform into ellipses under beam propagation, although they become tilted and change their proportions.

In order to visualize how the ellipse (5.3.21) propagates through some optical system, we firstly rewrite (5.3.21) in explicit form, i. e.

$$\Sigma_{\kappa\kappa}q^2 - 2\Sigma_{q\kappa}q\kappa + \Sigma_{qq}\kappa^2 = \varepsilon^2 \qquad (5.3.22)$$

Then, we list the geometrical properties of the ellipse (5.3.22), the proof being left to the reader (fig. 2)

- The axes of (5.3.22) make the angles ϑ and ϑ' with the q-axis, respectively given by

$$\vartheta = \frac{1}{2}\tan^{-1}\left(2\frac{\Sigma_{q\kappa}}{\Sigma_{qq} - \Sigma_{\kappa\kappa}}\right), \qquad \vartheta' = \vartheta + \frac{\pi}{2} \qquad (5.3.23)$$

- The points at which (5.3.22) intersects the q and κ axes are

$$q_I = \frac{\varepsilon}{\sqrt{\Sigma_{\kappa\kappa}}}, \qquad \kappa_I = \frac{\varepsilon}{\sqrt{\Sigma_{qq}}} \qquad (5.3.24)$$

- The maximum excursions along the coordinate axes are specified just by the variances (5.3.11), namely

$$q_M = \sqrt{\Sigma_{qq}}, \qquad \kappa_M = \sqrt{\Sigma_{\kappa\kappa}} \qquad (5.3.25)$$

- The area of the ellipse (5.3.22) turns to be π times the emittance ε, i.e.

$$A = \pi q_I \kappa_M = \pi q_M \kappa_I = \varepsilon \qquad (5.3.26)$$

which thereby takes this further meaning.

We visualize the effect on the light beam ellipse of the propagation through some optical system.

A diffraction region of length d, for instance, leaves unchanged $\Sigma_{\kappa\kappa}$:

[‡]Being $det\mathbf{\Sigma} > 0$, the conic (5.3.21) turns to be an ellipse.

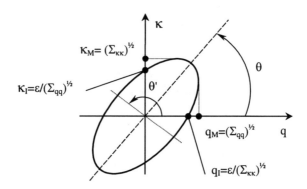

Figure 2   Phase-space ellipse based on the optical beam parameters: $\Sigma_{\kappa\kappa}$, $\Sigma_{q\kappa}$, Σ_{qq}

$$\Sigma_{\kappa\kappa}(z_o) = \Sigma_{\kappa\kappa}(z_i) \qquad (5.3.27)$$

and hence the beam ellipse rotates anticlockwise within the strip in phase plane limited by the straight line $\kappa = \pm\kappa_{max}$ and with the intersections $\pm q_{int}$ with the q-axis (fig. 3a).

Correspondingly, a thin lens f produces a tilt of the beam ellipse, clockwise if $f > 0$ and anticlockwise if $f < 0$, within the strip $q = \pm q_{max}$, leaving unchanged the intersections $\pm\kappa_{int}$ with the κ-axis (fig. 3b). The propagation of the phase space ellipse through a thin lens is indeed characterized by the invariance of Σ_{qq}:

$$\Sigma_{qq}(z_o) = \Sigma_{qq}(z_i) \qquad (5.3.28)$$

The propagation through a Fourier transformer and a magnifier is depicted in fig. 3c and fig.3d. The former interchanges the q and κ variances, carrying $\Sigma_{\kappa\kappa}(z_i)$ into $\Sigma_{\kappa\kappa}(z_o)$ and viceversa, apart from some multiplicative factors. On the other hand, a magnifier produces a squeeze of the ellipse along the q-axis and a clockwise rotation if $m > 1$, or a squeeze along the κ-axis and a counterclockwise rotation if $m < 1$.

The invariance of the phase space area -or equivalently of the brightness- associated with a source of radiation and determined by its spatial and angular extent has practical consequences on the coupling efficiency when transferring optical power from one component to another in a given optical configuration. It is evident that there is no optical system that can squeeze the phase space area of a given radiation

Moments of Wigner distribution function and phase space ellipse 163

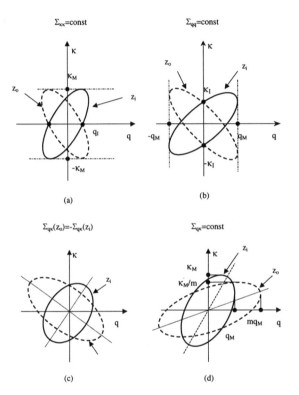

Figure 3 Propagation of the beam ellipse through (a) a diffraction region d, (b) a thin lens f (> 0), (c) a Fourier trasformer f (> 0) and (d) a magnifier m (> 1)

beam. As a consequence, if the phase space area representative of a given optical system -usually called the *acceptance*- is smaller than the phase space area of the radiation that must be coupled to it, some of the optical power will be lost. It is needless to say that the design of optical configuration, where radiation is coupled to some receiver, will greatly benefit from the evaluation of the maximum coupling efficiency achievable, which in turn is a problem of phase space areas (or volumes) and source brightness.

We give now our attention to the Gaussian signals, which play a significant role in optics both as real signals and convenient model signals. Gaussian beams may be introduced in several different ways. As the solution to the scalar paraxial wave equation in free space or as the fundamental modes of optical resonators [§] or as the coherent wavefields generated by the Gauss-Schell model sources.

Here, we briefly comment on Gaussian beams as solutions to the paraxial wave equation and characterize the law of propagation of the relevant parameters.

The simplest solution to the one-dimensional paraxial wave equation in free space takes the form

$$\varphi(q,z) = e^{-i\frac{\pi}{4}} \left(\frac{kQ_0}{\pi}\right)^{\frac{1}{4}} \frac{1}{\sqrt{Q(z)}} \exp\left\{-i\frac{k}{2Q(z)}(q-q_0)^2\right\} \quad (5.3.29)$$

with the constants being chosen in order to fullfil the normalization of a Gaussian beam (5.3.3). The parameter $Q(z)$, usually known as the complex curvature radius, fully characterize the optical properties of (5.3.29). Its z-dependence displays the simple form

$$Q(z) = Q_0 + z \quad (5.3.30)$$

where Q_0 denotes the value at $z = 0$ and is taken to be purely imaginary. It is evident that the real and imaginary parts of $\frac{1}{Q(z)}$ define the curvature radius of the phase fronts and the transverse extent of the beam. It is standard practice to write $\frac{1}{Q(z)}$ in the form

$$\frac{1}{Q(z)} = \frac{1}{R(z)} + i\frac{2}{kw^2(z)} \quad (5.3.31)$$

with $R(z)$ and $w(z)$ being reported as the phase-front radius of curvature and the beam spot size at z of a Gaussian beam. The law of propagation they obey can be easily inferred from the definition (5.3.31) and the rule (5.3.30). We will deduce them later when considering the second order moments of the brightness function.

[§]In this connection, Gaussian beams come out as the eigensolutions of the Huyghens integral equation (4.4.15) in which z_o is ideally separated from z_i by the resonator lenght L and the signal at z_o is required to be $\varphi(q, z_i + L) = \gamma\varphi(q, z_i)$ with the complex factor γ accounting for the diffraction losses and phase shifts.

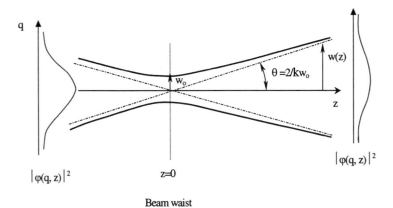

Figure 4 Variation of the beam spot size $w(z)$ of a Gaussian beam along the direction of propagation.

Now, we address the reader to figs. 4-5 to visualize the spreading out of the beam from the minimum size at the waist $z = 0$ and the corresponding evolution of $R(z)$, which shows that in the far zone the wave fronts are indistinguishable from those of a spherical wave diverging from the beam waist.

Turn now to work out the expression of the Wigner distribution function. Evaluating the integral (5.2.5) with the wave function (5.3.29), where for simplicity sake we put $q_0 = 0$, we end up with the following Gaussian form

$$B(q, \kappa, z) = \frac{k}{\pi} \exp\left[-2\frac{q^2}{w^2} - \frac{w^2}{2}(\kappa - \frac{k}{R}q)^2\right] \qquad (5.3.32)$$

which as expected is centered on $q = 0$ and for each point (q, z) the main spatial frequency component is $\kappa(q, z) = \frac{k}{R(z)}q$, specifying the transverse component of the wave vector $\mathbf{k}$ (see the geometry of the beam depicted in figs. 4-5).

Rewriting the Wigner function (5.3.32) in the more convenient model form (5.3.19), it is easy to specialize the matrix Σ of the second order moments as

$$\Sigma = \begin{pmatrix} \frac{w^2}{4} & k\frac{w^2}{4R} \\ k\frac{w^2}{4R} & \frac{1}{w^2} + k^2\frac{w^2}{4R^2} \end{pmatrix} \qquad (5.3.33)$$

which fixes the value of the emittance of a Gaussian beam as

$$\varepsilon_G = \frac{1}{2} \qquad (5.3.34)$$

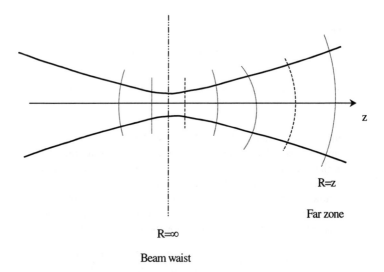

Figure 5 Variation of the radius of curvature of a Gaussian beam. At large distances the beam behaves as a spherical wave diverging from the beam waist.

the minimum achievable value according to (5.3.14).
The above is also in agreement with the general property expressed in (5.3.12); in fact we have

$$\Sigma_{qq}(z)\Sigma_{\kappa\kappa}(z) = \frac{1}{4}(1 + k^2\frac{w^4}{R^2}) \geq \frac{1}{4} \qquad (5.3.35)$$

the equality occurring when $R = \infty$, i.e. at the plane $z = 0$ where $R_0 = \infty$ and the beam spot size w_0 is minimum.

Exploiting the general relation (5.3.16) ruling the propagation of the matrix Σ, we can easily infer the law of propagation of the spot size $w^2(z)$ and the radius of curvature $R(z)$ in free space. Taking the reference frame at $z = 0$, where as already noted $R_0 = \infty$ we obtain the well known relations

$$w^2(z) = w_0^2(1 + \frac{z^2}{z_R^2}) \qquad (5.3.36)$$

$$R(z) = z(1 + \frac{z_R^2}{z^2})$$

where the above introduced parameter z_R

$$z_R \equiv k\frac{w_0^2}{2} \qquad (5.3.37)$$

known as the *Rayleigh distance*, is a measure of the directionality of the beam. It is taken as the borderline between the near zone: $z < z_R$ and the far zone: $z > z_R$, where both w and R increase linearly with the distance: $w \simeq \frac{2}{kw_0}z$ and $R \simeq z$.

As final comment let us note that the Wigner function of a Gaussian beam remain Gaussian after propagating through any arbitrary optical medium; the ellipse matrix Σ changes according to (5.3.16).

5.4 Optical characteristics of synchrotron radiation sources

We turn now to the main topic of the Chapter: to analyze the optical characteristics of synchrotron radiation sources. Therefore, we apply the considerations, developed in the previous sections, to synchrotron radiation sources. This can be accomplished in two steps.

First, we calculate the spectral brightness due to specific insertion devices - bending magnets, wigglers and undulators- according to the definitions (5.2.5), where $\varphi(\mathbf{q}, z)$ can be related to the electric field $E(\mathbf{q}, z)$ of the emitted radiation. Explicitly, we write in MKS units

$$\mathsf{B}(\mathbf{q},\boldsymbol{\kappa},z) = \left(\frac{k}{2\pi}\right)^2 \frac{2\varepsilon_0}{T} \int d\mathbf{q}' \left\langle E^*(\mathbf{q}-\frac{\mathbf{q}'}{2},z)E(\mathbf{q}+\frac{\mathbf{q}'}{2},z)\right\rangle e^{-i\boldsymbol{\kappa}\cdot\mathbf{q}'} \qquad (5.4.1)$$

where T is the time duration of the electric field and the ensemble average has been explicitly displayed to account for possible random fluctuations of the field.

The equivalent definition in terms of the Fourier transform $\widetilde{E}\left(\boldsymbol{\kappa},z\right)$ is straightforwardly written as

$$\mathsf{B}(\mathbf{q},\boldsymbol{\kappa},z) = \left(\frac{k}{2\pi}\right)^2 \frac{2\varepsilon_0}{T} \int d\boldsymbol{\kappa}' \left\langle \widetilde{E}^*\left(\boldsymbol{\kappa}-\frac{\boldsymbol{\kappa}'}{2},z\right) E\left(\boldsymbol{\kappa}+\frac{\boldsymbol{\kappa}'}{2},z\right) \right\rangle e^{-i\boldsymbol{\kappa}'\cdot\mathbf{q}} \qquad (5.4.2)$$

We recall that here, as throughout this Chapter, we will consider a narrow bandwidth $d\omega$ about a given frequency ω. Therefore, $E(\mathbf{q},z)$ in (5.4.1) should be more appropriately regarded as a spectral component of the emitted electric field at temporal frequency ω. As in (5.2.5), the brightness function in (5.4.1) is a spectral brightness although for notational simplicity we omit the subscript ω.

Furthermore, it is worth stressing that eq. (5.4.1) implicitly assumes that the electric field is a scalar quantity, i.e. with a single polarization component; this is consistent with the property of synchrotron radiation, since usually only horizontal component is important.

As second step, we must take account of the electron beam distribution in positions and angles. In this connection, we will derive on a rigorous basis a result, which is intuitively plausible as well: the source brightness is given by the convolution of the brightness due to the *ideal electron* in the beam and the e-beam distribution in size and angle.

Since it is the electron beam optics which determine the brightness as well as the intensity and the time structure of the photon source, we briefly review some of the geometry of electron beams stored into storage rings, already discussed and in more details, in the previous chapters.

It is common practice to utilize a coordinate system related to the *reference orbit*, the orbit fixed by the designed magnetic structure in correspondence with a nominal value of the electron energy and momentum. As previously remarked, few electrons move along the reference orbit as a consequence for instance of errors in electron position and momentum upon injection, of scattering processes and of emission of synchrotron light. The position of a particle moving near the ideal orbit is then specified by (x,y,s), where the curvilinear coordinate s is taken along the electron reference orbit whereas x and y are perpendicular to s and in the horizontal and vertical planes, respectively (fig. 6). The phase space appropriate to the electron beam motion is thereby the 4-dimensional linear space Γ_4 formed by the coordinates (x,y) and their derivatives (x',y'), which in the paraxial approximation turn to be the angles of the electron trajectory to the ideal orbit. The electron distribution in phase space, i.e. in position and angle, is governed by the characteristics of the ring - geometry and magnetic structure - and by the quantum fluctuations of the photon emission. Due to the stochastic character of the latter, the distributions of electrons in x and y position and x' and y' angles about the reference trajectory have Gaussian

Optical characteristics of synchrotron radiation sources

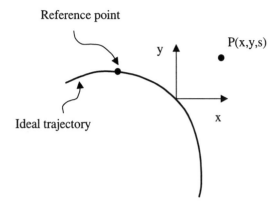

Figure 6 Coordinate system related to the electron reference trajectory, showing the curvilinear coordinate s, the radial coordinate x and the vertical y.

forms, with spatial r.m.s. widths σ_x and σ_y and angular r.m.s. widths $\sigma_{x'}$ and $\sigma_{y'}$. The horizontal and vertical emittances, ε_x and ε_y, define the areas of the projected ellipses in the $x - x'$ and $y - y'$ planes, respectively.

It is obvious that the optical characteristics of the photon source are strongly related to the spatial and angular distribution of the electrons in orbit as well as on the spatial character of the emission process itself. Therefore, in linear dimensions, width and height, the source size is simply fixed by the electron bunch sizes σ_x and σ_y : $\Sigma_x = \sigma_x$ and $\Sigma_y = \sigma_y$. On the other hand, the angular divergence of the optical source follows from the convolution of the electron angular distribution and the angular intensity dependence of synchrotron radiation from a single emitting electron. In particular, if one describes, as usual, the angular distribution of the emitted radiation for both σ- and π-components by a Gaussian with respect to the horizontal angle φ and the vertical ψ, with r.m.s. width σ_R, which is determined for the specific insertion device under consideration, then the angular divergence of the optical source is

$$\Sigma_{x'}^2 = \sigma_{x'}^2 + \sigma_R^2, \qquad \Sigma_{y'}^2 = \sigma_{y'}^2 + \sigma_R^2 \qquad (5.4.3)$$

in the horizontal and vertical planes respectively.

Thus, at some location along the reference orbit we can draw in the phase planes $x - x'$ and $y - y'$ the electron ellipse and the photon ellipse, having the same x and y dimensions but larger x' and y' extents (fig. 7).

As the electron beam proceeds around the orbit, the electron ellipses rotate in phase planes and change their proportions. Therefore, because in applications one

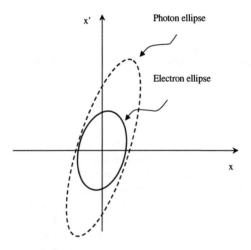

Figure 7 Horizontal phase space ellipses $(\sigma_x, \sigma_{x'})$ and $(\Sigma_x, \Sigma_{x'})$ for the electron beam and for the photons radiated by it at some orbit location s.

Optical characteristics of synchrotron radiation sources

generally collects radiations from a segments of orbit comprising some distance along the orbit, let us say Δs, one must account for the contributions from all the source elements, distributed along the orbit segment Δs and having differing geometrical characteristics. In other words, the light source has spatial and angular extent and it is extended longitudinally along the orbit direction; the appropriate optical phase space should be indeed 6-dimensional.

In order to make the problem somewhat manageable, one can transform the photon distributions from all the emitting elements along the orbit segment Δs to a single plane, thus getting an *effective planar source*, which can be described by a 4-dimensional phase space. This can be accomplished transforming, via the transfer matrix method, the photon ellipse at s to the corresponding ellipse at the center of the emitting Δs segment, for instance, and then summing up the contributions of all the constituent elements along Δs. For a more accurate discussion of this procedure, the reader is addressed to the above quoted references.

Here, we turn now to the brightness function based method and as already announced we will derive the addition theorem, which accounts for the effects of the electron beam size and angular divergence on the source brightness through a convolution over the electron phase space distribution of the brightness due to the reference electron, both taken at some fixed location s_0 along the orbit. This closely resembles the above procedure to transfer the source ellipse at s to the reference plane at s_0.

To establish this result, we will use the definition of B in terms of the Fourier transform $\widetilde{E}(\boldsymbol{\kappa}, z)$ of the electric field; according to the principle of optical reciprocity stated in section IV.4, $\widetilde{E}(\boldsymbol{\kappa}, z)$ relates to the far zone pattern of the radiated field: $E(\mathbf{q}, z + \ell)$ with $\ell \to \infty$.

It is worth noting that within the present context the ideal orbit plays the role as the optical axis in the previous sections. Thus the coordinate z is the same as the azimuth s. We retain here both symbols, the former being usual in light ray optics as the latter is in charged beam optics.

Let the reference plane be located along the reference orbit at $z = 0$. Then, we consider a random collection of electrons, distributed along the ideal orbit according to a Gaussian centered upon the reference electron, the ideal particle in the beam, which moves along the design trajectory with the nominal energy and momentum. In symbols, the longitudinal distribution of the electrons is

$$f(z) = \frac{1}{\sqrt{2\pi}\sigma_z} \exp\left(-\frac{z^2}{2\sigma_z^2}\right) \qquad (5.4.4)$$

where σ_z is the r.m.s. bunch length and the reference electron is at $z = 0$.

At a given time, say $t = 0$, the reference electron at $z = 0$ radiates the field $\widetilde{E}(\boldsymbol{\kappa}, 0)$. Let z_i be the longitudinal position of the i-th electron in the beam at that time. Furthermore, denote by x_i, y_i and x'_i, y'_i the transverse coordinates and angles of the i-th electron at the plane $z = 0$ relative to the reference electron trajectory. If different electrons are statistically independent and the variations of the magnetic guide field across the dimensions of the electron beam are negligible, the contribution of the i-th electron to the field radiated at $z = 0$ into the spatial frequency $\boldsymbol{\kappa}$ is the same as that due to the ideal electron at a frequency suitably corrected through the electron motion directions x'_i, y'_i, but weighted by the appropriate causality factor. In symbols, the electric field at the plane $z = 0$ due to the i-th electron is

$$\widetilde{E}_i\left(\boldsymbol{\kappa}, 0\right) = \exp\left[i\left(kz_i - \boldsymbol{\kappa}\cdot\mathbf{q}_i\right)\right]\widetilde{E}_0\left(\boldsymbol{\kappa} - \boldsymbol{\kappa}_i, 0\right) \qquad (5.4.5)$$

where $\widetilde{E}_0$ is the electric field due to the ideal electron, $\mathbf{q}_i$ is the electron position vector: $\mathbf{q}_i = (x_i, y_i)$ and $\boldsymbol{\kappa}_i = k(x'_i, y'_i)$, k being the wavenumber $k = \frac{2\pi}{\lambda}$..

The total electric field is obtained summing up the contributions (5.4.5), namely

$$\widetilde{E}\left(\boldsymbol{\kappa}, 0\right) = \sum_{i=1}^{N_e}\widetilde{E}_i\left(\boldsymbol{\kappa}, 0\right) \qquad (5.4.6)$$

N_e being the total number of electrons.

Having in mind eq. (5.4.2) we calculate the ensemble average

$$\left\langle\widetilde{E}^*\left(\boldsymbol{\kappa}_1, 0\right)\widetilde{E}\left(\boldsymbol{\kappa}_2, 0\right)\right\rangle =$$
$$\sum_{i=1}^{N_e}\left\langle\widetilde{E}_i^*\left(\boldsymbol{\kappa}_1, 0\right)\widetilde{E}_i\left(\boldsymbol{\kappa}_2, 0\right)\right\rangle + \sum_{i\neq j=1}^{N_e}\left\langle\widetilde{E}_i^*\left(\boldsymbol{\kappa}_1, 0\right)\widetilde{E}_j\left(\boldsymbol{\kappa}_2, 0\right)\right\rangle \qquad (5.4.7)$$

where the average involves the longitudinal and transverse (spatial and angular) distribution of the electrons.

Since there are no correlations between the longitudinal motion and the transverse, we can separately calculate the averages over the z- and (x, y, x', y')-distributions.

Thus, for the average over the longitudinal distribution in the second summation in (5.4.7) we obtain

$$\left\langle e^{-ikz_i}\right\rangle\left\langle e^{-ikz_j}\right\rangle = \left|\int e^{ikz}f(z)dz\right|^2 = \exp(-k^2\sigma_z^2) \qquad (5.4.8)$$

with the probability density (5.4.4). As it is usually the case that $\exp(-k^2\sigma_z^2) \ll \frac{1}{N_e}$, we can neglect the second summation in (5.4.7).

Furthermore, noting that the first term on the rhs of eq. (5.4.7) is the sum over identical contributions, we can write

$$\left\langle\widetilde{E}^*\left(\boldsymbol{\kappa}_1, 0\right)\widetilde{E}\left(\boldsymbol{\kappa}_2, 0\right)\right\rangle = N_e\left\langle e^{i(\boldsymbol{\kappa}_1-\boldsymbol{\kappa}_2)\cdot\mathbf{q}_e}\widetilde{E}_0^*\left(\boldsymbol{\kappa}_1 - \boldsymbol{\kappa}_e, 0\right)\widetilde{E}_0\left(\boldsymbol{\kappa}_2 - \boldsymbol{\kappa}_e, 0\right)\right\rangle \qquad (5.4.9)$$

where the average is understood over the transverse spatial and angular electron distribution, the electron variables being $\mathbf{q}_e = (x_e, y_e)$ and $\boldsymbol{\kappa}_e = k(x'_e, y'_e)$.

Optical characteristics of synchrotron radiation sources

We denote by $\rho_e(x, y, x', y', z)$ the electron transverse phase space distribution, which as noted is usually given by a Gaussian according to

$$\rho_e(x, y, x', y', z) = \frac{1}{2\pi\varepsilon_x\varepsilon_y} \exp\left\{-\frac{1}{2\varepsilon_x^2\varepsilon_y^2}\left[\mathbf{u}^{\mathsf{T}}\mathbf{\Sigma}^{-1}\mathbf{u}\right]\right\} \quad (5.4.10)$$

where $\mathbf{u}$ is the 4-dimensional vector $\mathbf{u} = \begin{pmatrix} x \\ y \\ x' \\ y' \end{pmatrix}$ and $\mathbf{\Sigma}$ is the matrix of second order moments and $\varepsilon_x, \varepsilon_y$ are the horizontal and vertical beam emittances.

Finally, inserting the expression (5.4.9) with $\kappa_1 = \kappa - \frac{\kappa'}{2}$ and $\kappa_2 = \kappa + \frac{\kappa'}{2}$ into the definition (5.4.2) we obtain

$$\mathsf{B}(\mathbf{q}, \boldsymbol{\kappa}, 0) = N_e \left\langle \mathsf{B}_0(\mathbf{q} - \mathbf{q}_e, \boldsymbol{\kappa} - \boldsymbol{\kappa}_e, 0)\right\rangle \quad (5.4.11)$$

which making the average explicit gives

$$\mathsf{B}(\mathbf{q}, \boldsymbol{\kappa}, 0) = N_e \int d\mathbf{q}' \int d\boldsymbol{\kappa}' \mathsf{B}_0(\mathbf{q} - \mathbf{q}', \boldsymbol{\kappa} - \boldsymbol{\kappa}', 0)\rho_e(\mathbf{q}', \boldsymbol{\kappa}') \quad (5.4.12)$$

The above expresses the important result that the total brightness at the reference plane $z = 0$ due to the electron beam is calculated as the convolution over the electron transverse phase space distribution function at the reference plane of the brightness due to a single electron.

It is worth noting that as the brightness relates to the phase space density distribution function of the radiation field, the expression (5.4.12) takes the meaning as the convolution between the phase space probability density of the electrons and the field radiated by the ideal electron.

As to the latter, we note that it can be inferred from the expression (2.2.3) for the far-field pattern at frequency ω on account of the optical reciprocity statement (4.4.23), thus getting

$$\widetilde{E}(\boldsymbol{\kappa}, 0) = \frac{1}{\sqrt{2\pi}} \frac{e}{4\pi\varepsilon_0 c} \int dt (\mathbf{n} \times (\mathbf{n} \times \boldsymbol{\beta})) \exp\left[i\omega(t - \frac{\mathbf{n}\cdot\mathbf{r}}{c})\right] \quad (5.4.13)$$

where the symbols are according to the geometry of fig. 4 and the spatial frequency $\boldsymbol{\kappa}$ is fixed by the transverse component of $\mathbf{n}$

$$\boldsymbol{\kappa} = k(n_x, n_y) \quad (5.4.14)$$

The result established in (5.4.12) provides a rigorous procedure to calculate the brightness function for synchrotron radiation once specified the electron beam phase space density distribution function $\rho_e(x, y, x', y', z)$ and the insertion device, and hence the electric field associated with a moving electron at the chosen reference plane.

This represents a complete characterization of the optical properties of the source. In fact, as showed in section 2, the brightness function propagates through an arbitrary optical medium according to the simple law (5.2.24). Moreover, integration of $B(q, \kappa, z)$ over q and κ provides the flux density and the angular distribution.

We compute now the brightness for undulator radiation. As the exact calculation of the brightness function due to a single electron is quite complicated, we exploit the approximate form for the angular distribution of the undulator radiation as a Gaussian with an appropriate r.m.s. width. Thus, we assume for the Fourier transform $\tilde{E}(\kappa, 0)$ of the electric field the Gaussian shape

$$\tilde{E}(\kappa, 0) \propto \exp(-\frac{\kappa^2}{k^2 \sigma_{R'}^2}) \tag{5.4.15}$$

where $\kappa^2 = \kappa_x^2 + \kappa_y^2$ and, for the first harmonic,

$$\sigma_{R'} = \sqrt{\frac{\lambda}{N\lambda_u}} \tag{5.4.16}$$

where λ_u is the undulator period and N the number of periods.

Accordingly, the electric field $E(q, 0)$ will have a Gaussian shape as well

$$E(q, 0) \propto \exp(-\frac{q^2}{\sigma_R^2}) \tag{5.4.17}$$

with spatial width

$$\sigma_R = \frac{2}{k\sigma_{R'}} \tag{5.4.18}$$

Using the definition (5.4.1) or (5.4.2) we can write down the brightness produced by a single electron at the plane $z = 0$, namely

$$B_0(q, \kappa, 0) = B_0 \exp\left[-2\left(\frac{q^2}{\sigma_R^2} + \frac{\kappa^2}{k^2 \sigma_{R'}^2}\right)\right] \tag{5.4.19}$$

where B_0 denotes the invariant value of the brightness at the phase space origin.

We resort now to the addition theorem (5.4.12) and calculate the convolution of the above derived brightness function with the electron phase space distribution function at $z = 0$. Assuming for the latter the product of two Gaussians, which specify the electron distribution in position and angle separately in the horizontal and vertical direction, i.e.

$$\rho_e(x, y, x', y', 0) = \rho_x(x, x')\rho_y(y, y') \tag{5.4.20}$$

and

Optical characteristics of synchrotron radiation sources

$$\rho_x(x, x') = \frac{1}{2\pi\varepsilon_x} \exp\left\{-\frac{1}{2}(\frac{x^2}{\sigma_x^2} + \frac{x'^2}{\sigma_{x'}^2})\right\} \qquad (5.4.21)$$

and similarly for $\rho_y(y, y')$. In the above, we have further supposed for simplicity that the $x - x'$ and $y - y'$ electron ellipses are oriented along the axes and hence the emittances are simply $\varepsilon_x = \sigma_x \sigma_{x'}$ and $\varepsilon_y = \sigma_y \sigma_{y'}$.

Combining therefore the expressions (5.4.5) and (5.4.20-5.4.21) as required by (5.4.12), we obtain the brightness due to the electron beam at $z = 0$ in the form

$$B_U(\mathbf{q}, \boldsymbol{\kappa}, 0) = B_U(0) \exp\left[-\frac{1}{2}\left(\frac{q_x^2}{\Sigma_x^2} + \frac{q_y^2}{\Sigma_y^2} + \frac{\kappa_x^2}{k^2 \Sigma_{x'}^2} + \frac{\kappa_y^2}{k^2 \Sigma_{y'}^2}\right)\right] \qquad (5.4.22)$$

where

$$\begin{aligned}\Sigma_x^2 &= \sigma_x^2 + \frac{\sigma_R^2}{4}, & \Sigma_{x'}^2 &= \sigma_{x'}^2 + \frac{\sigma_{R'}^2}{4} \\ \Sigma_y^2 &= \sigma_y^2 + \frac{\sigma_R^2}{4}, & \Sigma_{y'}^2 &= \sigma_{y'}^2 + \frac{\sigma_{R'}^2}{4}\end{aligned} \qquad (5.4.23)$$

The peak value $B_U(0)$ is obtained integrating (5.4.22) over all phase space domain and equating the final expression to the total radiant flux P, thus yielding

$$B_U(0) = \frac{P}{(2\pi)^2 \Sigma_x \Sigma_{x'} \Sigma_y \Sigma_{y'}} \qquad (5.4.24)$$

which thus completes the expression for the brightness of undulator radiation.

We specialize (5.4.22) to the limiting cases corresponding to specific relation between the widths - in position and angle- of the electron and photon beams. Then, we consider

1. *electron emittance dominated regime*, described by the relation

$$\sigma_x, \sigma_y \gg \sigma_R, \quad \text{and} \quad \sigma_{x'}, \sigma_{y'} \gg \sigma_{R'} \qquad (5.4.25)$$

In that case, the peak brightness is

$$B_U(0) = \frac{P}{(2\pi)^2 \varepsilon_x \varepsilon_y} \qquad (5.4.26)$$

and since P scales as the number N of undulator periods, so does $B_U(0)$ as well.

2. *diffraction limited regime*, described by the opposite relations

$$\sigma_x, \sigma_y \ll \sigma_R, \quad \text{and} \quad \sigma_{x'}, \sigma_{y'} \ll \sigma_{R'} \qquad (5.4.27)$$

which gives

$$B_U(0) = \frac{P}{\left(\frac{\pi}{2}\sigma_R \sigma_{R'}\right)^2} = \frac{P}{\left(\frac{\lambda}{2}\right)^2} \qquad (5.4.28)$$

176   Wigner distribution and synchrotron radiation sources

which again scales as N.

3. The intermediate regime is characterized by a mixture of the relations (5.4.25) and (5.4.27), namely

$$\sigma_x, \sigma_y \gg \sigma_{\mathsf{R}}, \quad \text{and} \quad \sigma_{x'}, \sigma_{y'} \ll \sigma_{\mathsf{R}'} \qquad (5.4.29)$$

Then, the peak brightness becomes

$$\mathsf{B}_{\mathsf{U}}(0) = \frac{P}{\pi^2 \sigma_x \sigma_y \sigma_{\mathsf{R}'}^2} \qquad (5.4.30)$$

which being $\sigma_{\mathsf{R}'} \propto N^{-\frac{1}{2}}$ goes with N^2.

As final comment, let us notice that in the case **2)** above, that is when the electron beam phase space volume is smaller than that of the undulator radiation produced by a single charge, the phase space volume associated with undulator radiation is nearly close to the minimum possible value. Thereby, in that case the undulator radiation is fully spatially coherent.

5.5 Problems

1. Exploiting the definition of $\langle \hat{p} \rangle$ given in Problem IV.5 and using the Fourier integral prove the relation (5.2.3).

2. Find the G-function for elementary optical systems as a free section of length d and a thin lens of focal length f and compound optical systems as a magnifier and a Fourier transformer (see also Problem 4).

3. Determine the radiant flux through an aperture with width $2a$ parallel to the source plane and symmetrically located around the z-axis at the plane z_o. Let the source be located at z_i.

4. Write down the general expression of the ray-spread function $\mathsf{G}(q, q\prime, \kappa, \kappa')$ relative to a cascade of optical systems in terms of the ray-spread function of each optical component in the cascade. (Apply the results to the second part of Problem 2.)

5. On account of the laws of transformations (5.3.17) and (5.3.30) find a formal expression of the parameter $Q(z)$ (or equivalently $\frac{1}{Q(z)}$) in terms of the second order moments Σ_{qq}, Σ_{pp} and Σ_{qp}. (There are two possible expressions.)

6. Deduce the equations of motion obeyed by the matrix Σ and hence by the second order moments of the brightness function.

7. Consider a pure phase modulation system, described by an input-output relation as

$$\varphi(q, z_o) = m(q)\varphi(q, z_i)$$

with $m(q) = \exp(i\gamma(q))$, $\gamma(q)$ real. Show that at the first order the propagation manifests itself through a shift of the frequency κ by the derivative $\frac{d\gamma}{dq}$.

8. From the results of Problem 6 deduce the equation ruling the variation of $\Sigma_{qq}(z)$ due to the propagation through a medium described by a z-independent Hamiltonian H. Then, infer the law (5.3.36) for the spot-size of a freely propagated gaussian beam.

Suggested bibliography

As to the Wigner distribution function we suggest the papers

1. E.P. Wigner "On the quantum correction for thermodynamic equilibrium" Phys. Rev. **40**, 749 (1932).
2. L. Cohen "Generalized phase-space distribution functions", J. Math. Phys **7**, 781-786 (1966).
3. M.D. Srinivas and E. Wolf "Some nonclassical features of phase-space representations of quantum mechanics" Phys. Rev. D **11**, 1477-1484 (1975).
4. R.F. O'Connell and E.P. Wigner "Quantum-mechanical distribution functions: conditions and uniqueness" Phys. Lett. A **83**, 145-147 (1981).
5. M.Millery, R.F. O'Connell, M.O. Scully and E.P. Wigner "Distribution functions in Physics: fundamentals" Phys. Rep. **106**, 121-167 (1984).

   The problems inherent to the applications of the Wigner distribution function formalism to optics and to the link between the optical Wigner distribution function and the brightness are treated in
6. A. Walther "Radiometry and Coherence" J. Opt. Soc. Am. **58**, 1256-1259 (1968).
7. A. Walther "Radiometry and Coherence" J. Opt. Soc. Am. **63**, 1622-1623 (1973).
8. E. Wolf "Coherence and Radiometry" J. Opt. Soc. Am. **68**, 6-17 (1977).
9. Ari T. Friberg "On the existence of a radiance function for finite planar source of arbitrary states of coherence" J. Opt. Soc. Am. **69**, 192-198 (1978).
10. M.J. Bastiaans "The Wigner distribution function applied to optical signals and systems" Opt. Comm. **25**, 26-30 (1978).
11. M.J. Bastiaans "Wigner distribution function and its application to first-order optics" J. Opt. Soc. Am. **69**, 1710-1716 (1979).
12. M.J. Bastiaans "The Wigner distribution function of partially coherent light" Optica Acta **28**, 1215-1224 (1981).
13. M.J. Bastiaans "Applications of the Wigner distribution function to partially coherent light" J. Opt. Soc. Am. A **3**, 1227-1238 (1986).

    The optics of synchrotron radiation is discussed in
14. G.K. Green "Spectra and optics of synchrotron radiation" BNL 50522 (1976).
15. S. Krinsky, M.L. Perlman and R.E. Watson "Characteristics of synchrotron radiation and its sources" in Handbook of Synchrotron Radiation, Vol. I, ed. E.E. Koch, North Holland, Amsterdam 1983.
16. K.-J. Kim "Brightness, coherence and propagation characteristics of synchrotron radiation", Nucl. Instr. and Meth. in Phys. Res. A **246**, 71-76 (1986), *where in particular the application of the Wigner distribution function based method is suggested and developed.*

Chapter 6
SYNCHROTRON RADIATION SOURCES, INSERTION DEVICES AND BEAM CURRENT LIMITATIONS

Summary

We discuss problems related to synchrotron radiation brightness, we analyze the role played by the e-beam emittance and examine various lattice schemes, proposed to optimize the emittance. We discuss the diffraction limited operation its physical meaning and explore the effect of the insertion device on the machine parameters. We present general considerations on the polarization properties of insertion devices and, finally, we analyze the effects which may limit the stored beam current.

6.1 Introduction

In chapter I we have discussed the optical properties of charged beam transport, we have introduced the concept of beam emittance and we have shown how this quantity in linked to the beam transverse dimensions and divergences. In chapter II we have discussed the dynamical interplay between beam emittance, synchrotron radiation power and machine parameters. In chapters IV, V we have studied the optical properties of the synchrotron radiation and it has been stressed that, within such a context a major role is played by the beam transverse dimensions and divergences which provide kind of quality parameters.

Since the emitting source is not point-like, the r.m.s. transverse dimensions of the photon beam, as well as its divergences, will be related to those of the electron beam, through the quadratic composition

$$\Sigma_\xi = \sqrt{\sigma_\xi^2 + \frac{\lambda L_u}{(4\pi)^2}} \qquad (6.1.1)$$

$$\Sigma'_\xi = \sqrt{\sigma'^2_\xi + \frac{\lambda}{L_u}}$$

as already remarked σ_ξ, σ'_ξ are relevant to the source part, while

$$\sigma_f = \frac{(\lambda L_u)^{1/2}}{4\pi} \quad , \quad \sigma'_f = \left(\frac{\lambda}{L_u}\right)^{1/2} \qquad (6.1.2a)$$

are the photon-beam diffraction limited transverse section and divergence respectively.

We can associate with 6.1.2a the "emittance"

$$\sigma_f \sigma'_f = \frac{\lambda}{4\pi} \qquad (6.1.2b)$$

and the Twiss parameters

$$\beta_f = \frac{L_u}{4\pi} \quad , \quad \gamma_f = \frac{4\pi}{L_u} \qquad (6.1.2c)$$

We have dwelt on the above points to emphasize the main thread of the history: the equilibrium beam dimensions depend on the radiated power and in turn are a crucial element to define the quality of the emitted radiation.

We have stressed the concept of quality factors of synchrotron radiation and we have formulated a quantitative definition, which is not provided by the photon flux, the real figure of merit being indeed the spectral brightness (in short brilliance), defined as the photon flux per unit solid angle and source area, emitted in a relative bandwidth (b.w.)

$$\mathfrak{B} = \frac{d^4 N}{dt\, d\Omega\, dS\, (d\lambda/\lambda)} \qquad (6.1.3)$$

which is usually expressed in

$$\frac{photons}{s\ mrad^2\ mm^2\ 0.1\%\ b.w.}$$

It is also evident that since dS is the source area, apart from diffraction effects $dS\, d\Omega \simeq \varepsilon_x \varepsilon_y$, therefore the smaller is the emittance the larger is the brightness.

An idea of the evolution of the X-ray brilliance during the years is provided by fig. 1.

We can illustrate more quantitatively the above point, by considering the case of the Brightness or peak brillance which in the case of the undulator can be defined as

$$\mathfrak{B} = \frac{\mathcal{F}}{(2\pi)^2 \Sigma'_x \Sigma'_y \Sigma_x \Sigma_y} \qquad (6.1.4)$$

where $\mathcal{F}$ is the photon flux introduced in chapter II.

Just to fix ideas and to specify some reference numbers we consider two limiting cases of the above definition

a) Beam-emittance dominated brightness

by assuming that ε_x, $\varepsilon_y \gg \frac{\lambda}{4\pi}$ we end up with

$$\mathfrak{B} = \frac{\mathcal{F}}{(2\pi)^2 \varepsilon_x \varepsilon_y} \qquad (6.1.5a)$$

Introduction

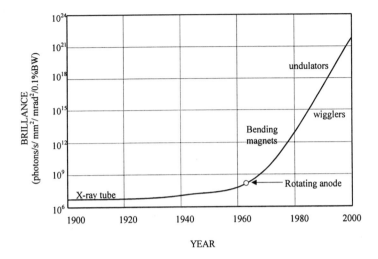

Figure 1   Increase of brillance over this century.

b) <u>Diffraction limited brightness</u>
in this case we assume point like sources and therefore

$$\mathcal{B} = \frac{4\mathcal{F}}{\lambda^2} \qquad (6.1.5b)$$

It is clear that orders of magnitudes may be the difference between the two limits. To give an example of best achieved performances we consider the case of ESRF source (January 1996):

$$\varepsilon_x \cong 4 \cdot 10^{-9}\, m \cdot rad \quad , \quad \varepsilon_y \cong 4 \cdot 10^{-11}\, m \cdot rad$$

which for photons of $12\, KeV$ $\left(1\, \text{Å}\right)$ yield

$$\frac{\varepsilon_x\, \varepsilon_y}{\varepsilon_p^2} \cong 16\, \pi^2$$

We expect therefore that the diffraction limited brightness be 2-orders of magnitudes larger than the best presently available value, which ranges around 10^{20}-10^{21}. The above numbers become more dramatic if confronted with the brightness of the first generation sources, of the order of 10^{13}.

Larger brightness means larger degree of coherence and thus the possibility of using synchrotron radiation for applications like phase-contrast imaging, diffraction topography etc..

After these general introductory remarks we are ready to discuss more specific and quantitative topics in the next sections.

6.2 Undulator brightness

The flux of a magnetic undulator has been derived in chapter II. The undulator spectrum consists of a series of narrows peaks centered at

$$\varepsilon_n [KeV] = \frac{0.95 \, n \, E^2 \, [GeV]}{\lambda_u [cm] \left(1 + \frac{k^2}{2}\right)} \tag{6.2.1}$$

with associated photon emittances given by

$$\varepsilon_{ph} [m \cdot rad] \simeq 1.039 \cdot \frac{10^{-10}}{n} \frac{\lambda_u [cm]}{E^2 [GeV]} \left(1 + \frac{k^2}{2}\right) \tag{6.2.2}$$

at low k the fundamental peak dominates and as one increases the k parameter, more and more harmonics appear and the fundamental is shifted towards lower frequencies

The brightness can therefore be written as

$$\mathfrak{B}_n = \frac{1.431 \cdot 10^{14} \, N \, Q_n \, I \, [A]}{(2\pi)^2 \, \Sigma_{x,n} \, \Sigma'_{x,n} \, \Sigma_{y,n} \, \Sigma'_{y,n}} \tag{6.2.3}$$

and we can more conveniently specialize Σ_ξ and Σ'_ξ as

$$\begin{aligned}
\Sigma_{\xi,n} &= \sigma_\xi \sqrt{1 + \left(\frac{\beta_{ph}}{\beta_\xi}\right)^2 \frac{\varepsilon_{ph}}{\varepsilon_\xi}} \\
\Sigma'_{\xi,n} &= \sigma'_\xi \sqrt{1 + \left(\frac{\beta_\xi}{\beta_{ph}}\right)^2 \frac{\varepsilon_{ph}}{\varepsilon_\xi}}
\end{aligned} \tag{6.2.4}$$

The dependence on n of the optical beam section and divergences is due to the dependence on the harmonic number contained in the photon emittance, see equation 6.2.2.

The term in the square root may just be a negligible correction as in the $1^{\underline{st}}$ generation source or the dominating part as in the diffraction limited dominated source.

In fig. 2 we have reported the undulator brightness vs. the photon energy, for three different values of the emittance, differing by two orders of magnitudes.

The three curves can be representative of the $1^{\underline{st}}$, $2^{\underline{nd}}$, $3^{\underline{rd}}$ generation sources. The case with larger brightness corresponds to the ESRF parameters.

It is now important to appreciate the difference with the diffraction limited case. To this aim we note that the photon emittance in practical units is given by equation 6.2.2, so that the diffraction-limited brightness writes

Wiggler and bending magnets

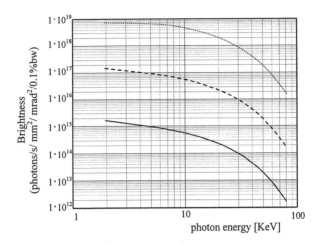

Figure 2 Comparison between achievable brightness in the 1$^{\underline{st}}$, 2$^{\underline{nd}}$, 3$^{\underline{rd}}$ generation sources. The case with larger brightness corresponds to the ESRF parameters.

$$\mathfrak{B}_n \left[\#photons/s/0.1\%\, b.w./mm^2/mrad^2 \right] = 3.356 \cdot 10^{20} \frac{N Q_n E^4 \left[GeV\right] I\left[A\right]}{\lambda_u \left[cm\right]^2 \left(1 + \frac{k^2}{2}\right)^2} \quad (6.2.5)$$

In this case by keeping constant λ_u, k, N the brightness scales as the fourth power of the energy, for further remarks on this point see the concluding section to this chapter. Fig. 3 provides a comparison between a third generation source and the diffraction limited case,
it is evident that this limit may provide brightness values three orders of magnitude larger than the presently achievable best performances.

6.3 Wiggler and bending magnets

As already remarked in chapter II, the spectral properties of the emitted radiation changes with increasing k. In particular, when k becomes very large the change is dramatic. The wiggler spectrum can indeed be viewed as the incoherent sum of bending magnet spectra originated from $2N$ point sources.

The approximation of the flux given in the previous section holds in the case in which small angular divergence and small irradiated area are required. According to chapter II the flux distribution of wiggler magnet radiation is given by $2N$ (with N being the number of magnet periods) the appropriate formula for bending magnets. Accordingly we get

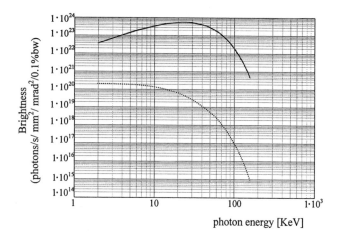

Figure 3 Comparison between brightness achievable in the diffraction limit case (continuous line) and with the present performances in third generation machines (dotted line).

$$\left[\frac{d\mathcal{F}}{d\Omega\, d\omega/\omega}\right]_{\psi=0} = 1.325 \cdot 10^{13} E^2 \,[GeV]\, I\,[A]\, 2N \left(\frac{\varepsilon}{\varepsilon_c}\right) K_{2/3}^2 \left(\frac{\varepsilon}{2\,\varepsilon_c}\right), \quad (6.3.1)$$

$$\varepsilon_c\,[KeV] = 0.665 \cdot E^2\,[GeV]\, B\,[T]$$

which represents the on axis radiated flux reported in fig. 4 for the ESRF source, the lower curve yields the same quantity for bending magnet operation.

In fig. 5 we have also reported the relevant brightness. It is evident that for a large range of photon energies undulators may provide the largest brightness values.

Before concluding this section we want to mention two points

a) the computation of the modified Bessel function $K_\nu(x)$ can be easily simplified by the use of the following algorithm

$$K_\nu(x) \cong h\left[\frac{e^{-x/2}}{2} + \sum_{r=1}^M \exp\left(-x\cosh(r\,h)\right)\cdot\cosh(\nu\,r\,h)\right] \quad (6.3.2a)$$

which also yields

$$\int_x^\infty K_\nu(\eta)\,d\eta \cong h\left[\frac{e^{-x/2}}{2} + \sum_{r=1}^M \frac{\exp\left(-x\cosh(r\,h)\right)}{\cosh(r\,h)}\cdot\cosh(\nu\,r\,h)\right] \quad (6.3.2b)$$

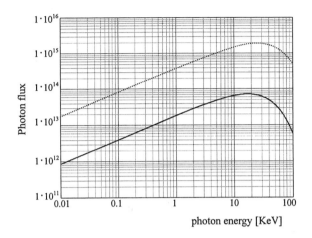

Figure 4 Wiggler magnet photon flux vs. photon energy. The dotted line is calculated for the ESRF parameters. The continuous curve refers to a bending magnet photon flux.

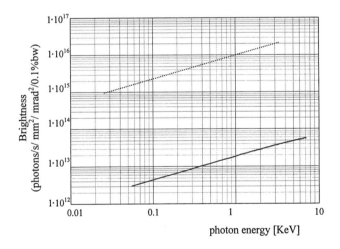

Figure 5 Wiggler magnet brightness vs. photon energy. The dotted line is calculated for the ESRF parameters. The continuous curve refers to a bending magnet photon flux.

for most practical case h can be chosen around 0.5 and M around 30.

b) We have not mentioned an other quantity which plays a significant role, namely the flux generated per unit horizontal angle (obtained by integrating equation 6.3.1 over the vertical angle) we get

$$\frac{d\mathcal{F}}{d\vartheta} = 2.457 \cdot 10^{13} E\,[GeV]\,I\,[A]\,2N\,S\left(\frac{\varepsilon}{\varepsilon_c(\vartheta)}\right) \tag{6.3.3a}$$

where

$$\varepsilon_c(\vartheta) = \varepsilon_c \left[1 - \frac{\gamma\vartheta}{k}\right]^{1/2} \tag{6.3.3b}$$

The dependence of the critical frequency on ϑ is due to the fact that the magnetic field must be taken at the point of electron's trajectory tangent to the direction of observation.

An idea of the dependence of 6.3.3a, namely of the flux generated by a wiggler per unit horizontal angle, on the photon energy is given in fig. 5.

6.4 Polarization characteristics of insertion devices

In chapter II we have marginally touched on the problems associated with the polarization of .

As already remarked, planar undulators generate linearly polarized radiation, but in applications, different types of polarizations may be required. Helical, crossed and variably polarized undulators have been proposed and exploited to generate intense radiation with "exotic" polarization properties. In this section, we give the basic tools, to treat the polarization aspects of synchrotron radiation and then we apply these concepts to a specific example.

We recall here a few basic concepts about Jones vectors and matrices, a deeper discussion can be found in the suggested bibliography at the end of the chapter.

The general polarization state can be specified by means of a column vector $\underline{v}$ of the form

$$\underline{v} = \begin{pmatrix} a_x e^{i\delta_x} \\ a_y e^{i\delta_y} \end{pmatrix} \tag{6.4.1}$$

The elements of the vector are projection on the x and y axes of a given reference frame of the vector representing the electromagnetic field. The positive quantities $a_{x,y}$ are the wave amplitudes while the real quantities $\delta_{x,y}$ are the initial phases at the considered point. The polarization state is not affected by the multiplication of $a_{x,y}$ by a common factor. The normalization can be therefore made by setting equal to one, in suitable units, the quantity $|a_x|^2 + |a_y|^2$ which is proportional to the wave intensity. The polarization does not change if the same contribution to the phase is added to $\delta_{x,y}$. Accordingly, we could always set δ_x equal to zero and

Polarization characteristics of insertion devices

make the corresponding adjustment to the phase δ_y. Linear polarization states are characterized by $\delta_x - \delta_y = 0$ or π.

Denoting by ϑ the angle between the direction of polarization and the x-axis we define, as Jones vector, the quantity

$$\underline{l}_\vartheta = \begin{pmatrix} \cos \vartheta \\ \sin \vartheta \end{pmatrix} \qquad (6.4.2)$$

($\vartheta = 0$ and $\vartheta = \frac{\pi}{2}$ to x, y polarizations)

We consider as particular cases $\vartheta = \frac{\pi}{4}$ and $\vartheta = -\frac{\pi}{4}$, which leads to

$$\underline{l}_{\frac{\pi}{4}} = \frac{1}{\sqrt{2}} \begin{pmatrix} 1 \\ 1 \end{pmatrix} \quad , \quad \underline{l}_{-\frac{\pi}{4}} = \frac{1}{\sqrt{2}} \begin{pmatrix} 1 \\ -1 \end{pmatrix} \qquad (6.4.3)$$

The circular polarization states are characterized by x and y components with same amplitudes and phase-difference $\delta_x - \delta_y = \pm \frac{\pi}{2}$, which yields right and left circular polarizations

$$\underline{l}_r = \frac{1}{\sqrt{2}} \begin{pmatrix} 1 \\ i \end{pmatrix} \quad , \quad \underline{l}_l = \frac{1}{\sqrt{2}} \begin{pmatrix} 1 \\ -i \end{pmatrix} \qquad (6.4.4)$$

the effect of an optical component acting on the polarization state can be described by means of a suitable 2x2 Jones matrix $\widehat{A}$, namely

$$\underline{v}' = \widehat{A}\,\underline{v} \qquad (6.4.5a)$$

i.e.

$$\begin{pmatrix} a'_x e^{i\delta'_x} \\ a'_y e^{i\delta'_y} \end{pmatrix} = \begin{pmatrix} \zeta_1 & \zeta_2 \\ \zeta_3 & \zeta_4 \end{pmatrix} \begin{pmatrix} a_x e^{i\delta_x} \\ a_y e^{i\delta_y} \end{pmatrix} \qquad (6.4.5b)$$

where ζ_j are the complex entries of $\widehat{A}$. An optical component which does not change the intensity of $\underline{v}$ is characterized by a matrix with

$$\left| \det \widehat{A} \right| = 1 \qquad (6.4.6a)$$

where $\det \widehat{A}$ denotes the determinant of $\widehat{A}$ and $\left| \det \widehat{A} \right|$ its modulus. If 6.4.6a is satisfied $\widehat{A}$ is said to provide a unitary transformation, if $\det \widehat{A} = 1$ it is said special unitary. In the second case $\widehat{A}$ can be parametrized as

$$\widehat{A} = \begin{pmatrix} \zeta_1 & \zeta_2 \\ -\zeta_2^* & \zeta_1^* \end{pmatrix} \quad , \quad |\zeta_1|^2 + |\zeta_2|^2 = 1 \qquad (6.4.6b)$$

Several components are able to change the polarization without altering the intensity, the most important examples are the wave plates

$$\widehat{M}_0(\varphi) = \begin{pmatrix} 1 & 0 \\ 0 & e^{i\varphi} \end{pmatrix} \quad (6.4.7)$$

and rotators

$$\widehat{R}(\alpha) = \begin{pmatrix} \cos\alpha & -\sin\alpha \\ \sin\alpha & \cos\alpha \end{pmatrix} \quad (6.4.8)$$

It is worth noting that

$$\widehat{M}_0\left(\frac{\pi}{2}\right) \underline{l}_{\frac{\pi}{4}} = \underline{c}_r \quad (6.4.9)$$
$$\widehat{M}_0\left(\frac{\pi}{2}\right) \underline{l}_{-\frac{\pi}{4}} = \underline{c}_l$$

i.e. linear polarization at an angle $\pm\frac{\pi}{4}$, is transformed into right (left) circular polarization by a wave-plate matrix at $\frac{\pi}{2}$, conversely we have

$$\widehat{M}_0\left(-\frac{\pi}{2}\right) \underline{c}_r = \underline{l}_{\frac{\pi}{4}} \quad (6.4.10)$$
$$\widehat{M}_0\left(-\frac{\pi}{2}\right) \underline{c}_l = \underline{l}_{-\frac{\pi}{4}}$$

i.e. a wave-plate matrix at $-\frac{\pi}{2}$ transforms circularly into linearly polarized light.

In general the Jones vectors can be used to describe the polarization basis, an other important concept associated with the polarization vectors is the polarization rate (P.R.), defined as the difference between the intensities of two orthogonal directions divided by the total intensity. In the case of the vector 6.4.1 we will define the polarization rate

$$\mathfrak{p}_v = \frac{|a_x|^2 - |a_y|^2}{|a_x|^2 + |a_y|^2} \quad (6.4.11)$$

More in general we introduce the P.R., $\mathfrak{p}_v$ for the x,y polarization, $\mathfrak{p}_T$ for the $\underline{l}_{\frac{\pi}{4},-\frac{\pi}{4}}$ case and $\mathfrak{p}_H$ for the circularly polarized states. Along with the above rates, we should introduce also $\mathfrak{p}_D$, namely the "natural" or unpolarized component of the light. Completely unpolarized light has the property that the intensity is the same in any axis of an arbitrary orthogonal direction. P.R.s are dimensionless quantities between -1 and 1, are not independent and related to each other through

$$\mathfrak{p}_v^2 + \mathfrak{p}_H^2 + \mathfrak{p}_T^2 + \mathfrak{p}_D^2 = 1 \quad (6.4.12)$$

The use of the above concepts allows a more clear and rigorous treatment of the polarization properties of synchrotron radiation.

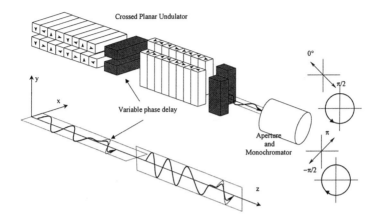

Figure 6 Scheme of a beam line with a crossed planar undulator and photon polarization properties.

The case of the crossed undulator may provide a genuine example. In this case a pair of planar undulators are oriented at right angles to each other (as in fig. 6). The amplitude of the radiation of this device is a linear superposition of two parts one along the x-direction and the other along y, namely

$$\underline{A}_c = \widehat{M}_0(\varphi)\,\underline{l}_{\frac{\pi}{4}} \tag{6.4.13}$$

Expressed in the circular basis the field amplitude of the crossed undulator writes

$$\underline{A}_c = \frac{1}{2}\left[\left(i + e^{i\varphi}\right)\underline{c}_r - \left(i - e^{i\varphi}\right)\underline{c}_l\right] \tag{6.4.14a}$$

and in the $\underline{l}_{\frac{\pi}{4},-\frac{\pi}{4}}$ basis

$$\underline{A}_c = \frac{1}{2}\left[\left(i + e^{i\varphi}\right)\underline{l}_{\frac{\pi}{4}} - \left(i - e^{i\varphi}\right)\underline{l}_{-\frac{\pi}{4}}\right] \tag{6.4.14b}$$

Thus getting, in the first case a polarization rate of

$$\mathfrak{p}_H = \sin\varphi \tag{6.4.15a}$$

and in the second

$$\mathfrak{p}_T = \cos\varphi \tag{6.4.15b}$$

A further important example is that provided by the variably polarization undulator. The on axis field of the variably-polarized undulator proposed by Sasaki is provided by

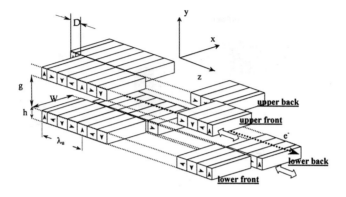

Figure 7 The Sasaki variably-polarizing undulator.

$$B_x = a B_0 \sin\left(\frac{\delta}{2}\right) \cos\left(\frac{2\pi}{\lambda_u} z + \frac{\delta}{2}\right) \quad (6.4.16)$$
$$B_y = b B_0 \cos\left(\frac{\delta}{2}\right) \sin\left(\frac{2\pi}{\lambda_u} z + \frac{\delta}{2}\right)$$
$$B_z = 0$$

where δ is an adjustable phase difference linked to the shift $\mathcal{D}$ shown in fig. 7, $a B_0$ and $b B_0$ are the horizontal and vertical components of the peak magnetic field generated by the undulator.

The undulator radiation intensity can be evaluated by using the methods outlined in chapter II (for further comments see also appendix A), thus finding that the radiation spectrum consists of a series of harmonics whose peak is centered at

$$\omega_m = \frac{2 m \gamma^2 \omega_u}{1 + \frac{k^{*2}}{2} + \gamma^2 \vartheta^2} \quad (6.4.17)$$

$$k^* = k \sqrt{\left[a \sin\left(\frac{\delta}{2}\right)\right]^2 + \left[b \cos\left(\frac{\delta}{2}\right)\right]^2}$$

It is evident that when $\delta = \frac{\pi}{2}$, the magnetic field 6.4.16 reduces to that of an helical undulator, while for $\delta = 0$ or π to a linearly polarized field with magnetic field along y or x-axis respectively. We expect, therefore that the emitted radiation reflects the polarization properties of the undulator, we find that on axis ($\psi = 0$) the spectrum per unit solid angle can be written as

$$\left[\frac{d^2\mathcal{I}}{d\Omega d\omega}\right]_{\psi=0} = \frac{e^2\gamma^2 T^2}{4\pi^2 c}\sum_{m=0}^{\infty}\sin c^2\left[(\omega H - m\omega_u)\frac{T}{2}\right]\cdot\left[|P_m^x|^2 + |P_m^y|^2\right] \quad (6.4.18)$$

$$H = \frac{1}{2\gamma^2}\left[1 + \frac{k^{*2}}{2} + \gamma^2\vartheta^2\right]$$

with x and y denoting the polarization directions and

$$|P_m^x|^2 = \frac{m^2\omega_u^2\left[k\,b\cos\left(\frac{\delta}{2}\right)\right]^2}{\left\{1 + \frac{k^2}{2}\left[\left(a\sin\left(\frac{\delta}{2}\right)\right)^2 + \left(b\cos\left(\frac{\delta}{2}\right)\right)^2\right]\right\}^2}\cdot\left[J_{\frac{m-1}{2}}\left(\widetilde{\xi}_\omega\right) + J_{\frac{m+1}{2}}\left(\widetilde{\xi}_\omega\right)\right]^2$$

$$|P_m^y|^2 = \frac{m^2\omega_u^2\left[k\,a\sin\left(\frac{\delta}{2}\right)\right]^2}{\left\{1 + \frac{k^2}{2}\left[\left(a\sin\left(\frac{\delta}{2}\right)\right)^2 + \left(b\cos\left(\frac{\delta}{2}\right)\right)^2\right]\right\}^2}\cdot$$

$$\cdot\left[J_{\frac{m-1}{2}}\left(\widetilde{\xi}_\omega\right) - J_{\frac{m+1}{2}}\left(\widetilde{\xi}_\omega\right)\right]^2 \quad (6.4.19a)$$

where

$$\widetilde{\xi}_\omega = m\frac{k^2}{4}\frac{\left[a\sin\left(\frac{\delta}{2}\right)\right]^2 - \left[b\cos\left(\frac{\delta}{2}\right)\right]^2}{1 + \frac{k^2}{2}\left[\left(a\sin\left(\frac{\delta}{2}\right)\right)^2 + \left(b\cos\left(\frac{\delta}{2}\right)\right)^2\right]} \quad (6.4.19b)$$

It is evident that P.R. can be varied by properly adjusting δ; for $\delta = 0$ and π it can be easily checked that 6.4.19a reduces to the analogous relation for plane undulators reported in chapter II.

It is also worth noting that for $\delta = \frac{\pi}{4}$ and $a = b$, P^x and P^y combine to yield circular polarization, $\widetilde{\xi}_\omega = 0$ and the harmonic $m = 1$ only survives on axis.

6.5 Emittance optimization

In the previous section we have seen that the main goal of any machine designed as synchrotron radiation source is that of providing a beam with small emittances. In chapter II we have shown that the equilibrium horizontal emittance for an isomagnetic ring can be written as

$$\varepsilon_x = \frac{C_q\gamma^2\langle H\rangle}{J_x\rho} \quad (6.5.1)$$

where

$$\langle H\rangle = \frac{1}{2\pi\rho}\int\left[\gamma_x D^2 + 2\alpha_x DD' + \beta_x D'^2\right]ds \quad (6.5.2)$$

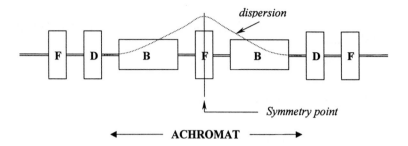

Figure 8 Double focusing achromat (DFA) lattice.

It is therefore evident that by assuming J_x constant, the minimization of the emittance can be achieved by minimizing the dispersion function D. This means that in bending magnets, when photons are emitted, D must be kept low and β functions must be optimized.

The integral in equation 6.5.2 can be evaluated e.g. for a single bending magnet of length L and radius ρ. At second order in ρ/L we get e.g.

$$I \cong \left(\gamma_0 D_0^2 + 2\alpha_0 D_0 D_0' + \beta_0 D_0'^2\right) L + (\alpha_0 D_0 + \beta_0 D_0') \frac{L^2}{\rho} - \quad (6.5.3)$$

$$- (\gamma_0 D_0 + \alpha_0 D_0') \frac{L^3}{3\rho} + \left(\frac{\beta_0 L}{3} - \frac{\alpha_0 L^2}{4} + \frac{\gamma_0 L^3}{20}\right) \frac{L^2}{\rho^2}$$

where the subscript $_0$ denotes the value at the magnet entrance. Equation 6.5.3 can be exploited to derive optimum values of the Twiss parameters yielding the minimum emittance.

Let us however consider a specific example provided by the double focusing achromat (DFA) lattice or basic Chasman-Green, which is shown in fig.8.

A focusing quadrupole magnet is inserted between two dipoles. By adjusting the strength of the quadrupole, the dispersion generated by the first magnet can be cancelled by passing through the second dipole. The emittance of the lattice can be therefore evaluated and by assuming $D_0 = D_0' = 0$ at the entrance of the magnet we get (see also equation 6.5.3)

$$\varepsilon_x = \frac{C_q \gamma^2}{J_x} \vartheta^3 \left(\frac{\beta_0}{3L} - \frac{\alpha_0}{4} + \frac{\gamma_0 L}{20}\right) \quad (6.5.4)$$

where ϑ is the bending angle, which is inversely proportional to the total number of dipoles inside the machine. The emittance can be minimized by a proper adjustment of the optical parameters and in the present case the minimum of 6.5.4 is achieved

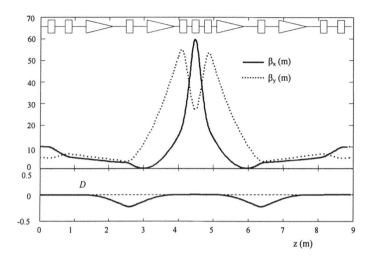

Figure 9   Optical functions β_x, β_y and off-energy function D behaviours in a combination of double focusing achromat (DFA) and triplet achromat (TAL) lattices.

when the minimum of β function, within the dipoles, occurs at distance $s = \frac{3}{8}L$, from the magnet entrance and its value is

$$\beta = L\frac{\sqrt{3}}{8\sqrt{5}} \qquad (6.5.5)$$

thus getting

$$\varepsilon_x = \frac{C_q \gamma^2}{J_x}\vartheta^3 \frac{1}{4\sqrt{15}} \qquad (6.5.6)$$

The basic structure can be made more flexible by adding defocussing quadrupoles upstream and down-stream the focusing one in order to ensure focusing in both planes. An example of behaviour of the optical functions in a Chasman-Green type lattice is provided in fig. 9.

A further example of lattice is the triplet achromat (TAL) shown in fig.10, in this case no quadrupoles are inside the insertion straight section and the emittance write

$$\varepsilon_x = \frac{C_q \gamma^2}{J_x}\vartheta^3 \frac{2}{3}\left(\frac{\beta}{L}\right)^* \qquad (6.5.7)$$

where $\left(\frac{\beta}{L}\right)^*$ is the optimum value of β in the middle of the insertion section

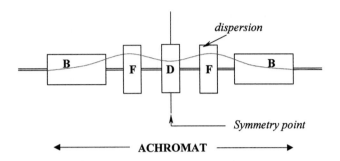

Figure 10   Triplet achromat (TAL) lattice.

$$\left(\frac{\beta}{L}\right)^* = \frac{\sqrt{3}}{2}\left[\frac{1}{5} + \frac{L_1}{L} + \frac{4}{3}\left(\frac{L_1}{L}\right)^2\right]^{1/2} \quad (6.5.8)$$

with L_1 being the length of the insertion. A clear disadvantage of this type of lattice is the dependence of the emittance on the optimum Twiss function inside the insertion.

For completeness sake we report an example of behaviour of optical functions as shown in fig. 11.

As final examples we mention the triple bend achromat (TBA) (fig. 12) FODO (fig.13) lattices. In the first case the basic cell consists of one focusing quadrupole in between the bending magnets. The relevant minimum emittance writes

$$\varepsilon_x = \frac{C_q \gamma^2}{J_x} \vartheta^3 \frac{7}{36\sqrt{15}} \quad (6.5.9)$$

The FODO structure is a sequence of alternating focusing and defocussing quadrupoles, separated by bending magnets. The emittance of this lattice reads

$$\varepsilon_x = 4\frac{C_q \gamma^2}{J_x} \vartheta^3 F(\varphi_c) \quad (6.5.10)$$

where φ_c is the betatron phase-advance per cell. The minimum of F is obtained for φ_c ranging between 100^0 and 135^0 where the function is flat and takes values around 0.

The general conclusion one may draw from the above considerations is that

$$\varepsilon_x [mm \cdot mrad] \sim 10^{-6} \frac{E^2 [GeV]}{J_x N_d^3} \cdot \Phi \quad (6.5.11)$$

The values reached by the function Φ, for the various lattice configuration are given in Tab VI.1

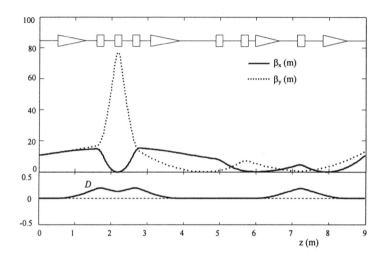

Figure 11 Behaviour of the optical functions in the TAL lattices.

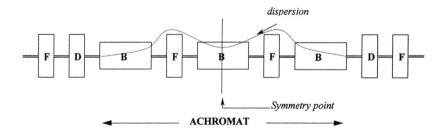

Figure 12 Triple bend achromat (TBA) lattice.

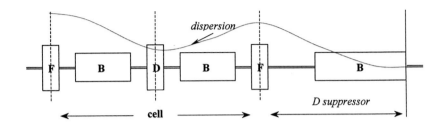

Figure 13 FODO lattice.

The real emittance values are usually larger than the minimum achievable values for compromise reasons of machine stability, which will be discussed in the concluding section.

Table VI.1

	F_{min}	examples	Energy (GeV)	N_d	F
DFA	2.35 E-05	ESRF	6	64	5.02 E-05
TAL	2.43 E-05(*)	ACO	0.536	8	
TBA	1.83 E-05	ALS(**)	1.5	36	1.14 E-04
FODO(***)	4.52 E-04				

(*) to be multiplied by $(\beta/L)_{opt}$
(*) with gradient in the dipoles
(***) regular cells without dispersion suppressor

By comparing 6.5.11 with the photon emittance we get

$$\frac{\varepsilon_x}{\varepsilon_{ph}} \sim 9.62 \cdot 10^9 \frac{n\, E^4\, [GeV]}{J_x\, N_d^3\, \lambda_u\, [cm]\, \left(1 + \frac{k^2}{2}\right)} \cdot \Phi \qquad (6.5.12a)$$

By assuming as typical values $J_x = 2$, $N_d = 30$, $\lambda_u\, [cm] = 10$, $k \simeq 2$, $\Phi \simeq 2 \cdot 10^{-5}$ we get that to reach radial emittance values close to the diffraction limit, the energy should scale as

$$E\, [GeV] \simeq \frac{1.7}{\sqrt[4]{n}} \qquad (6.5.12b)$$

By including the vertical emittance and by requiring that

$$\varepsilon_x\, \varepsilon_y \sim \varepsilon_{ph}^2 \qquad (6.5.13)$$

under the same assumption we get

$$E\, [GeV] \simeq \frac{1.7}{\sqrt[4]{n}} \cdot \left(\frac{1-k}{k}\right)^{1/4} \qquad (6.5.14)$$

where k is the radial coupling factor. It is evident that the photon emittance dominated brightness can hardly be reached for sources operating above $5\,GeV$ in the KeV region.

Before concluding the section, for completeness sake, let as consider the case of beam emittance dominated brightness, which writes

$$\mathfrak{B}_n = \frac{k}{(1+k^2)^2} \frac{1}{\Phi^2} \frac{J_x N_d^3}{E^4\,[GeV]} N\, Q_n\, I\,[A] \qquad (6.5.15)$$

Apart from unessential numerical factors, the beam emittance dominated regime, differs from the diffraction limited case because the brightness scales as E^{-4} and not as E^4.

In this section we have seen that, to achieve a high degree of brightness for sources operating in the hard X-ray region, small emittances are needed.

Short wavelengths require high beam energies, small emittances demand for low energies. One part of the problem lies in these conflicting requirements.

We have considered the problem of emittance optimization, without considering what it implies for the whole of the machine itself. This aspect of the problem will be touched on in the next section.

6.6 Effect of the insertion device on the beam parameters

The time a beam can be kept in a Storage Ring is a measure of the beam life time. Limitation to the life-time can be due to various effects (scattering by residual gas, quantum fluctuation of radiation, intrabeam scattering etc.).

Within the present context the most severe limitations come from the transverse motion. In the linear treatment of betatron oscillation the particle is lost when its amplitude exceeds the aperture of the vacuum chamber. However, the unavoidable non linearities, present in the accelerator, cause a limitation of the maximum betatron amplitude.

Fig. 14 shows the relation between initial amplitude and maximum final amplitude due to non linear effects. We define as *Dynamical Aperture* the initial amplitude for which the final one approaches infinity.

After these remarks let us go back to the problem of emittance minimization, which implies the necessity of minimizing the dispersion functions and the betatron function in the bending magnets.

Strong focusing is therefore required and thus large quadrupole gradient. According to the discussion of chapter I this induces large chromaticity and individual particle will experience tune shift due to non-zero momentum deviation, we remind that

$$\xi_{x,y} = \frac{\Delta \nu_{x,y}}{(\Delta p/p)} = -\frac{1}{4\pi} \int \beta_{x,y}(s)\, K_{x,y}(s)\, ds \qquad (6.6.1)$$

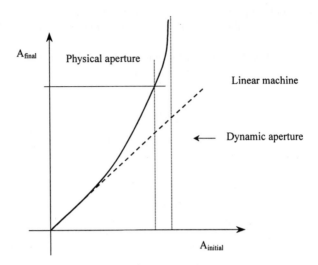

Figure 14   Dynamical aperture reductions due to non-linearities. The increased betatron amplitude by non-linearities is drawn as a function of the initial amplitude.

off-momentum particles may be lost. As also remarked in chapter I, chromaticity can be corrected by using sextupoles*. The strength of these sextupoles must be very strong since the dispersion generated by the bending magnets is small. As a consequence non-linear effects become important and thus severe limitations on the dynamic aperture arise. If this quantity is not adequate, amplitude oscillation, induced at the injection by Coulomb scattering, cannot be accommodated within the machine aperture and thus the injection efficiency becomes very poor. Complicated sextupole arrangement are then necessary to get a reasonable amplitude.

Before closing this chapter, we must clarify a final point, relevant to the effect of the insertion device on the beam qualities and on the machine stability. In chapter II we have seen that equilibrium emittance and energy spread can be written in terms of the radiation integrals. Denoting by I_m^{ID} the $m^{\underline{th}}$ radiation integral of the insertion device, we can write either emittance and energy spread by replacing I_m with $I_m + I_m^{ID}$ (see chapter II). Alternatively one can write

$$\varepsilon_x = \varepsilon_x^0 \frac{1 + \frac{I_5^{ID}}{I_5^0}}{1 + \frac{I_2^{ID} - I_4^{ID}}{I_2^0 - I_4^0}} \qquad (6.6.2)$$

---

*We should also remind that to prevent instabilities of the head-tile type demands for slightly positive chromaticity values.

Effect of the insertion device on the beam parameters

$$\sigma_\varepsilon = \sigma_\varepsilon^0 \left[\frac{1 + \frac{I_3^{ID}}{I_3^0}}{1 + \frac{2 I_2^{ID} + I_4^{ID}}{2 I_2^0 + I_4^0}} \right]^{1/2}$$

where the upper "^{0}" denotes the values calculated without the contribution of the ID-part.

To get explicitly the equilibrium emittance and energy spread with the inclusion of the ID effects, we must evaluate the relevant radiation integrals.

We assume that the ID is characterized by a sinusoidal field variation along the axis of propagation and by a curvature radius ρ_q, period length λ_q and number of periods N_q. Within these assumptions one gets

$$I_2^{ID} = \frac{\lambda_q N_q}{2 \rho_q^2} \quad , \quad I_3^{ID} = \frac{4 \lambda_q N_q}{3\pi \rho_q^2} \quad , \quad I_4^{ID} = -\frac{3 \lambda_q^3 N_q}{32\pi^2 \rho_q^2}, \quad (6.6.3)$$

$$I_5^{ID} = \frac{\lambda_q^4}{4\pi^2 \rho_q^5} \left[\frac{3}{5\pi} + \frac{3}{16} \right] \langle \gamma \rangle L_{ID} - \frac{9 \lambda_q^3}{40\pi^4 \rho_q^5} \langle \alpha \rangle L_{ID} + \frac{\lambda_q^2}{15\pi^3 \rho_q^5} \langle \beta \rangle L_{ID}$$

where $\langle \rangle$ denotes average of the Twiss parameters in the insertion of length L_{ID}.

To these values one should add two further integrals contributing to I_4 and I_5 due to non-vanishing dispersion functions in the dispersive section[†]

$$i_4^{ID} = \frac{7 \lambda_q^2 D_0'}{9\pi^2 \rho_q^3} \quad (6.6.4)$$

$$i_5^{ID} = \frac{4 \lambda_q N_q}{3\pi \rho_q^3} \left[\gamma_0 D_0' + \beta_0 D_0'^2 \right]$$

and the subscript $_0$ stands for value at the entrance of the device. The integral i_4^{ID} contributes directly to I_4^{ID} while for i_5^{ID} and I_5^{ID} the contribution is more complicated. By assuming that in bur machines the I_4 contribution can be neglected with respect to I_2, we end up with

---

[†]the effect is complicated by the fact that the self - ID generated dispersion should be combined to that present in the straight section without the ID. The ID generated dispersion is given by

$$\sigma(s) = \left(\frac{\lambda_q}{2\pi}\right)^2 \frac{1}{\rho_{ID}} \cos\left(\frac{2\pi s}{\lambda_q}\right)$$

$$\sigma'(s) = -\left(\frac{\lambda_q}{2\pi}\right) \frac{1}{\rho_{ID}} \sin\left(\frac{2\pi s}{\lambda_q}\right)$$

and equations 6.6.3,6.6.4 yields an idea of the net result which in general should be evaluated numerically.

$$\overline{\sigma}_\varepsilon = \sigma_\varepsilon \left[\frac{1 + \frac{4}{3\pi} \frac{L_{ID}}{2\pi \rho} \left(\frac{\rho}{\rho_q}\right)^3}{1 + \frac{1}{2} \frac{L_{ID}}{2\pi \rho} \left(\frac{\rho}{\rho_q}\right)^2} \right] \qquad (6.6.5)$$

From the above relation it follows that if

$$\left(\frac{\rho}{\rho_q}\right) < \frac{3\pi}{8} \qquad (6.6.6)$$

there is a reduction of energy spread. As to the emittance since the dominant part of I_5^{ID} is provided by the $\langle \beta \rangle$ dependent term, we get

$$\overline{\varepsilon_x} = \varepsilon_x \frac{1 + \frac{\lambda_q^2}{15\pi^3 \rho_q^5} \frac{\langle \beta \rangle L_{ID}}{I_5^{ID}}}{1 + \frac{1}{2} \frac{L_{ID}}{2\pi \rho} \left(\frac{\rho}{\rho_q}\right)^2} \qquad (6.6.7)$$

From which one also derive a practical condition under which the beam emittance is unperturbed or reduced with respect to the bur machine case[‡]

$$\lambda_q^2 \, [cm] \leq 5.87 \cdot 10^9 \, \frac{E \, [GeV]}{B^3 \, [T]} \, \frac{\varepsilon_x \, [m \cdot rad]}{\langle \beta \, [m] \rangle} \qquad (6.6.8)$$

The discussion developed in this chapter is aimed at giving a preliminary idea of the problems one faces with in the design of an accelerator designed as synchrotron radiation source. We have kept the discussion at a qualitative level, by privileging the physical aspects. The reader interested in more specific details is addressed to the bibliography at the end of the chapter.

6.7 Current limitations

We have seen that the brightness of a S.R. source depends on the e-beam current and we did not comment on the problems which may lead to current limitations in modern Storage Rings.

Roughly speaking these effects can be divided in two main categories
a) Incoherent collective effects
b) coherent collective effects.

Most of them although well understood played a secondary role in the Storage Rings generation not designed for S.R. purposes, the circulating current was indeed not so intense as in the case of second or third generation devices, so that previously considered small perturbing effects have started to play a mayor role.

The incoherent collective effects (I.C.E.) are those in which any individual circulating particle affects the motion of the other particles, without involving the "memory" of the previous evolution. Examples of I.C.E. are

---

[‡]Note that $I_5^0 \cong \frac{I_2^0 \varepsilon_x}{C_q \gamma^2}$

- Beam-gas interaction,
- Ion-trapping process,
- Intra-beam scattering.

We will briefly discuss the above effects and the treatment will be necessary short and heuristic.

6.7.1 Beam-gas interaction

This type of effect includes four process
 a) Elastic scattering on nuclei
 b) Bremsstrahlung on nuclei
 c) Elastic scattering on electrons
 d) Inelastic scattering on electrons

Each one of the above process is characterized by a cross-section, the total cross-section, σ_z i.e. the sum of the individual cross-sections, specifies the beam life-time due to beam-gas interaction

$$\tau = \frac{c}{\sigma_z \beta n} \qquad (6.7.1)$$

where β is the reduced particle speed and n is the density of the residual gas related to the gas pressure by

$$n\left[m^{-3}\right] = 3.217 \cdot 10^{22} \, P \, [Torr] \qquad (6.7.2)$$

To give an idea of the order of magnitudes involved in, we specify the parametrization of the various cross sections.

in the case a) the process leads to an angular kick for the betatron motion, it is evident that, if the induced amplitude exceeds the vacuum chamber, the particle is lost. The cross-section of the process, in practical units reads

$$\sigma_a \left[cm^2\right] \cong 1.3 \cdot 10^{-32} \frac{\mathcal{Z}^2}{E^2 \, [GeV]} \left(\frac{\langle \beta_y \rangle}{a^2}\right) \qquad (6.7.3)$$

where $\mathcal{Z}$ is the atomic number of the residual gas components, a the half-chamber aperture and $\langle \beta_y \rangle$ the average envelope of the beta function along the vertical direction along which the loss is assumed to occur. In the case b) the process leads to an energy loss for the circulating particles. The electron emits, indeed, a photon and the nucleus is left unexcited, the electron is lost if the energy variation induced by the process exceeds the radio frequency acceptance ε_{RF} of the ring. The cross section of the process reads

$$\sigma_b \left[cm^2\right] \simeq 3.1 \cdot 10^{-27} \cdot \left[\ln\left(\varepsilon_{RF}^{-1}\right) - \frac{5}{8}\right] \cdot \ln\left(\frac{183}{\mathcal{Z}^{1/3}}\right) \qquad (6.7.4)$$

In the process c) the circulating electrons may transfer energy to the electrons of the residual gas, in this case too the particles are lost when the energy deviation is larger than ε_{RF} and the cross-section is given by

$$\sigma_c \left[cm^2 \right] \simeq 2.5 \cdot 10^{-28} \frac{Z}{E \left[GeV \right]} \frac{1}{\varepsilon_{RF}} \tag{6.7.5}$$

Finally, in the process d) the scattering occurs for an electron of the atom and the momentum transfer leaves the atom excited. the cross-section writes

$$\sigma_d \left[cm^2 \right] \simeq \frac{Z \, \sigma_b \left[cm^2 \right]}{\ln \left(\frac{183}{Z^{1/3}} \right)} \cdot \left[3 \ln \left(\frac{48.9 \, E \left[GeV \right]}{\varepsilon_{RF}} \right) - 1.4 \right] \tag{6.7.6}$$

In most practical cases the first two processes only play a significant role and the bremsstrahlung is more significant at higher energy.

Before closing these few remarks on beam gas interaction, we stress that the gas pressure depends on the beam current itself. It consists indeed of two contributions, the static pressure P_0 present in the vacuum chamber also in absence of circulating beam and a dynamic pressure, proportional to the current. This last contribution can be understood as follows. Emitted synchrotron-radiation photons hit the vacuum chamber walls, the produced photo-electrons, emitted on a large solid angle, hit the cavity walls too and drive the molecules out. Typical pressure values to reach beam life times of the order of 10 hours are $P \simeq 5 \cdot 10^{-10} \, Torr$.

6.7.2 Ion trapping processes

A characteristic phenomenon of the electron Storage Rings is the so called ion trapping which causes a local enhancement of the gas pressure inside the volume of the circulating beam.

The ions, created by the interaction of the electrons with the rest gas, are affected by the electromagnetic forces of the circulating bunches and experience a transverse focusing with transverse angular displacements specified by

$$\Delta x' = a_x \cdot x \tag{6.7.7}$$
$$\Delta y' = a_y \cdot y$$

The parameters $a_{x,y}$ are provided by

$$a_{x,y} = \frac{2 \, r_p \, c \, \mathcal{N}_e}{\mathcal{M} \, \sigma_{x,y} \cdot (\sigma_x + \sigma_y)} \tag{6.7.8}$$

where r_p is the proton classical radius, $\mathcal{M}$ the mass number of the ion and $\mathcal{N}_e$ the number of electrons in the bunch. It is evident that a beam provided by a series of equally spaced bunches, with identical population, can be viewed as a periodic structure, acting on the ion as a kicker followed by a drift section, namely

$$\widehat{M} = \begin{pmatrix} 1 & \delta_B \\ 0 & 1 \end{pmatrix} \begin{pmatrix} 1 & 0 \\ -a & 1 \end{pmatrix} \tag{6.7.9}$$

Current limitations

where δ_B is the time separation between two adjacent bunches and a is the focusing parameter. According to the discussion of chapter I, the motion is stable, i.e. ion are trapped if

$$-2 < Tr\left(\widehat{M}\right) = 2 - a\,\delta_B < 2 \qquad (6.7.10)$$

which yields a critical ion mass number provided by

$$(\mathcal{M}_c)_{x,y} = \frac{r_p \mathcal{N}_e \Delta_B}{2\,\sigma_{x,y} \cdot (\sigma_x + \sigma_y)} \quad , \quad \Delta_B = c \cdot \delta_B \qquad (6.7.11)$$

The trapping process continues until defocusing due to ion space-charge in the drift region is strong enough to establish the equilibrium. In this last case the drift matrix should be replaced by the defocusing matrix provided by

$$\widehat{M}_{ion} = \begin{pmatrix} \cosh(g\,t) & \frac{1}{g}\sinh(g\,t) \\ g\sinh(g\,t) & \cosh(g\,t) \end{pmatrix} \qquad (6.7.12)$$

where

$$g^2 = \frac{e}{\mathcal{M}\,m_p}\frac{\partial E_y}{\partial y} = \frac{4\pi\,r_p}{\mathcal{M}}c^2\,\frac{d_i}{1+\frac{\sigma_y}{\sigma_x}} \qquad (6.7.13)$$

where d_i is the ion-density. By expanding 6.7.12 up to the second order in $g\,t$ and by applying the stability criterion to one period, we get the condition for trapping

$$\frac{1}{\mathcal{M}}\,(\mathcal{M}_c - k\,d_i) < 1 \quad , \qquad (6.7.14a)$$

$$k = \frac{\pi\,r_p\,\Delta_B^2}{1+\frac{\sigma_y}{\sigma_x}}$$

In conclusion for zero initial density only masses with $\mathcal{M} > \mathcal{M}_c$ can be trapped, if $d_i \neq 0$ the threshold decreases and

$$\mathcal{M} > \mathcal{M}_c - k\,d_i \qquad (6.7.14b)$$

The accumulation limit for a bunched beam is reached when the total number of trapped ions corresponds to the total number of electrons in the machine. the maximum ion density is therefore given by

$$d_i = \frac{A_c}{k} = \frac{W}{2\pi R}\frac{1}{2\pi\,\sigma_x\sigma_y} \quad , \quad \Delta_B = 2\pi R \qquad (6.7.15)$$

where W is the total number of electron in the bunches.

The increase of the local density will provide an increase of local pressure provided by the relation

$$p = p_0 \left(1 + \frac{d_i}{n\,[m^{-3}]}\right) \qquad (6.7.16)$$

where $n\,[m^{-3}]$ is given by equation 6.7.2.

It is perhaps worth to spend a few more words on the evaluation of the pressure rise due to ion trapping. Let us therefore remark again that the ionization process is a by product of the beam-gas interaction. Within this context critical parameters are the ionization cross-sections and production rate. Denoting by σ_i the ion-production cross-section, we can write the rate for ion production as

$$\frac{dN^+}{dt} = \sigma_i\, c\, n \quad \text{per electron} \qquad (6.7.17a)$$

and in practical units we find that a current I lasting for a time τ will provide an ion density

$$N^+\left[m^{-1}\right] \cong 1.875 \cdot 10^{30}\, \sigma\left[m^2\right]\, I\,[A]\, \tau\,[s] \qquad (6.7.17b)$$

According to the previous relation we get for a round beam a pressure increase provided by

$$P_i\,[Torr] = \frac{n^+\,[m^{-1}]}{4\pi\, n\,[m^{-3}]\,(\sigma^0\,[m])^2} \qquad (6.7.18)$$

Just to give a feeling on the involved order of magnitudes we note that the rate of production of CO^+ at $1 \cdot 10^{-9}\,Torr$ is $1.45\,CO^+ \cdot electron^{-1} \cdot s^{-1}$. Therefore $100\,mA$ will lead, after $1\,s$, to a linear density of ions of $3 \cdot 10^{10}\,ions/m$ and for a round beam with $\sigma_x = \sigma_y = 1\,mm$ the pressure rises from $1 \cdot 10^{-9}$ to $8.4 \cdot 10^{-9}\,Torr$.

6.7.3 Intra-beam scattering

Particles inside the bunch perform betatron oscillations around the closed equilibrium orbits. In the moving frame of the bunch the motion is essentially transverse (apart for the slow synchrotron motion). Coulomb scattering may occur between two particles and their transverse momenta can be transformed into longitudinal momentum. When transforming back to the laboratory system the variation of the longitudinal energy is multiplied by a factor γ, it can be therefore quite large and if it exceeds the R.F. acceptance ε_{RF} the particles are lost and this process, known as Touschek effect, may limit the beam life time. We will derive the Touschek life time by following the treatment given in classical textbooks, quoted at the end of the chapter. the geometry of the scattering process is shown in fig.15, it refers to a reference frame where the motion is non relativistic and takes into account two particles with equal but opposite momenta. The differential cross-section of the process is given by the Möller formula

$$\frac{d\sigma}{d\Omega} = \frac{4\,r_0^2}{(v/c)^2}\left[\frac{4}{(\sin\vartheta)^4} - \frac{3}{(\sin\vartheta)^2}\right] \qquad (6.7.19)$$

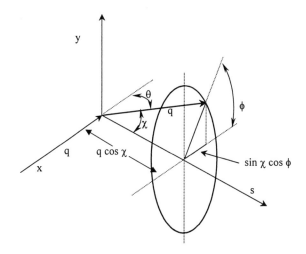

Figure 15 Geometry of the Touschek scattering.

where v is the relative velocity in the center of mass of the system and ϑ is the scattering angle.

For later convenience we introduce the dimensionless momentum

$$q = \frac{p_x}{m_0 c} = \frac{1}{2}\left(\frac{v}{c}\right) \qquad (6.7.20)$$

where p_x is the component of the momentum in the horizontal direction and v the relative velocity in the center of mass. After an elastic collision between two particles in the center of mass and in the horizontal plane, the components of the particles momentum in the longitudinal direction are $q\,m_0\,c\,\cos\chi$. Going back to the laboratory system we find

$$p_s = \gamma\left[p'_s + \frac{\beta}{c}E'\right] \qquad (6.7.21)$$

Since we have assumed non relativistic motion we can neglect the E' dependent term and write

$$p_s \cong \gamma\,q\,m_0\,c \qquad (6.7.22)$$

The particles are lost if the momentum component along the longitudinal direction is larger than the momentum acceptance of the radio-frequency, i.e.

$$|q\cos\chi| > \frac{\Delta p_{RF}}{\gamma} \qquad (6.7.23)$$

$$|\cos\chi| > \frac{\Delta p_{RF}}{|q|\gamma} = \mu$$

where Δp_{RF} is the maximum-momentum acceptance of the radio-frequency. By referring again to fig.15 we get

$$\cos\vartheta = \sin\chi\,\cos\varphi \qquad (6.7.24)$$
$$d\Omega = \sin\chi\,d\chi\,d\varphi$$

we get for the total in the center of mass

$$\sigma = \frac{4\,r_0^2}{(v/c)^2}\int_0^{\cos^{-1}(\mu)}\sin\chi\,d\chi\int_{-\pi}^{\pi}d\varphi\left[\frac{4}{\left(1-(\sin\chi\,\cos\varphi)^2\right)^2} - \frac{3}{\left(1-(\sin\chi\,\cos\varphi)^2\right)}\right] \qquad (6.7.25)$$

which after integration yields

$$\sigma = \frac{8\pi\,r_0^2}{(v/c)^2}\left[\frac{1}{\mu^2} - 1 + \ln\mu\right] \qquad (6.7.26)$$

After the evaluation of the cross-section the next step is the derivation of loss rate and life time.

Referring to fig.16 we can write the loss rate of the particles as

$$d\left(\frac{dN}{dt}\right) = (\sigma\,\nu)\,\rho\,dN \qquad (6.7.27)$$

Since the flowing in and scattered particles belong to the same ensemble, namely $dN = \rho\,dV$, we may conclude

$$d\left(\frac{dN}{dt}\right) = (\sigma\,\nu)\,\rho^2\,dV \qquad (6.7.28)$$

thus getting for the total loss rate in the center of mass

$$\frac{dN}{dt} = \overline{\sigma\,\nu}\int\rho^2\,dV \qquad (6.7.29)$$

where it has been assumed that the product $\sigma\,\nu$ is independent of the beam coordinates. The integral can be worked out more easily in the laboratory system, where we assume a decoupled Gaussian distribution, i.e.

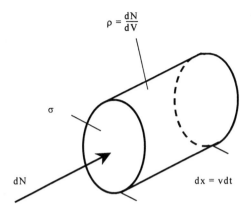

Figure 16 Sketch for the derivation of the differential loss rate.

$$\rho(x,y,s) = \frac{N}{(2\pi)^{3/2} \sigma_x \sigma_y \sigma_s} \cdot \exp\left[-\frac{1}{2}\left(\frac{x^2}{\sigma_x^2} + \frac{y^2}{\sigma_y^2} + \frac{s^2}{\sigma_s^2}\right)\right] \quad (6.7.30)$$

The integration yields

$$\int \rho^2 \, dV = \frac{N^2}{8 \pi^{3/2} \sigma_x \sigma_y \sigma_s} \quad (6.7.31)$$

and the average $\overline{\sigma \nu}$ should be performed with regard to the change in relative moments leading to particle loss, i.e.

$$\overline{\sigma \nu} = 2 \int_{\frac{\Delta_{PRF}}{\gamma}}^{x} g(q)(\sigma \nu) \, dq \quad (6.7.32)$$

where

$$g(q) = \frac{1}{\sqrt{2\pi}\sigma_q} \cdot \exp\left[-\frac{1}{2}\frac{q^2}{\sigma_q^2}\right] \quad , \quad \sigma_q = \frac{\gamma \sigma'_x}{\sqrt{2}} \quad (6.7.33)$$

By exploiting the above relation we can define the inverse of the beam life time as $1/\tau = \frac{1}{N}\frac{dN}{dt}$, thus getting after some algebra§

$$\frac{1}{\tau} = \frac{\sqrt{\pi} r_0^2 c}{\gamma^3} \frac{\mathcal{N}_e}{v_B \sigma'_x} \frac{1}{\varepsilon_{RF}^2} \cdot \left\{ \varepsilon \int_\varepsilon^\infty \frac{1}{n^2}\left[\frac{n}{\xi} - \frac{1}{2}\ln\left(\frac{n}{\xi}\right) - 1\right] e^{-n} \, dn \right\} \quad (6.7.34)$$

---
§The reader is invited to derive eq. 6.7.34.

where $v_B = (4\pi)^{3/2} \sigma_x \sigma_y \sigma_s$ is the bunch volume and $\xi = \left(\frac{\Delta p_{RF}}{\gamma \sigma_{p_x}}\right)^2$.
The assumption that the collision occurs in the horizontal plane amounts to the approximation of flat beam. Eq. 6.7.34 should be considered valid in this approximation. In the case of round beam we find

$$\frac{1}{\tau} = \frac{2\pi r_0^2 c}{\gamma^4} \frac{\mathcal{N}_e}{v_B \sigma'_x \sigma'_y} \frac{\mathcal{D}(\varepsilon)}{\varepsilon_{RF}} \tag{6.7.35a}$$

where

$$\mathcal{D}(\varepsilon) = \sqrt{\varepsilon} \int_\varepsilon^\infty \frac{1}{n^{3/2}} \left[\frac{n}{\xi} - \frac{1}{2}\ln\left(\frac{n}{\xi}\right) - 1\right] e^{-n} \, dn \tag{6.7.35b}$$

It is worth noting that the Touschek effect may provide significant reductions of the beam life-time for low-energy $\mathcal{S.R.}$. Along with the Touschek effect we should also include the treatment of the multiple Touschek effect which may lead not to particle losses but may manifest itself as a noise source for the particle motion yielding additional energy spread. The treatment is however beyond the scope of this chapter and the interested reader is addressed to the bibliography at the end of the chapter. Regarding the coherent effects the only one we will treat is that associated to the so called microwave instability which provides the most serious electron beam brightness limitation in modern Storage Rings.

We will just touch on the problem which will be reconsidered in the next chapter.

We have mentioned that the microwave instability is one of the main limitations of the e-beam brightness in modern Storage Rings. It manifests itself through an anomalous increase of the energy spread and of the bunch length along with a particularly noisy evolution of the whole system. The effect depends on the amount of stored current in the bunch itself but not in the other bunches (single bunch effect). It is primarily due to the interaction of the beam with the surroundings (vacuum pipe, discontinuities in the walls, electrodes, etc.). A simple but effective way to model this instability is provided by the following set of equations, describing the evolution of the longitudinal coordinates of particle circulating in the ring $[(\bar{\varepsilon}, \bar{z}) \equiv (\varepsilon_{n+1}, z_{n+1})$ and $(\varepsilon, z) \equiv (\varepsilon_n, z_n)]$

$$\bar{z} = z + c\alpha_c T_0 \bar{\varepsilon} \ ,$$
$$\bar{\varepsilon} = \varepsilon - \frac{\omega_s^2 T_0}{c\alpha_c} z - 2\frac{T_0}{T_s}\varepsilon + 2\sigma_\varepsilon^0 \sqrt{\frac{T_0}{T_s}} r +$$
$$+\rho\omega_s T_0 \sigma_\varepsilon^0 \frac{1}{N} \sum_{k=1}^N \exp\left[-\frac{\mu \sigma_p}{\sigma_z^0}(z - z_k)\right] \vartheta(z - z_k) \tag{6.7.36}$$

Most of the symbols have already been specified in the text. We remind that z is the longitudinal coordinate and $\varepsilon = \frac{E - E_0}{E_0}$ is the relative energy shift from

Current limitations

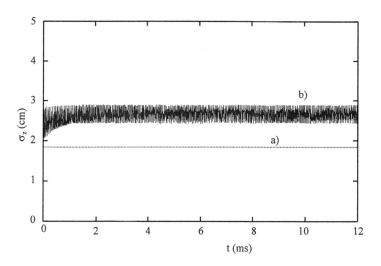

Figure 17   Bunch length vs time: a) natural regime, b) microwave dominated regime ($\rho = 50$).

the synchronous energy. We recognize that the last framed term account for a new contribution added to the ordinary radio-frequency, damping and quantum noise effects (r is a random number from 0 to 1) (see chapter II). The physical meaning of the new term, leading to the microwave instability can be understood as a force term due to a near field, excited by all the electrons before the jth electron and experienced by the jth electron itself. This field is characterized by a decay constant $\frac{\mu \sigma_p}{\sigma_z^0}$ and by a constant ρ, which can be associated with the bunch current and in the present context plays the role of a quantity characterizing the strength of the instability.

The evolution of electron bunch length σ_z under the influence of the microwave instability ($\rho = 50$) is given in fig. 17. It is evident that the relevant values are significatively larger than the case without microwave instability contribution ($\rho = 0$, natural case).

An important criterion known as Boussard's criterion, fixes the following relation between average current and energy spread when the microwave instability is active

$$I = \frac{\alpha_c^2 \; cE/e}{\omega_s \frac{z_n}{n} R} \sigma_\varepsilon^3 \qquad (6.7.37)$$

we have denoted by z_n the longitudinal coupling impedance at the nth revolution harmonic. The sources of longitudinal coupling impedance which determines the threshold of the microwave instability include the beam pipe discontinuities and cavities, free space impedances from bending magnets and from the wigglers. The best achieved values range from 0.1 to 0.5 Ω. The above equation can be exploited in different ways as we will see in the next chapter, but it can be used to infer a threshold for the appearance of the instability and the threshold is indeed fixed by

$$I_{th} \sim \frac{\alpha_c^2 \, cE/e}{\omega_s \frac{z_n}{n} R} \sigma_{\varepsilon,n}^3 \qquad (6.7.38)$$

where $\sigma_{\varepsilon,n}$ is the natural energy spread. When the beam current is larger then I_{th} the energy spread and the bunch length start to grow following approximatively the scaling 6.7.37.

By recalling that $\sigma_z = \frac{c\alpha_c}{\omega_s} \sigma_\varepsilon$ and $\hat{I} = \frac{2\pi R}{\sigma_z} I$ we can derive also the following relation from 3.3.6 regarding the peak current values

$$\hat{I} = \frac{2\pi \, \alpha_c E/e}{\frac{z_n}{n}} \sigma_\varepsilon^2 \qquad (6.7.39)$$

By exploiting typical values we infer from 6.7.39 that the threshold peak current is of the order of *Ampers*.

Before concluding this chapter we will discuss a further instability which is a different manifestation of the anomalous energy spread, affects the high intensity storage rings and is characterized by the behavior shown in fig.18, where we have reported the energy spread vs. time. Simulations indicate that e-beam environment coupling impedances of purely inductive nature determine a type dynamics while a pure resistive impedance gives rise to a noisy behavior of the type shown in fig.19.

The mechanisms underlying the behavior of fig.18 are particularly interesting and will be discussed here, because of their similarity with the FEL-$S\mathcal{R}$ physics presented in Chap.III. The argument we will give is crude but effective.

The interaction of a $S\mathcal{R}$ bunch with the machine environment creates a short range field, which does not cumulate turn after turn. Such a field is experienced by the bunch itself and the conditions are provided for the on set of the instability, which determines the anomalous growth of the energy spread and of the associated bunch length. When the bunch characteristics are modified, i.e. the peak current is reduced by the induced bunch lengthening, the conditions to support the instability no more hold, it is switched off, the initial beam conditions are restored, because of the ordinary damping mechanism and the process may start again.

Both figs.18,19 are the result of a substantive computational effort, based on multiparticle codes, which include as much Physics is possible and require a considerable amount of computer time. A simplified model capable of reproducing the Saw-Tooth instability, with a significant degree of accuracy, has been, however, proposed. This model is based on a couple of non linear equations accounting for the evolution of the instability growth rate and of the induced energy spread, derived under the assumption that the instability growth is controlled by a term due to the wake

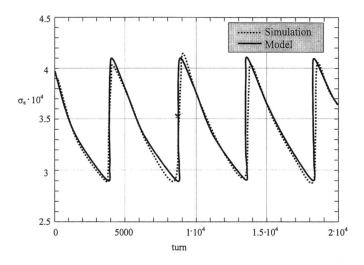

Figure 18   Energy spread vs number of turns. Comparison between simulations and model equations above microwave instability with pure inductive impedance.

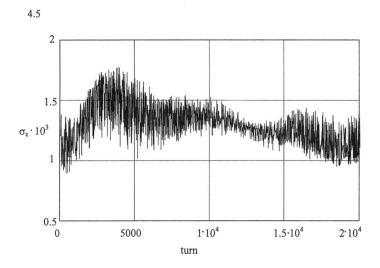

Figure 19   Noisy behavior due to pure resistive impedance.

field counteracted by the Landau damping, associated with a spread in the oscillation frequencies. The equations are reported below for the sake of completeness

$$\frac{d}{dt}a = \left[\frac{A}{(1+\sigma^2)^{\frac{1}{4}}} - B \cdot (1+\sigma^2)^{\frac{1}{2}}\right] \cdot a, \qquad (6.7.40)$$

$$\frac{d}{dt}\sigma^2 = \left(\alpha - \frac{2}{\tau_s}\right) \cdot \sigma^2$$

where α is the instability growth rate and σ is the ratio between the induced and natural energy spread $\sigma_{\epsilon,n}$. We must also underline that eqs.6.7.40 implicitely contain the assumption that the induced energy spread combines quadratically with the natural part. The coefficients A and B, in terms of machine parameters write

$$A = \frac{n}{T_0}\sqrt{\frac{(2\pi)^{\frac{3}{2}}I_0 \cdot \nu_s \cdot \left|\frac{Z_n}{n}\right|}{\frac{E_0}{e} \cdot \sigma_{\epsilon,n}}}, \qquad (6.7.41)$$

$$B = \frac{n}{T_0}(2\pi) \cdot \alpha_c \cdot \sigma_{\epsilon,n},$$

and their dimensions are s^{-1}. Examples of evolution provided by eqs.6.7.41 are given in figs.20 where we have plotted the instability growth rate and the induced energy spread.

We have quoted the microwave and the Saw-Tooth instabilities for the obvious reason that they are responsible for a degradation of the beam qualities which must be accounted for in the analysis of the performances of insertion devices exploiting accelerators, affected by these types of instability. A second less obvious reason is due to a very interesting effect that it has been noted that their interplay of the FEL produces a mechanism inducing the switching off of the instability itself, this new and unexpected phenomenology will be discussed in the forthcoming Chapter.

Current limitations

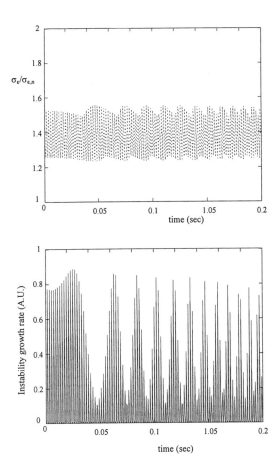

Figure 20 Induced energy spread and instability growth rate.

Suggested bibliography

For the first three sections the reader is addressed to the bibliography of chapters II, IV.
For the high brilliance lattices and the effects of insertion devices, see e.g.

1. A. Ropert, CAS (CERN Accelerator School), " Synchrotron Radiation and Free Electron Lasers" (CERN 90-93, 2 April 1990).
2. H. Wiedeman " An Ultra Low Emittance Mode for PEP Using Damping Wigglers " Nucl. Instr. Meth. A266, 24 (1988).
3. R.Chasman and K. Green " Preliminary Design of a Dedicated Synchrotron radiation Facility ".
4. D.Einfeld and G.Mulhhaupt " Choice of the Principal Parameters and Lattice of BESSY " Nucl.Instr.Meth.172 (1980).
5. M. Sommer " Optimisation de l'Emittance d'une Machine pour Rayonnement Synchrotron " DCI/NI/20/81 (1981).
6. H. Wiedeman " Linear Theory of the ESRP Lattice " ESRP-IRM/9/83 (1983).
7. A. Jackson " A Comparison of Chasman-Green and Triple Bend Achromat Lattices " Particle Accelerators 22, 111 (1987).
8. G. Vignola " The Use of Gradient Magnets in Low Emittance Storage Rings " Nucl. Instrum. Meth. A 246 (1986).
9. A. Wrulich " Study of FODO structures for a synchrotron light source " Particle Accelerators 22 257 (1988).
10. L. Smith " Effects of Wigglers and Undulators on Beam Dynamics" LBL-ESG Tech.Note-24 (1986).
11. M. Katoh and Y. Kamiya " Effects of Insertion Devices on Beam Parameters " IEEE Particle Accelerator Conference 437 (1987).
12. A. Wrulich " Effects of Insertion Devices on Beam Dynamics" ST/M-87/18 (1987).
13. M. Katoh " The Effects of Insertion Devices on Betatron Functions and thir Corrections in the low Emittance of the Photon Factory Storage Ring, KEK Report 86-12 (1987).
14. P. J. Briant " Design of a Ring Lattice ", CAS (CERN Accelearator School, Fourth General Accelerator Physics Course) CERN 91-04 May 1991.
For the Current Limitations see e.g.
15. M. Cornacchia " Requirements and Limitations of Beam Quality in Synchrotron Radiation Sources" CAS (CERN Accelerator School), " Synchrotron Radiation and Free Electron Lasers" (CERN 90-93, 2 April 1990).p. 53.
16. J. Haissinski " These Laboratoire de l'Accelerateur Lineaire, Orsay (1965).
17. G. Brianti " The Stability of Ions in Bunched-Beam Machines " Proc. of the CERN Accelerator School, CERN 84-15 (1984).

18. C. Bernardini et al. " Lifetime and Beam Size in Electron storage Rings" , Proc. Int. Conf. on High Energy Accelerators, Dubna (1963) p.411-415.
19. H. Bruck " Accelerateur Circulaires de Particles " Presses Universitaires de France, Paris (1966).
20. H. Bruck and J. Le Duff " Beam Enlargement in Storage Ring by Multiple Coulomb Scattering " , Proc. 5th Int. Conf. on High Energy Accel. , Frascati (1965).
21. J. LeDuff "Single and Multiple Touschek Effect " Proc.Int.Workshop on Synchrotron Radiation Instrumentation, Taipei, Taiwan, Republic of China, February 22-27 (1988).
22. J. LeDuff " Current and density Limitations in Existing Electron Storage Rings " Nucl. Instr. Meth. A 239, 83 (1985).
23. Y. Miyahara " A New formula for the Lifetime of a Round Beam Caused by the Touschek Effect in an Electron Storage Ring " Japanese Journal of Applied Physics 24 (1985).

For a more general treatment see also

24. J. D. Bjorken and S. Mitingawa Particle Accelerators 13,115 (1983).
25. A. Wrulich " Single-Beam Life-Time " Proc. of the CERN Accelerator, CERN 94-01 (1994).

For the Instabilities see e.g.

26. A. W. Chao " Physics of Collective Beams in High Energy Accelerators " John Wiley and Sons New York (1986) and references therein.
27. K.J.Wille " Storage Rings and Longitudinal Instabilities" Nucl. Instr. Meth.A 393, 18 (1997).
28. D. Boussard CERN LABII/INT/75-2 (1975).

For the Saw-Tooth Instability see

29. S. Heifets proc. of the 14th $ICFA$ workshop, Frascati (1998) to be published.
30. G. Dattoli, L.Mezi, M. Migliorati and L. Palumbo " A simple Model of Saw-Tooth instability in Storage Rings " Il Nuovo Cimento A, to be published.

Chapter 7
CONSTRUCTING AND MEASURING INSERTION DEVICES

summary

In this chapter different type of technology and materials to realize IDs are considered. Measurement techniques both to characterize blocks of rare hearth permanent magnet and magnetic field behaviour in IDs are described.

7.1 Introduction

As already remarked insertion devices are those magnetic devices not strictly related to the machine optics but inserted as source of synchrotron or coherent radiation. The most common ID are undulator (UM) and wiggler (WM) magnets. As stressed in section 2.6, both UMs and WMs are a sequence of magnets with alternating polarity as sketched in fig.10. Electrons crossing the ID experience an alternating magnetic field and are accelerated in the transverse direction with respect to that of propagation. The alternating magnetic field can be realized by conventional or superconducting electromagnets or by using permanent magnets in pure or hybrid configurations. The conventional electromagnets are now completely replaced by permanent magnet devices. The superconducting solution is limited to the cases in which very high on axis magnetic field intensity is required and is usually limited to devices with few magnetic poles. The difference between UM and WM depends on the amplitude of the e-beam oscillation with respect to the aperture of the emission cone.

There are two main geometries for an undulator: planar or helical. An ideal planar, or linearly polarized, undulator generates a sinusoidal magnetic field behavior in one of the transverse directions and no field in the other (see eq. 2.6.8) while an ideal helical, or helically polarized, undulator generates a sinusoidal magnetic field in both transverse directions, the e- beam follows an helical path around the undulator axis. In planar UM the radiation emitted is linearly polarized while in the helical case it is circularly polarized.

The on axis magnetic field, generated in a planar conventional electromagnet is given by the expression

$$B_0 = \frac{B_p}{\cosh(\xi)} \cdot \left(\frac{1 - \frac{\sinh(\xi)}{3\sinh(3\xi)}}{1 - \frac{\tanh(\xi)}{3\tanh(3\xi)}}\right) \quad , \quad \xi = \frac{\pi\lambda_u}{g} \qquad (7.1.1)$$

where infinitely wide poles with infinite permeability has been assumed; λ_u is the period, g the gap and

$$B_p\,[T] = \frac{32\,\mu_0\,N\,I\,[A]}{\sqrt{2}\,\pi\,\lambda_u\,[m]} \cdot \left(\frac{\cosh\left(\frac{\pi g}{\lambda_u}\right)}{\sinh(\xi)} - \frac{\cosh\left(3\frac{\pi g}{\lambda_u}\right)}{3\sinh(3\xi)}\right) \quad , \qquad (7.1.2)$$

NI is the number of ampere-turns per coil.

To give an example of the magnetic field generated in an helical geometry we will consider the case of two coaxial multi-wire solenoids carrying current in opposing direction in order to cancel the axial field component. The surviving transverse components reads

$$B_x = B_0 \sum_{n=0}^{\mathcal{N}-1} \cos\left[\frac{2\pi}{\lambda_u}(z-z_n)\right] \qquad (7.1.3)$$

$$B_y = B_0 \sum_{n=0}^{\mathcal{N}-1} \sin\left[\frac{2\pi}{\lambda_u}(z-z_n)\right] \quad ,$$

with $\mathcal{N}$ being the number of wires per coil

$$B_0\,[T] = \frac{8\pi \cdot 10^{-7}\,I\,[A]}{\lambda_u\,[m]} \cdot \left[\frac{2\pi}{\lambda_u}a \cdot K_0\left(\frac{2\pi}{\lambda_u}a\right) + K_1\left(\frac{2\pi}{\lambda_u}a\right)\right] \quad , \qquad (7.1.4)$$

a the coil radius and $K_{0,1}$ modified Bessel functions.

In the previous chapters we have discussed different types of IDs, including Optical Klystrons, crossed and variably polarization undulators. These types of "exotic" devices have been proposed for different reasons: to enhance the Free Electron Laser gain in Storage Rings or to get elliptical polarizations. It is however clear that the basic element of any of these IDs is a plane undulator configuration. For this reason, in this context, we will limit our attention to a description of the main criteria for the construction and magnetic field characterization of the basic structure of a permanent magnet insertion device in pure and hybrid configuration.

A pure planar UM is a sequence of permanent magnets usually arranged in the so called Halbach configuration as shown in fig. 4.a. The field amplitude for infinitely wide blocks can be calculated analytically and reads

$$B_0 = 2\,B_r \cdot \frac{\sin\left(\frac{\varepsilon\,\pi}{M}\right)}{\frac{\pi}{M}} \cdot \left(1 - e^{-k_u h}\right) \cdot e^{-\frac{k_u g}{2}} \quad , \quad k_u = \frac{2\pi}{\lambda_u} \qquad (7.1.5)$$

where B_r is the remanent field, M the number of blocks per period, h the block height and ε the fill factor to take account of spaces between the blocks.

Rare hearth permanent magnets

In the hybrid configuration magnetic field strength is induced in ferromagnetic poles by permanent magnet bocks disposed as in fig. 4.b. The on-axis field amplitude, assuming infinite permeability in the steel, can be evaluated using the following formula

$$B_0 = \frac{4 B_r}{\sqrt{2}\,\pi} \cdot \left(\frac{1}{\sinh(\xi)} - \frac{1}{3 \sinh(3\xi)} \right) \quad , \quad \xi = \frac{\lambda_u}{g} \tag{7.1.6}$$

In this introductory section we have summarized the characteristics of the on axis field of electromagnet and PM undulators. In the following sections we will enter more deeply in the description of the characteristics of rare hearth magnets and the techniques for realizing PM blocks. We will also describe the most frequently used techniques for the characterization of PM blocks and for the magnetic field measurement of UM and WM.

7.2 Rare hearth permanent magnets

Sinterized rare hearth magnets exhibit very hard magnetic characteristics and very high residual fields.

Inside permanent magnets the induction vector and magnetic field strength have opposite directions. With reference to fig. 1 the relevant working region is the fourth quadrant of the B-H diagram. By assuming that the induction vector is constant in the magnet, the working point is that fixed by the condition $B \cdot H$ be a maximum. This condition, called Evershed's criterion, corresponds to the maximum in energy density localized in the PM or to the volume minimum of PM material for given condition in the air gap.

The total magnetic field is provided by the following expression

$$\mathbf{B} = \mu_0\,\mu\,\mathbf{H} + \mathbf{B}_r \tag{7.2.7}$$

where $\mathbf{B}_r$ is the remanent field, whose meaning can be clarified by splitting the magnetic induction vector $\mathbf{B}$ into the parallel and perpendicular components to the magnetization direction, namely

$$\begin{aligned} B_\| &= \mu_0\,\mu_\|\,H_\| + B_r \\ B_\perp &= \mu_0\,\mu_\perp\,H_\perp \end{aligned} \tag{7.2.8}$$

The remanent field B_r is therefore the value of the parallel component of $\mathbf{B}$, $B_\|$, at $H_\| = 0$ in the B-H diagram, $\mu_\|$ and $\mu_\perp$ are the values of permeability in the direction parallel and perpendicular to the magnetization direction respectively.

The parallel component of the magnetic strength is $H_\|$ give by

$$H_\| = \frac{1}{\mu_0\,\mu_\|} B_\| - H_c \tag{7.2.9}$$

where H_c is the cohercitivity magnetic strength. The large value of H_c for these materials is due both to the crystallographic structure and the sintering technique. From Ampere's law we get

$$\nabla \times \frac{1}{\mu_0 \mu} \mathbf{B} = \mathbf{J} = \nabla \times \mathbf{H}_c \qquad (7.2.10)$$

To understand the reasons of RE-PM properties we will briefly describe the producing technique. Grains of the order of few μm of materials are obtained from a molten mixture of the RE-PM compounds solidified by rapid cooling and subjected to a crunching and milling process. The grains are magnetically highly anisotropic. The powder is exposed to a strong magnetic field and to high pressure, the grains physically rotate and dispose their magnetically preferred axis (easy axis). The aligned grains are then sintered, i.e. embedded in a single structure. The process ends by exposing the material to a very high magnetic filed in order to align practically all magnetic moment along the direction of magnetization (easy axis). As a consequence RE-PM exhibits a very high intrinsic cohercitivity magnetic strength $_I H_c$ and are not limited in performance by the induced cohercitivity magnetic strength that we are simply denoted by H_c (see fig. 1). These materials present discrete directions of easy magnetization (easy axis) separated by hard magnetization directions.

Before the introduction of sinterized rare hearth compounds ferrite-based or AL-NI-CO materials were widely used to realize permanent magnets. The limits of these materials are a lower residual field for ferrites and a very low cohercitivity magnetic strength for AL-NI-CO, as a consequence, AL-NI-CO, in spite of the higher residual field with respect ferrites, exhibits a drastic reduction of the induction B with increasing H. The material used in rare hearth PM are Samarium-Cobalt or Neodymium-Iron-borum. In fig. 2 we have reported the hysteresis diagram for $SmCO_5$ in comparison with Barium Ferrite and AL-NI-CO.

SmCo is less sensitive to temperature variations (higher Curie temperature) than NdFeBr. NdFeBr is actually preferred for the higher residual field. NdFeBr is also mechanically harder but the presence of iron imposes specific treatment of the blocks to avoid oxidation.

Expression 7.2.10 allows to conclude that a PM block with faces parallel to each other and homogeneously magnetized orthogonally to one face generates a magnetic field equivalent to that a current sheet of intensity (see fig. 3a)

$$I = \frac{B_r}{\mu_0} \cdot h \qquad (7.2.11)$$

where h is the block thickness. Furthermore according to the Ampere's equivalence law, the magnetic field of the PM block is equivalent, at sufficient distance, to that a dipole moment $\vec{m}$ directed orthogonally to the block face.

This fact is useful in the case of non homogeneously magnetized blocks which can be characterized using the amplitudes and directions of the relevant dipole mo-

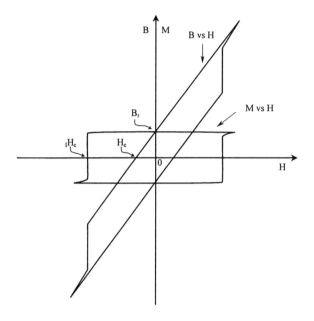

Figure 1 $B\text{-}H$ diagram.

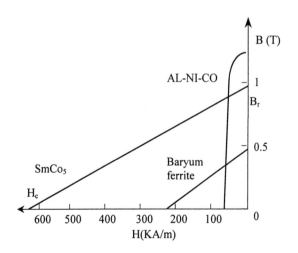

Figure 2 Hysteresis diagram for $SmCO_5$ in comparison with Barium Ferrite and AL-NI-CO.

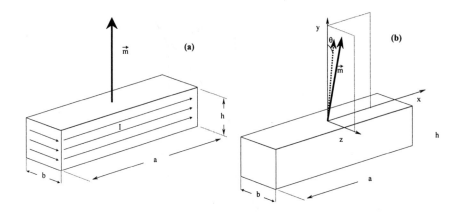

Figure 3   Homogeneously (a) and non homogeneously (b) magnetized REC blocks.

ments. (see fig. 3b).

The magnetic field strength generated at a certain point by a block of the type shown in fig.3 can be calculated analytically and reads

$$B_i = \frac{\mu_0}{4\pi} \sum_{\substack{j=1 \\ k \neq i \neq j}}^{3} \left[M_i \tan^{-1}\left(\frac{r_i r_k}{r_j r}\right) \right]_{r_{i1} r_{j1} r_{k1}}^{r_{i2} r_{j2} r_{k2}} + M_j \left[\ln(r_k + r) \right]_{r_{i1} r_{j1} r_{k1}}^{r_{i2} r_{j2} r_{k2}} \qquad (7.2.12)$$

where $1, 2, 3$ refers to the x, y, z components of the vector $r = \sqrt{r_x^2 + r_y^2 + r_z^2}$ respectively. The meaning of the above expression should be understood as that of a volume integral. The indices denote the contributions at each edge of the block, from the lower to the upper vertex, as shown in fig.4

7.3   Measuring techniques for PM blocks

Different methods have been adopted to measure the PM blocks magnetic characteristics. Before the use of schimming technique to correct the inhomogeneities of the magnetic devices (see the next section), the characterization of the single PM block was a crucial point to realize an undulator, in particular in pure configuration.

In this context we will limit ourselves to describe the following techniques:
a) rotating coil
b) hall probe
c) Helmholtz coil

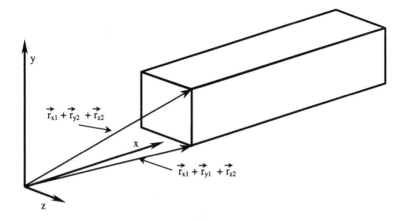

Figure 4 Contribution to the magnetic field in the origin of the reference frame, calculated at an edge of the block.

Rotating coil technique

This method allows the evaluation of the magnetic moment angles with respect to the normal to the block face and the relative magnetic strength with respect a reference magnet suitably chosen in the array of blocks to be measured. In fig. 5 the scheme of a modification of this method is shown. The block to be measured and the reference one are inserted in a rotating holder with polarization directions parallel or antiparallel as clarified below. The magnetic flux variation induces a current in a coil surrounding the rotating holder. Amplitude and phase displacement with respect to a reference signal are analyzed. For a pure dipole field the output voltage signal is sinusoidal, whilst, due to the finite size of the blocks, multipole contributions appear. Two measurement are carried out for each block: one with $\vec{m}$ and $\vec{m_r}$ parallel (sum measurement), the other with $\vec{m}$ and $\vec{m_r}$ antiparallel (difference measurement), $\vec{m_r}$ being the magnetic moment of the reference block.

With the blocks disposed as in fig. 5 the $m_{y,z}$ components can be evaluated, to evaluate m_x the rotating holder must be spinned around the x-axis.

The relative magnetization strength and polarization angle are defined as

$$\varepsilon = \frac{|\vec{m_r}| - |\vec{m}|}{|\vec{m_r}|} \quad , \quad \tilde{\vartheta} = \frac{\vec{m_r} \cdot \vec{m}}{|\vec{m_r}||\vec{m}|} \qquad (7.3.13)$$

and can be obtained using the following relations

$$\varepsilon = \eta \cos \psi \quad , \quad \tilde{\vartheta} = \eta \sin \psi \qquad (7.3.14)$$

224   Constructing and measuring insertion devices

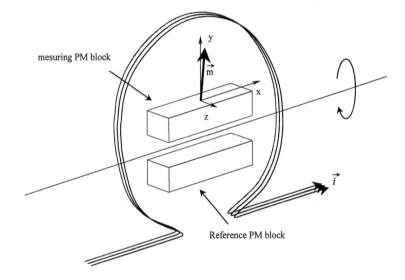

Figure 5   Sketch of the rotating holder equipment with the blocks arranged for the measurement of magnetic moment $\vec{m}$ in the plane yz.

where

$$\eta = 2\frac{V_+}{V_-} \quad , \quad \psi = \psi_- - \psi_+ \qquad (7.3.15)$$

are the voltage amplitude ratio and phase difference of the output signal relative to the momenta difference (−) and sum (+) measurements respectively. The angle ϑ, formed by $\vec{m}$ with the normal to the block face in the plane yz, is finally given by

$$\vartheta = \tilde{\vartheta} + \vartheta_r \quad . \qquad (7.3.16)$$

ϑ_r is obtained repeating the same procedure, one time for all, with the same block rotated around the y-axis of 180°. It is straightforward to verify that

$$\vartheta_r = -\frac{1}{2}\left(\tilde{\vartheta}_0 + \tilde{\vartheta}_\pi\right) \qquad (7.3.17)$$

where $\tilde{\vartheta}_0$ and $\tilde{\vartheta}_\pi$ refers to the measurement after and before the 180° block rotation respectively.

Hall probe technique

This method consists in measuring the magnetic field strength components at a certain number of points around the block. For example the measure of B_y taken with an Hall probe at the points indicated in fig.6 enables to determine the magnetization components and the block center offsets by means the following expressions

$$M_z = \frac{B_1 - B_3 - B_6 + B_8}{4C_1} \quad M_x = \frac{B_2 + B_7}{2C_2} \quad M_y = \frac{B_4 - B_5 - B_9 + B_{10}}{4C_3} \qquad (7.3.18)$$
$$\Delta_z = \frac{B_1 - B_3 + B_6 - B_8}{4\,M_y\,C_4} \quad \Delta_x = \frac{B_2 - B_7}{2\,M_y\,C_5} \quad M_y = \frac{B_4 - B_5 + B_9 - B_{10}}{4\,M_y\,C_6}$$

the coefficients $C_{1,6}$ can be found easily in any given case by calculating the field produced at the relevant points by blocks with the magnetization vector aligned along the x,y or z direction using eq.7.2.12.

The blocks must be located accurately in position and angle with respect to the probe.

Helmholtz coil technique

Helmholtz coil is generally used to generate an high uniform magnetic field in a certain region of space and consists of two coaxial circular current loops or coils of n turns having a mean radius r and placed at a distance r apart on their axes as shown in fig.7. The magnetic field at the center axial point 0 is

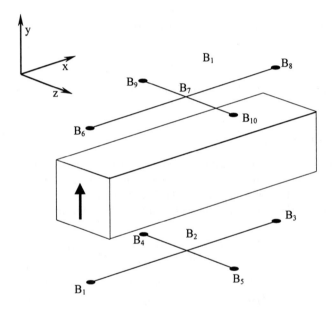

Figure 6 Measurement of PM block with the Hall probe technique. In the figure the points where the probe takes the field measurements are indicated.

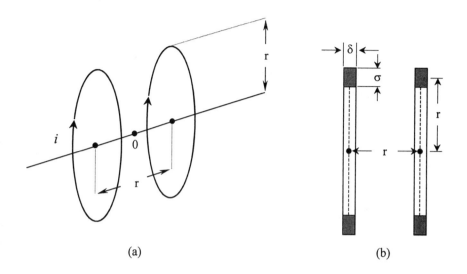

(a)   (b)

Figure 7 Helmholtz coil sketch. r is the mean radius, σ and δ are the thickness and height of the coil respectively.

$$H = 0.0286 \frac{\pi\, i\, n}{r} \qquad (7.3.19)$$

where H is in oersteds, r is in cm and i in $Amperes$. The maximum field uniformity is obtained when the ratio between the cross section dimensions (see fig.7b) reads $\frac{\sigma}{\delta} = 1.0776$.

Measuring PM-blocks no current is carried in the coils, the block to be measured is put inside an Helmholtz coil with its easy-axis parallel to the coil axis and is then turned by 180°. The flux changes according to the following expression

$$\Delta\phi = 2\,C\,|\vec{m}| \qquad (7.3.20)$$

and induces current in the coils. The coil constant C can be determined using a calibration block or by the practical formula

$$C = 0.7155\, \frac{n}{r} \qquad . \qquad (7.3.21)$$

To completely characterize the block, two possible techniques are available: we can rotate the block around main axis or translate and stop the block in the homogeneous region.

The advantage of the Helmholtz coil is the independence from shape and position of the block when placed in the homogeneous region that is the previously defined region, in witch this specific geometry ensures an high uniform magnetic field when current is carried in the coils. A more exhaustive description of the Helmholtz coil technique can be found in the bibliography at the end of this chapter.

7.4 Magnetic field correcting techniques

Before the introduction of schimming correction techniques, in PM UM and WM, the hybrid configuration only allowed the possibility of an efficient correction of field inhomogeneities after the assembling of the structure. Owing to the presence of ferromagnetic poles both electromagnetic or mechanical corrections are possible in a hybrid PM structure, but, as counterpart, the presence of ferromagnetic material itself limits the correction to a specific operating gap. Electromagnetic corrections by coils presents problems both for compactness as well as for the temperature sensitivity of PM materials. Mechanical corrections are preferred and realized with the so called "tuning studs" as shown in fig.8.

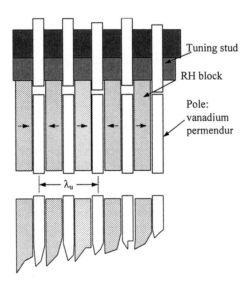

Figure 8   Scheme of an hybrid undulator with tuning studs corrections.

By adjusting the distance between each stud and the upper face of the iron pole, it is possible to change locally the magnetic flux through each poles. By changing the operating gap the effect changes, due to the finite permeability of ferromagnetic materials.

The schimming technique consists in positioning iron shims on poles to subtract a small part of magnetic flux and correct, by strength field subtraction, local

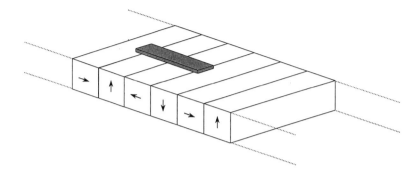

Figure 9 Example of schimming correction.

inhomogeneities. The correction holds at any operating gap, the reason is that the schims are completely saturated and become uninfluent with respect to any magnetic circuit modifications. An example of the modality of positioning is shown in fig.9.

To realize an efficient procedure to correct the field inhomogeneities it is necessary to have a fast magnetic measurement system to evaluate the behavior of the magnetic field and the of the first and second field integrals. After the positioning of the schims we must check the corrections and, eventually, adjust the disposition of the schims. Computer codes able to simulate the effect of schims, starting from a measured magnetic field behaviour, can predict with sufficient accuracy the optimum schim distributions to minimize the field inhomogeneities, in particular in pure PM configuration. In the next section we will discuss different magnetic measurement systems. A deeper treatment of the schimming technique can be found in the bibliography at the end of this chapter.

7.5 Magnetic field measurements

Different techniques are used to measure the magnetic field characteristics of Undulators and Wigglers. In the following we will discuss some of these techniques with particular reference to those which, in our opinion, allow the best characterization of the magnetic field with minimal amount of time.

Magnetic field mapping

To map the field in the volume gap of the device to be measured, an Hall probe mounted on a motorized and computer controlled xyz-positioning system can be used. A whole mapping with sufficient spatial definition requires a prohibitively large amount of time to store all the data. To avoid this problem we can limit the field characterization to second order in the transverse position and express the field as

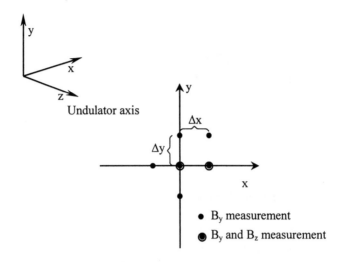

Figure 10 A possible scheme of the measurements to carry out in order to map the magnetic field in undulators and wigglers.

$$B_{x,y,z} = \sum_{\substack{n,m \\ n+m \leq 2}} b_{x,y,z}^{n,m}(z)\, x^n y^m \qquad (7.5.22)$$

It is therefore sufficient to measure the magnetic field components along a minimum of 5 axis around the undulator axis plus the measurements of $B_{y,z}$ on the undulator axis itself, according to the scheme of fig.10. To avoid losses of precision in calculating the coefficients b in eq.7.5.22, a better accuracy is obtained by increasing the number of axis up to 9. This method allows a whole information on the magnetic field behaviour in the device and the evaluation of the field integrals, it needs however large time and is unsuitable for shimming compensation techniques. The direct measurement of the field with Hall probe is limited to the measure of the on axis behaviour.

Many different methods have been used to evaluate quickly the magnetic characteristics, in particular the field integrals. We will describe the rotating and flipping coil methods and the stretched and pulsed wire techniques.

The first and second integrals of the transverse field components are defined as

$$\begin{aligned} I_x &= \int B_x\, dz & I_y &= \int B_y\, dz \\ I_{xx} &= -\int z B_x\, dz & I_{yy} &= -\int z B_y\, dz \end{aligned} \qquad (7.5.23)$$

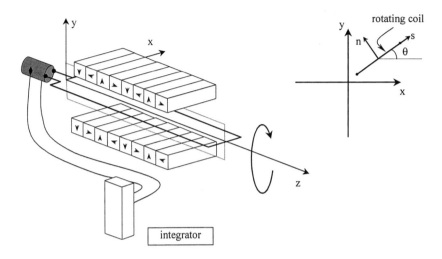

Figure 11   Sketch of rotating or flipping coil measurement system.

In an ideal device the first and second integrals are zero in both planes. In real cases these quantities determine a global change of the electron beam angle and position that, for a relativistic e-beam reads

$$\Delta x' = -\frac{e}{\gamma mc}I_y, \quad \Delta y' = -\frac{e}{\gamma mc}I_x, \quad \Delta x = -\frac{e}{\gamma mc}I_{yy}, \quad \Delta y = -\frac{e}{\gamma mc}I_{xx} \quad (7.5.24)$$

Rotating and flipping coil measurement system

This method uses a coil rotating around the axis of the device to be measured, as shown in fig.11. The same scheme can be used to perform a flipping coil measurement. In this case the coil is rapidly rotated by 90° around the magnetic axis. Because the methods described in the following yields the same information of flipping coil with some advantages in constructing the apparatus and in versatility, we will limit ourself to describe the rotating coil technique.

The integrated voltage induced in the coil of length L, width h and surface $S = L h$ is

$$\int_{t_1(\theta_1)}^{t_2(\theta_2)} V\, dt = N\,\Delta\Phi = N \iint \vec{B} \cdot \vec{n}\, dS = N \int_0^L \int_{-h/2}^{h/2} \vec{B} \cdot \vec{n}\, dz\, ds$$

where s correspond to the coordinate of a generic point in the transverse reference frame joined with the rotating coil, as sketched in fig.11.

Stretched wire measurement system

A N turns coil is realized from a multistrand wire stretched along the magnetic axis of the UM or WM to be measured, the coil is closed outside the device as described in fig.12. The translation of the wire by a quantity Δx induces a voltage in the coil proportional to the change of the magnetic flux Φ

$$\int V\,dt = N\,\Delta\Phi \qquad (7.5.25)$$

by assuming Δx small enough to consider B_y independent of x, $\Delta\Phi$ can be expressed as follows

$$\Delta\Phi = \iint B_y\,dxdz \approx \Delta x \int B_y dz = \Delta x\, I_y \qquad (7.5.26)$$

Similar considerations hold in the case of vertical translation of the wire. From eqs.7.5.25,7.5.26 we can deduce that

$$I_x = \frac{\int V\,dt}{N\Delta y}\quad,\quad I_y = \frac{\int V\,dt}{N\Delta x} \qquad (7.5.27)$$

The method to measure the second integrals is similar. In this case the two ends of the wire are translated in opposite directions. If the length of the wire is L, the integrated voltage induced in the coil is

$$\int V\,dt = \int_{-L/2}^{L/2} dz \int_{-x(z)}^{x(z)} B_y\,dx \qquad (7.5.28)$$

where $x(z) = \frac{\Delta x\,z}{L}$. By assuming again Δx small enough to consider B_y independent of x, we obtain

$$\int V\,dt = \frac{2N\Delta x}{L} \int_{-L/2}^{L/2} z\,B_y\,dz \qquad (7.5.29)$$

A similar procedure for vertical translation finally yields

$$I_{xx} = -\frac{L}{2N\Delta y}\int V\,dt\quad,\quad I_{yy} = -\frac{L}{2N\Delta x}\int V\,dt \qquad (7.5.30)$$

This method allows also the derivation of multipole field components in a way similar to the rotating coil. A specific treatment of this argument can be found in the bibliography at the end of the chapter.

Pulsed wire measurement system

This method exploits the excitation of a vibration wave produced by a current pulse on a tungsten wire threaded through the magnetic device to be measured. To measure the magnetic field characteristics in both x and y directions, orthogonal optical

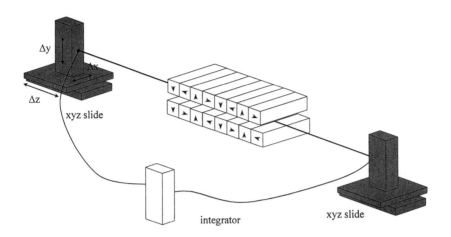

Figure 12   Sketch of Stretched wire measuring system.

detectors formed by a laser-photodiode pair are mounted on translation stages and moved relative to the wire's equilibrium position until the sensor output voltage is a linear function of the wire's displacement d, that occurs when the output voltage is approximatively one-half of the maximum value (see fig.13). This method allows a fast evaluation of the on axis field and of the first and second integral of the field. The first integral is obtained by exiting the wire with a short current pulse ($\sim 20\,\mu s$). The second integral is recorded by using a long current pulse (few ms). The behaviour of the on-axis magnetic field can be obtained by numerical differentiation of the first integral field signal.

The evaluation of the first and second integral of the field is a primary step for the correct insertion of the device in Storage Rings.

The mapping of the field plays a crucial role in correcting the field inhomogeneities in order to optimize the synchrotron radiation spectrum and to minimize inhomogeneous broadening in the spectral emission lines in UM in particular for FEL operation.

For a detailed description of the measurement systems briefly described in this section, the reader is addressed to the bibliography of this chapter.

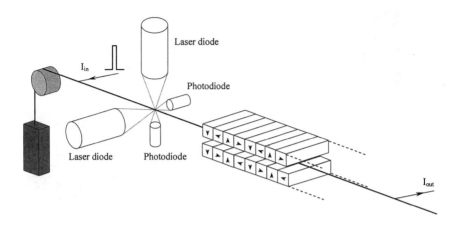

Figure 13 Sketch of pulsed wire measuring system.

Suggested bibliography

For a general treatment of magnetism and of the characteristics of magnetic materials the reader is addressed to:

1. B.D. Cullity, *Introduction to Magnetic Materials*, Addison-Wesley publishing company.
2. *Magnetism and Metallurgy*, edited by A.E. Berkowitz, E. Kneller, Academic Press.
3. S. Chikazumi, *Physics of Magnetism*, J.Wiley & sons ed.
   For a more specific treatment of rare hearth permanent magnets properties see e.g.
4. M.G. Benz, D.L. Martin, Appl. Phys. Letters,17, No. 4, 176 (1970).
5. J.B.Y. Tsui, K.J. Strnat, Appl. Phys. Letters,18, No. 4, 107 (1971).
6. J.B.Y. Tsui, D.J. Iden, K.J. Strnat, A.J. Evers, IEEE Trans. Magn. Mag-8, No. 2, 188 (1972).
7. K. Halbach, Nucl. Instrum. Methods 169, 1-10 (1980).
8. R. Pauthenet, Journal de Physique, colloque C1, 45, 285 (1984).
   For a more detailed description of the permanent magnet blocks measuring techniques see:
9. F.Ciocci, E. Fiorentino, A. Renieri, E. Sabia, Proc. Int. Conf. on "Insertion Devices for Synchrotron Sources" R.Tatchyn, I. Lindau eds, SPIE 582, 27 (1986).
10. M.W. Poole, R.P. Walker, IEEE Trans. Magn. Mag-17, No. 5, 1978 (1981).

11. D.H. Nelson, P.J. Barale, M.I. Green, D.A. Vandyke, IEEE Trans. Magn. Mag-24, No. 1, 1098 (1988).
    An exhaustive description of the schmming technique can be found in the following ESRF reports:
12. J. Chavanne, E. Chinchio, P. Elleaume, ESRF89CH21T - J. Chavanne, P. Elleaume, F. Revol, ESRF89CH26T.
    For a detailed description of the magnetic field measurement systems see:
13. D. Zangrando, R.P. Walker, Nucl. Instrum. Methods A376, 275 (1996).
14. R.W. Warren, Nucl. Instrum. Methods A272, 257(1988).

Chapter 8
FREE ELECTRON LASERS AS INSERTION DEVICES

Summary

We discuss the brightness of FEL devices operating in various configurations.
We consider low energy oscillators, $S\mathcal{R}$ Optical Klystron FELs, self amplified spontaneous emission FELs. Various mechanisms of coherent harmonic generation are also analyzed.
Finally we comment on the role of fourth generation synchrotron radiation sources.

8.1 Introduction

In the previous chapter we have shown that high current, low emittance e-beam passing through undulators, placed in a straight section of the ring, produces high brightness radiation, about five or six orders of magnitude larger than that in bending magnets.

The undulator radiation can be trapped between two mirrors, forming an optical resonator, and if its length is chosen in such a way that the reflected optical bunch overlaps a fresh electron bunch at the undulator entrance and if the necessary resonance conditions are satisfied, the signal can be amplified and the device may operate as a FEL oscillator. We have also seen that e-beam with very high brightness (large current and low transverse and longitudinal emittances) may amplify the spontaneous radiation emitted in the beginning part of he device. This is the so called self amplified spontaneous emission (SASE) mechanism. SASE based FELs offer the very interesting possibility of providing the "fourth generation" light sources. They are indeed the only known generators (albeit without any experimental evidence) of coherent radiation, down to the X-ray spectral region, with brightness exceeding by orders of magnitude that of the third generation.

In this chapter we will reconsider the FEL devices from the point of view of a synchrotron light source and we will try to understood the role of the transition to the "fourth generation".

Regarding this last point, times are perhaps premature to draw any conclusion. The crucial question to be focused on should be: *Storage Rings* or *Linacs*? is the *latter* or the *first better suited for the SASE operation*? The next sections are an attempt of discussing the FEL from the point of view of insertion devices.

We cannot however view an FEL device as an insertion device. It is Authors opinion that a real insertion device is an element capable of affecting an e-beam with an electric or a magnetic field determined " *a priori*" *with not induced fields included*.

We will not limit ourselves to the short wave-length range and we will discuss the problem by including infra-red oscillators.

8.2 FEL oscillator brightness

In chapter III we have described the physical mechanisms underlying the gain and saturation processes of FELs. We have noted that, within such a context, a quantity of paramount importance is the saturation intensity, namely

$$I_s \left[\frac{W}{cm^2}\right] \cong 10^{22} \frac{E^4 [GeV]}{N^4} \frac{1}{[\lambda_u [cm] \, k \, f_b(k)]^2} \qquad (8.2.1)$$

We can associate to 8.2.1 a corresponding brightness value. Which can be exploited as reference number. By assuming that the operating FEL is diffraction limited i.e. $(\lambda \simeq \lambda_R)$

$$\sigma_L = \frac{(\lambda L_c)^{1/2}}{4\pi} \quad , \quad \sigma'_L = \left(\frac{\lambda}{L_c}\right)^{1/2} \qquad (8.2.2)$$

and that the laser bandwidth is

$$\left(\frac{\Delta \omega}{\omega}\right)_L \sim \frac{\lambda}{2\pi \sigma_z} \qquad (8.2.3)$$

We can define the saturation brightness as

$$\mathfrak{B}_s \cong 3.977 \cdot 10^{42} \left(\frac{E[GeV]}{N}\right)^4 \frac{\sigma_z [mm]}{(\lambda_u [cm] \cdot [k \, f_b(k)])^2} \qquad (8.2.4)$$

which can be used as a reference value for the peak intracavity brightness.

By recalling that the optimum outcoupled laser power density is linked to the saturation intensity and to the peak e-beam density by

$$I_L = \frac{1}{2} g_0 I_s = \frac{1}{4N} P_E \qquad (8.2.5)$$

we can associate with 8.2.5 the peak laser brightness

$$\tilde{\mathfrak{B}}_L \simeq 6.4 \cdot 10^{37} \, \frac{\hat{I}[A]}{N} \, \frac{E[GeV]^3}{\lambda_u [cm] \cdot L_c [cm]} \, \frac{\sigma_z [mm]}{1 + \frac{k^2}{2}} \qquad (8.2.6)$$

To derive 8.2.6 we have assumed that the laser is diffraction limited and that the e-beam current density is

FEL oscillator brightness

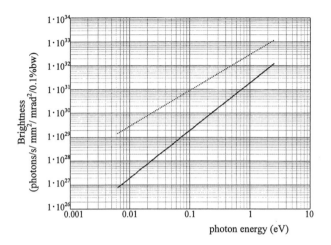

Figure 1 $\mathfrak{B}_s$ (continuous line) and $\mathfrak{B}_L$ (dotted line) vs $\varepsilon\,[eV]$ for a FEL with $N = 30$, $\lambda_u = 5\,cm$, $k = 2$, $\sigma_z = 1\,mm$, $\hat{I} = 10\,A$, $L_c = 10\,m$.

$$\hat{J}\left[A/m^2\right] = \frac{\hat{I}\,[A]}{\pi\left(\frac{\lambda L_c}{16\pi^2}\right)} \qquad (8.2.7)$$

where L_c is the cavity length. Equation 8.2.7 implies that e-beam and optical beams have the same phase space area, i.e.

$$\varepsilon_{x,y} = \frac{\lambda}{4\pi} \qquad (8.2.8a)$$

and that their sections are transversely matched, i.e.

$$\beta_{x,y} = \frac{1}{2} L_c \qquad (8.2.8b)$$

The units of the brightness $\mathfrak{B}_s$ and $\mathfrak{B}_L$ are the same defined in the previous chapter.

We must underline that the definition given in this section are appropriate for a FEL oscillator operating with a pulsed e-beam, the continuous case will not be considered in this book.

An idea of the achievable values is given in fig. 1. The differences between $\mathfrak{B}_s$ and $\mathfrak{B}_L$ are due to the fact that the first refers to the intracavity value.

The equilibrium intracavity brightness is linked to $\mathfrak{B}_s$ by (see equation 3.3.17)

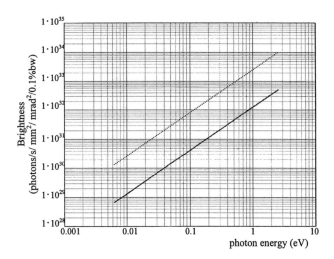

Figure 2 Equilibrium intracavity brightness $\hat{B}_E$ (continuous line) and output laser brightness (dotted line) vs $\epsilon\,[eV]$, same parameters of fig.1, and $\eta = 10\%, \eta_A = 5\%$.

$$\hat{\mathcal{B}}_E \simeq \hat{\mathcal{B}}_s\, F_c\,[G^*, \eta] \qquad (8.2.9a)$$

where

$$F_c\,[G^*, \eta] \simeq \frac{2}{\pi} \frac{1-\eta}{\eta}\left[1 - e^{-h(G^*,\eta)}\right] \qquad (8.2.9b)$$

The output laser brightness, given by 8.2.6, is the maximum achievable value. A more correct expression, including the cavity losses, should be provided by

$$\hat{\mathcal{B}}_L^* \cong \hat{\mathcal{B}}_L\, \overline{F}_c\,[G^*, \eta_A, \eta_P] \qquad (8.2.10a)$$

where

$$\overline{F}_c\,[G^*, \eta_A, \eta_P] = \eta_A\, F_c\,[G^*, \eta_A, \eta_P] \quad ; \quad \eta = \eta_A + \eta_P \qquad (8.2.10b)$$

we have denoted by η_A and η_P the active and passive losses respectively.

An idea of the corrections due to the losses-depending functions F_c and $\overline{F}_c$ is given by fig. 2 where we have reported $\hat{\mathcal{B}}_E$ and $\overline{\mathcal{B}}_L^*$.

Average and peak brightness are linked by

$$\overline{\mathcal{B}}_L \simeq \zeta\,(\delta_c, \tau_M, \tau_R, \nu_R)\, \hat{\mathcal{B}}_L^* \qquad (8.2.11a)$$

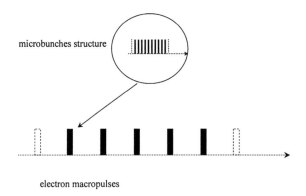

Figure 3   e-bunch structure

where

$$\zeta(\delta_c, \tau_M, \tau_R, \nu_R) = \delta_c \cdot (\tau_M - \tau_R) \cdot \nu_R \qquad (8.2.11b)$$

We have denoted by δ_c and τ_M the machine duty-cycle and the macropulse duration, respectively (see fig. 3).

The repetition frequency is specified by ν_R, while τ_R is the signal rise time, i.e. the time necessary to reach e^{-1} of the equilibrium intracavity power.

This quantity can be deduced from the logistic equation, introduced in chapter III, from which we get that the number of round trips necessary to get $1/e\,x_e$ (with x_e being the dimensionless equilibrium intracavity power)

$$n^* = \frac{\ln\left[\frac{1}{e-1}\frac{X_e - X_0}{X_0}\right]}{\ln\left[(1-\eta)(1+G)\right]} \qquad (8.2.12a)$$

which for $x_e \gg x_0$ and for low gain reduces to

$$n^* = \frac{\ln\left[\frac{X_e}{X_0}\right]}{(1-\eta)G - \eta} \qquad (8.2.12b)$$

The rise time is linked to the cavity length and to n^* by

$$\tau_R\,[\mu s] \simeq \frac{2}{3} 10^{-2} L_c\,[m] \cdot n^* \qquad (8.2.13)$$

and an idea of the dependence of τ_R on gain and cavity losses is given by fig. 4. By assuming $\delta \sim 5\%$, $\nu_R \sim 10\,Hz$, $\tau_R \sim 10\,\mu s$ and τ_R negligible, we find $\zeta \sim 5 \cdot 10^{-6}$.

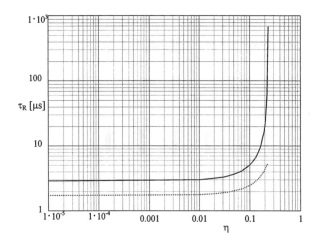

Figure 4 Rise time vs cavity losses, $L_c = 10m, G = 0.5$ (continuous line) $G = 0.3$ (dotted line)

We have mentioned that the relative bandwidth of an FEL operating with e-bunches of length σ_z is provided by equation 8.2.2. This implies that at the saturation, the length of the optical pulse should be equivalent to σ_z.

A simple example supporting this statement, is provided by the use of the logistic equation, which can be rewritten in the form

$$x(z,n) = x_0(z) \frac{[(1-\eta)(1+G(z))]^n}{1 + \frac{x_0(z)}{x_e(z)}[[(1-\eta)(1+G(z))]^n - 1]} \qquad (8.2.14)$$

to include the longitudinal coordinate dependence. The gain function $G(z)$ is defined as

$$G(z) = 0.85 \cdot g_0 f(z) \qquad (8.2.15)$$
$$f(z) = \exp\left(-\frac{z^2}{2\sigma_z^2}\right)$$

The function $f(z)$ represents the current longitudinal profile.

The z-dependence of the equilibrium intracavity dimensionless intensity can be derived from

$$G(z) \frac{1 - \exp(-\beta X_e(z))}{\beta X_e(z)} = \frac{1-\eta}{\eta} \qquad (8.2.16)$$

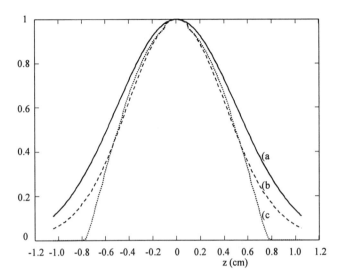

Figure 5 Evolution of the normalized laser pulse longitudinal profile: a) $n = 0$, b) $n = 10$, c) $n = 10^3$.

An example of evolution of spatial slope is provided by fig. 5. For small n the spatial width begins to narrow, while, when saturation occurs, it broadens to reach eventually the same width of the electron pulse.

Complete saturation, i.e. optical pulse length equating that of the electron bunch, is reached after a large number of round trips and for small cavity losses.

8.3 FEL Storage Ring brightness

In chapter III we have shown (see equation 3.4.8a) that the output power of a $\mathcal{SR}$ FEL is

$$I_{out} = \frac{1}{4N} \chi P_s \qquad (8.3.1)$$

where χ is the efficiency function.

According to the super-mode theory the $\mathcal{SR}$ FEL bandwidth and longitudinal length are given by

$$\left(\frac{\Delta\omega}{\omega}\right)_L \simeq \frac{1}{\pi}\sqrt{\frac{\lambda}{N\sigma_z}} \quad , \quad \sigma_E \simeq \frac{1}{2}\sqrt{N\lambda\sigma_z} \qquad (8.3.2)$$

and by recalling that

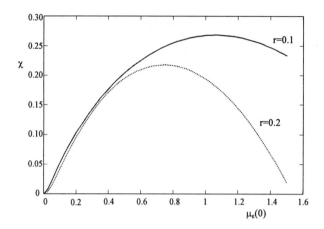

Figure 6 $\mathcal{SR}$ FEL efficiency function χ vs. $\mu_\varepsilon(0)$ for different r values.

$$I_{out} = \frac{1}{2}\chi g_0 I_s \frac{T}{\tau_s} \tag{8.3.3a}$$

we can conclude that the brightness of a $\mathcal{SR}$ FEL is

$$\mathfrak{B}_{\mathcal{SRF}} \cong \frac{1}{2}\chi\, \mathfrak{B}_L\, \sqrt{\mu_c}\,\frac{T}{\tau_s} \tag{8.3.3b}$$

The problem of a correct evaluation of the $\mathcal{SR}$ FEL output power and of the relevant brightness are essentially linked to the evaluation of the efficiency function χ. It contains all the informations relevant to the gain, energy spread, losses, detuning parameter, etc..

The evaluation of χ, in general terms, is rather cumbersome and requires a substantive computational effort. The method outlined in chapter III, according to which χ is defined as

$$\chi = 1.422\, r\, \tilde{x}_e\, \mu_\varepsilon^2(0) \tag{8.3.4a}$$

where $\tilde{x}_e$ is provided by the equilibrium equation

$$(1+\tilde{x}_e)\left[1 + 1.7\,\mu_\varepsilon^2(0)\,(1+\tilde{x}_e)\right]^2 = \frac{1}{r^2} \tag{8.3.4b}$$

It is evident that (see also figs. 13 and 14) that reasonable values of χ can be taken around 0.2 and 0.3 (see fig.6) . Our estimation is consistent with analogous estimations based only on heuristic argument.

Most of the operating $\mathcal{SR}$ FEL devices exploits an O.K. configuration to enhance the gain. A sufficiently accurate description of the $\mathcal{SR}$ O.K. FEL dynamics can be accomplished by using the same procedure, described in chapter III for the $\mathcal{SR}$ FEL, the only difference being that we must take into account the scaling laws, discussed for the O.K. FELs (see chapter III section 5). We will parametrize the O.K. FEL induced energy spread using the relation

$$\sigma_i(x_\delta) \cong \frac{0.365}{\sqrt{2\beta_\delta}\, N\,(1+\delta)} \exp\left(-\frac{\beta_\delta x_\delta}{4}\right) \left[\frac{\beta_\delta x_\delta}{1-\exp(-\beta_\delta x_\delta)} - 1\right]^{1/2}, \quad \beta_\delta \cong 0.53\frac{\pi}{2} \quad (8.3.5)$$

which is accurate for $x_\delta \lesssim 1$ and for $x_\delta \ll 1$ can be approximated as

$$\sigma_i(x_\delta) \cong \frac{0.365}{2N(1+\delta)}\sqrt{x_\delta} \quad (8.3.6)$$

The equations describing the intracavity intensity and indeed energy spread can be therefore written as (see sections III.4 and III.5)

$$\frac{d\tilde{x}_\delta}{dt} = \frac{\tilde{x}_\delta}{T_\delta}\left[\frac{1}{\sqrt{1+\tilde{\sigma}^2}\exp[\zeta(\tilde{\sigma}^2,\delta)]} - r_\delta\right] \quad (8.3.7a)$$

$$\frac{d\tilde{\sigma}^2}{dt} = -\frac{2}{\tau_s}\left(\tilde{\sigma}^2 - \tilde{x}_\delta\right)$$

where

$$\tilde{x}_\delta = \left[\frac{0.365}{2N(1+\delta)}\right]^2 \frac{1}{2\sigma_\varepsilon^2(0)}\frac{\tau_s}{T}\frac{I}{I_{s,\delta}} \quad (8.3.7b)$$

$$\zeta(\tilde{\sigma}^2,\delta) = 1.288\left[8N\sigma_\varepsilon^2(0)(1+\delta)\left(1+\tilde{\sigma}^2\right)^{0.5}\right]^2$$

$$T_\delta = \frac{T}{G^*_{O.K.}}, \quad r_\delta = \frac{\eta}{G^*_{O.K.}}, \quad G^*_{O.K.} = 0.85\,[8\,g_0\,(1+0.915\,\delta)]$$

We remind that $I_{s,\delta}$ is the saturation intensity of the O.K. FEL, albeit we have used less accurate formulae that those derived in III.5, we stress that the approximation made are adequate for the present purposes. The equilibrium and output intensities can be evaluated by following the same procedure we exploited for the conventional $\mathcal{SR}$ FEL. The first step is the derivation of the dimensionless equilibrium intensity $\tilde{x}_{\delta,E}$, obtained from the condition

$$(1 + \tilde{x}_{\delta,E})\exp[\zeta(\tilde{x}_\delta,\delta)] = \frac{1}{r_\delta^2} \quad (8.3.8)$$

The outcoupled power can be then written as

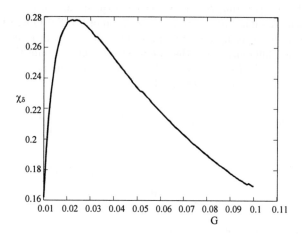

Figure 7 O.K. $\mathcal{SR}$ FEL efficiency function vs G. $\delta = 3$, $\eta_A = 10\%$, $\eta_p = 1\%$, $N = 21$, $\sigma_\varepsilon = 8 \cdot 10^{-4}$.

$$I_0 \cong \frac{1}{4N\,(1+\delta)} \chi_\delta \, P_s \qquad (8.3.9)$$

where the efficiency factor χ_δ writes

$$\chi_\delta = 0.468\,\mu_\varepsilon^2\,(1+\delta)\ r\,\tilde{x}_{\delta,E} \quad , \quad r = \frac{\eta}{0.85\,g_0} \qquad (8.3.10)$$

We remind that g_0 is the small signal gain coefficient relevant to the one undulator section. Typical values of the attainable O.K. $\mathcal{SR}$ FEL efficiency are shown in fig. 7;

being χ_δ and χ comparable we can conclude that the brightness of FEL oscillators operating in the O.K. or conventional configuration are therefore comparable.

8.4 SASE FEL brightness

The previous sections have been devoted to FEL oscillators, in which the use of mirrors, and thus of optical cavities, is crucial. A FEL can also be exploited as an amplifier of a seed laser. The advantage in this case is that mirrors are no more necessary. In the case of a high quality electron beam, with large current, the single pass gain can be so large that the spontaneous emitted radiation, in the beginning part of the undulator, can be amplified to an intense, quasi-coherent radiation. This is the so called mechanism of self-amplified-spontaneous-emission (SASE). The reason why SASE is attracting so a great deal of attention, will be clarified in the following sections. According to the discussions developed in section III.2 various performances of SASE can be expressed by a single parameter ρ. For the sake of completeness we report the logistic equation, reproducing the SASE evolution up to the

$$I(z) = \frac{1}{9} I_0 \frac{e^{4\pi\sqrt{3}\rho z/\lambda_u}}{1 + \frac{I_0}{9\rho P_E}\left[e^{4\pi\sqrt{3}\rho z/\lambda_u} - 1\right]} \tag{8.4.1}$$

It is evident from 8.4.1 that most of the SASE dynamics can be expressed through the parameter ρ: the e-folding length is about $\frac{1}{4\pi\rho}$ the undulator periods, the SASE saturates in about ρ^{-1} periods, the saturation power is about ρ times the electron beam power.

·To evaluate the brightness we proceed as follows. We remind that the laser power is linked to the small signal gain coefficient and to the saturation intensity by

$$I_L \cong \frac{1}{2} g_0 I_s \tag{8.4.2}$$

we also remind that the small signal gain coefficient and the saturation intensity satisfy the identity

$$g_0 I_s = \frac{1}{2N} P_E \tag{8.4.3}$$

where P_E is the e-beam power density.

In the case of high gain devices the laser power is linked to the e-beam power by

$$I_L \cong \rho P_E \quad , \quad \rho = \frac{1}{4\pi}\left(\frac{\pi g_0}{N^3}\right)^{1/3} \tag{8.4.4}$$

which according to 8.4.3 can be rewritten as

$$I_L \cong 2N \rho g_0 I_s \tag{8.4.5}$$

As already remarked optimized high gain devices require $N\rho \sim 1$. This condition allows to fix the intrinsic small signal gain and the final power associated with an optimized high gain FEL device, from eqs. 8.4.4 and 8.4.5 we find indeed

$$g_0 \cong \frac{(4\pi)^3}{\pi} \quad , \quad I_L \cong \frac{2}{\pi}(4\pi)^3 I_s \tag{8.4.6}$$

This last result represents a noticeable result. It states indeed that the final power of an optimized high-gain device with constant parameter is about $1.26 \cdot 10^3$ times the intrinsic saturation intensity. This means that the brightness of a high gain FEL device is just

$$\mathfrak{B}_{H.G.F.} \simeq 1.26 \cdot 10^3 \, \mathfrak{B}_s \, . \tag{8.4.7}$$

It must be understood that this last result hods in a 1-dimensional analysis and does not contain any informations about pulse propagation and transverse mode effects. It is however worth noting that by exploiting the e-beam - optical beam phase-space matching condition

$$\varepsilon \sim \frac{\lambda}{4\pi} \quad (\lambda = \frac{\lambda_u}{2\gamma^2}\left(1 + \frac{k^2}{2}\right)) \tag{8.4.8}$$

we can rewrite the saturation intensity in the form

$$I_s\left[\frac{MW}{cm^2}\right] \cong F(k) \frac{1}{N^4 \, (\varepsilon\,[m \cdot rad])^2} \;, \quad F(k) \cong 1.097 \cdot 10^{-4} \left(\frac{1+\frac{k^2}{2}}{k\,f_b(k)}\right)^2 \tag{8.4.9}$$

which clearly indicates that the saturation intensity is an intrinsic measure of the laser beam brightness. By recalling that (see eq. 3.2.1)

$$g_0 = \frac{16\,\pi}{\gamma} \lambda\,[m]\; L_u\,[m] \; \frac{|J\,[A/m^2]|\,N^2}{1.7 \cdot 10^4}\, \xi\, f_b^2\,(\xi) \tag{8.4.10}$$

we can use equation 8.4.6 to derive the following condition on the current density

$$J\left[A/m^2\right] \cong G(k) \frac{1}{N^3\,\sqrt{\lambda_u\,[m]}} \frac{1}{(\lambda\,[m])^{3/2}} \tag{8.4.11}$$

$$G(k) \cong 19.28\,\pi \cdot 10^4 \left[\frac{\left(1+\frac{k^2}{2}\right)^{3/4}}{k\,f_b(k)}\right]^2$$

These last relations may be exploited to understand the expected output power of a diffraction limited high gain FEL and the required current density levels (see fig. 8).

We can conclude that

$$P_L \cong \frac{2}{\pi}\,(4\pi)^3\, I_s$$

i.e. the maximum power attainable in a high gain FEL is about $1.26 \cdot 10^3$ times the saturation intensity.

8.5 Emission by a prebunched e-beam

Coherent emission at the fundamental harmonic occurs, after that the e-beam induced energy modulation transforms into spatial bunching. When the power increases, substantive bunching occurs and along with the fundamental, higher order harmonics are generated. The process of higher harmonic generation is extremely sensitive to

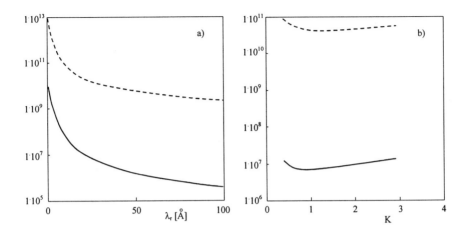

Figure 8 Diffraction limited high gain FEL: a) maximum power (continuous line) and e.beam current density (dashed line) vs wavelength for $k = 1$, $\lambda_u = 2\,cm$, $\rho = 10^{-3}$; b) same vs k for $\lambda = 30\,\text{Å}$.

the e-beam quality, therefore the induced energy spread, intrinsically linked to the saturation mechanism, is the naturally limiting factor of the self- induced harmonic generation (SIHG).

To discuss the emission process by a prebunched e-beam, we go back to section III.3 and rederive the small signal equations by assuming an initial non-uniform phase distribution, i.e.

$$f(\zeta_0) = \sum_{n=-\infty}^{+\infty} b_n e^{in\zeta_0}, \quad \left(\int_0^{2\pi} f(\zeta_0)\, d\zeta_0 = b_0 = 2\pi\right) \quad (8.5.1)$$

The coefficient b_n specify the harmonic content of the bunching and thus we will have

$$\left\langle e^{-i\zeta_0}\right\rangle_{\zeta_0} = \frac{1}{2\pi}\int_0^{2\pi} f(\zeta_0) e^{-i\zeta_0}\, d\zeta_0 = b_1 \quad (8.5.2)$$

$$\left\langle e^{-i2\zeta_0}\right\rangle_{\zeta_0} = \frac{1}{2\pi}\int_0^{2\pi} f(\zeta_0) e^{-i2\zeta_0}\, d\zeta_0 = b_2$$

With these assumptions the small signal equation becomes (see equations 3.3.3-3.3.6)

$$\begin{aligned}\frac{da}{d\tau} &= -2\pi g_0 b_1 + i\pi g_0 \int_0^{\tau} d\tau' (\tau - \tau')\, a(\tau')\, e^{-i\nu_0(\tau-\tau')} + \\ &\quad + i\pi g_0 b_2 \int_0^{\tau} d\tau' (\tau - \tau')\, a^*(\tau')\, e^{-i\nu_0(\tau+\tau')}\end{aligned} \quad (8.5.3)$$

By making the assumption that bunching is weak enough that the coefficients with $n \geq 2$ can be neglected, it is straightforward to prove that 8.5.3 can be cast in the same form 3.2.12a, with initial conditions

$$a(0) = a_0 \qquad (8.5.4)$$
$$\left(\frac{da}{d\tau}\right)_{\tau=0} = -2\pi g_0 b_1$$
$$\left(\frac{d^2 a}{d\tau^2}\right)_{\tau=0} = -2\pi i \nu_0 g_0 b_1$$

The solution can be therefore found in the form

$$a(\tau) = \sum_{j=1}^{3} a_j e^{-i(\nu + \delta \nu_j)\tau} \qquad (8.5.5a)$$

with

$$\sum_{j=1}^{3} a_j = a_0 \quad , \quad \sum_{j=1}^{3} a_j \delta \nu_j = -2\pi i g_0 b_1 - a_0 \nu \quad , \quad \sum_{j=1}^{3} a_j \delta \nu_j^2 = 2\pi i \nu_0 g_0 b_1 + a_0 \nu^2$$

(8.5.5b)

It is however more interesting to get the evolution of a in regions of special interest
a) Low gain regime ($g_0 \leq 0.3$)
b) Intermediate gain regime ($0.3 \leq g_0 \leq 10$)
c) High gain regime ($g_0 \geq 10$).
In the first case we get

$$a(\tau) = a_0 \left[1 + \pi g_0 \frac{2\left(1 - e^{-i\nu_0 \tau}\right) - i\nu_0 \tau \left(1 + e^{-i\nu_0 \tau}\right)}{\nu_0^3}\right] - 2\pi g_0 b_1 \left(\frac{\sin\left(\frac{\nu_0 \tau}{2}\right)}{\frac{\nu_0}{2}}\right) e^{-i\nu_0 \tau}$$

(8.5.6)

The most important conclusion one can draw from the previous equation is that the field may grow even for a vanishing input seed a_0. For $a_0 = 0$, $\nu_0 = 0$ and $\tau = 1$ we find that the coherently generated power is

$$I_c \cong \frac{5 g_0}{\pi^2} |b_1|^2 \cdot g_0 I_s \qquad (8.5.7)$$

which can provide a substantive fraction of the optimum outcoupled laser power, in fact (see equation 8.2.5)

$$\frac{I_c}{I_L} \cong \frac{10 g_0}{\pi^2} |b_1|^2 \qquad (8.5.8)$$

Emission by a prebunched e-beam   251

The brightness of the coherently generated power can be evaluated from equation 8.2.6; by keeping into account equation 8.5.8 and the fact that the relative bandwidth is just $\frac{1}{2N}$, we get

$$\mathfrak{B}_c \cong \frac{10\, g_0}{\pi^2} |b_1|^2 \mu_c\, \mathfrak{B}_c \qquad (8.5.9)$$

By assuming e.g. $b_1 \sim 0.1$, $g_0 \simeq 0.3$, $N = 50$, $\hat{I}\,[A] = 10$, $E\,[GeV] = 0.1$, $\lambda_u\,[cm] = 5$ we get $\mathfrak{B}_c \cong 5.76 \cdot 10^{24}$. When g_0 exceeds 0.3 we can exploit a perturbative procedure (see section 3.2) to obtain the higher order g_0 corrections. By taking $a_0 = 0$, we find, at the second order in g_0,

$$a(\tau) \cong -2\pi\, g_0\, b_1 \left\{ \frac{\sin\left(\frac{\nu_0 \tau}{2}\right)}{\frac{\nu_0}{2}} e^{-i\nu_0 \tau / 2} - \frac{1}{2}\pi\, g_0 \frac{(i\nu_0^2 \tau^2 + 4\nu_0 \tau - 6i)\, e^{-i\nu_0 \tau} + 6i + 2\nu_0 \tau}{\nu_0^4} \right\} \qquad (8.5.10)$$

The effect of the second order corrections in g_0 is provided by fig. 9 and at $\nu_0 = 0$, $\tau = 1$ we get

$$\frac{I_c}{I_L} \cong \frac{5\, g_0}{\pi^2} |b_1|^2 \cdot \left(1 + \frac{\pi^2}{24^2} g_0^2 \right) \qquad (8.5.11)$$

In this case too the brightness can be evaluated by following the same procedure leading to equation 8.5.9

In the case of the high gain regime we find that for $a_0 = 0$ and $\nu_0 = 0$, the field amplitude grows as

$$a = \frac{2i\, b_1\, (\pi g_0)^{2/3}}{3} \{ -e^{-i(\pi g_0)^{1/3}\tau} + e^{+i\frac{\pi}{3}} e^{\frac{\sqrt{3}}{2}(\pi g_0)^{1/3}\tau} e^{+\frac{i}{2}(\pi g_0)^{1/3}\tau} + \qquad (8.5.12a)$$
$$+ e^{-i\frac{\pi}{3}} e^{-\frac{\sqrt{3}}{2}(\pi g_0)^{1/3}\tau} e^{+\frac{i}{2}(\pi g_0)^{1/3}\tau} \}$$

and the relevant square modulus reads

$$|a|^2 = 4\frac{|b_1|^2 (\pi g_0)^{4/3}}{9} \{ -e^{-\sqrt{3}(\pi g_0)^{1/3}\tau} + e^{\sqrt{3}(\pi g_0)^{1/3}\tau} - \qquad (8.5.12b)$$
$$-2[e^{-\frac{\sqrt{3}}{2}(\pi g_0)^{1/3}\tau} \cdot \cos\left(\frac{\pi}{3} - \frac{3}{2}(\pi g_0)^{1/3}\tau\right) +$$
$$+ e^{+\frac{\sqrt{3}}{2}(\pi g_0)^{1/3}\tau} \cdot \cos\left(\frac{\pi}{3} + \frac{3}{2}(\pi g_0)^{1/3}\tau\right)]\}$$

It is worth noting that for small times ($\tau \ll 1$) the field grows quadratically with τ according to the simple relation

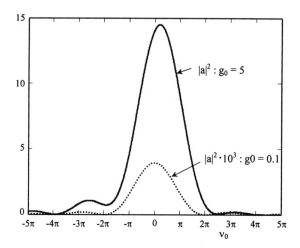

Figure 9 $|a|^2$ vs ν_0 at $\tau = 1$ for different g_0 values.

$$|a|^2 \simeq 4\pi^2 |b_1|^2 (g_0\tau)^2 \qquad (8.5.13)$$

which in practical units writes

$$I\left[\frac{MW}{cm^2}\right] \simeq 4.8 \cdot 10^{-10}\pi^2 |b_1|^2 \left(\frac{k}{\gamma}\right)^2 (J_0(\xi) - J_1(\xi))^2 \cdot \left|J\left[A/m^2\right]\right|^2 z\,[m]^2 \qquad (8.5.14)$$

We have assumed that the prebunched beam with current density J and reduced relativistic energy γ is injected into a linear undulator, z refers to the longitudinal coordinate of propagation inside the undulator ($\tau = z/L_u$).

Introducing ρ parameter, equation 8.5.12b can be written in practical units as

$$I\left[\frac{MW}{cm^2}\right] \simeq 1.15 \cdot 10^{-6}\pi^2 \gamma \rho |b_1|^2 \left|J\left[A/m^2\right]\right|^2 \cdot \qquad (8.5.15)$$
$$\cdot \left\{ \cosh\left(4\sqrt{3}\pi\rho\frac{z}{\lambda_u}\right) - \left[\begin{array}{c} e^{-2\sqrt{3}\pi\rho\frac{z}{\lambda_u}} \cos\left(\frac{\pi}{3} - 6\pi\rho\frac{z}{\lambda_u}\right) + \\ +e^{2\sqrt{3}\pi\rho\frac{z}{\lambda_u}} \cos\left(\frac{\pi}{3} + 6\pi\rho\frac{z}{\lambda_u}\right) \end{array}\right] \right\}$$

The above analysis is independent of the condition of low or high gain regime. In the latter case the τ^2 (or equivalently the z^2) growth precedes the exponential growth, dominated by the fastest root. An idea of the intensity evolution vs. τ is offered by fig. 10.

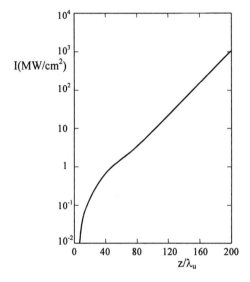

Figure 10 FEL signal evolution vs $\frac{z}{\lambda_u}$ in the case of prebunched e.beam, $|b_1| = 1.32 \cdot 10^{-3}$, $\rho = 2.28 \cdot 10^{-3}$.

Regarding the exponential growth, it is interesting to notice that, when the fastest root dominates, equation 8.5.15 writes

$$I\left[\frac{MW}{cm^2}\right] \simeq I_0 \left[\frac{MW}{cm^2}\right] \exp\left(4\sqrt{3}\pi\rho\frac{z}{\lambda_u}\right) \qquad (8.5.16)$$

where

$$I_0 \left[\frac{MW}{cm^2}\right] = 5.77 \cdot 10^{-7} \pi^2 \gamma \rho \, |b_1|^2 \, \left|J\left[A/m^2\right]\right|^2 \qquad (8.5.17)$$

As is well known the efficiency of a constant parameters FEL amplifier is ρ, it is therefore possible to link saturation length and bunching coefficient imposing the condition

$$I_0 \left[\frac{MW}{cm^2}\right] \cdot \exp\left(4\sqrt{3}\pi\rho\frac{z_s}{\lambda_u}\right) \simeq 0.511 \cdot 10^{-4} \gamma \rho \, \left|J\left[A/m^2\right]\right|^2 \qquad (8.5.18)$$

According to 8.5.16 we also get

$$|b_1| \simeq 3\, e^{-2\sqrt{3}\pi\rho N_s} \quad , \quad N_s = \frac{z_s}{\lambda_u} \qquad (8.5.19)$$

The above equation provides an interesting relation between the beam bunching coefficient and the number of periods necessary to reach the onset of saturation.

To gain a deeper insight into the details of the problem it is worth discussing the topic from the numerical point of view. Expanding the r.h.s. of 8.5.5a, containing the root dependent part, we get

$$a(\tau) = e^{-i\nu_0 \tau} \sum_{m=0}^{\infty} \frac{(-i)^m}{m!} \alpha_m \tau^m \qquad (8.5.20)$$

$$\alpha_m = \sum_{j=1}^{3} \delta \, v_j^m \, a_j$$

Furthermore according to the initial conditions one finds that the coefficients α_m are specified by the following recurrence relations

$$\alpha_m = \pi \, g_o \, \alpha_{m-3} - \nu_0 \, \alpha_{m-1} \qquad (m \geq 3) \qquad (8.5.21)$$
$$\alpha_0 = a_0 \quad , \quad \alpha_1 = -2\pi i \, g_0 \, b_1 - \nu_0 \, a_0 \quad , \quad \alpha_2 = 2\pi i \, \nu_0 \, g_0 \, b_1 - \nu_0^2 \, a_0$$

According to the above equations the numerical solution of equation 8.5.3 with $b_2 = 0$ can be easily obtained.

It is important to add a few words of comment, better clarifying the physical content of the above results. Referring to fig. 10, we notice that the evolution vs. τ of an optical signal induced by a prebunched beam is characterized by three phases

Self-induced higher order generation

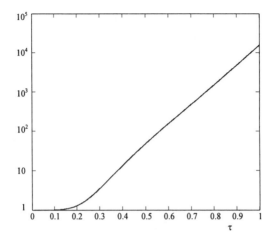

Figure 11 FEL signal evolution vs τ as induced by an input seed.

a) quadratic growth
b) preexponential growth
c) exponential growth.

In phase a) the evolution is dominated by the non homogeneous part of equation 8.5.3. In the second phase the field reaches a sufficiently large value, allowing the second term to play a non negligible role and coherence to develop. In the third step, when full coherence is established, the system grows exponentially. One can get a further physical insight, comparing the different behaviour between the field evolution induced by prebunched beam or by an input seed (see fig. 11).

In the latter case the phases a) and b) are replaced by a lethargy length in which the coherence develops.

8.6 Self-induced higher order generation

In the previous section we have remarked that the mechanism of emission from a prebunched beam is a further tool to generate coherent radiation. This mechanism can be exploited in many flexible ways.

We have remarked that the SASE mechanism requires an e-beam with very large brightness to amplify, up to the saturation, the radiation emitted in the initial part of the undulator. Whenever available the use of a seed laser can be exploited to reduce the saturation length, we have also seen that the effect of beam prebunching can be viewed as an equivalent seed to overcome the problems due to the signal growth from the "vacuum". As already remarked the FEL itself is one of the most suited tools to produce e-beam bunching.

To understand the SIHG mechanism in FEL we note that the FEL pendulum equation can be derived from the Hamiltonian

$$H = \frac{1}{2}\nu^2 - \frac{1}{2i}\left[a e^{-i\zeta} - a^* e^{i\zeta}\right] \tag{8.6.1}$$

where ν and ζ are canonically conjugated variables. The (ν, ζ) phase-space distribution can be derived from the Liouville equation

$$\frac{\partial}{\partial \tau} f(\nu,\zeta) = -\nu \frac{\partial}{\partial \zeta} f(\nu,\zeta) + \frac{1}{2}\left[a e^{-i\zeta} - a^* e^{i\zeta}\right] \frac{\partial}{\partial \nu} f(\nu,\zeta) \tag{8.6.2}$$

By expanding $f(\nu,\zeta)$ in Fourier series, we get

$$\begin{aligned}
f(\nu,\zeta) &= \sum_{n=-\infty}^{+\infty} b_n(\nu) e^{in\zeta} \\
f(\nu,0) &= b_n(\nu) \delta_{n,0} \\
b_0(\nu) &= \frac{1}{\sqrt{2\pi}\pi\mu_\varepsilon} \exp\left[-\frac{(\nu-\nu_0)^2}{2(\pi\mu_\varepsilon)^2}\right]
\end{aligned} \tag{8.6.3}$$

where b_n are the already introduced bunching coefficients, which contains explicitly the energy dependence. The higher harmonic dimensionless field a_n can be evaluated from b_n according to the relation

$$\frac{d}{d\tau} a_n = -2\pi g_n \int_{-\infty}^{+\infty} b_n(\nu,\tau)\, d\nu \tag{8.6.4}$$

and g_n is the gain coefficient of the $n^{\underline{th}}$ harmonic.

The analytical evaluation of the a_n including saturation effects is cumbersome. As in the case of the fundamental, we can use a Padé like approximation to get e.g. for the low gain case

$$a_3(\tau) \cong 2\pi i\, g_3\, a_0^3\, a_{3,0}(\tau) \cdot \left[1 + |a_0|^2 \left(\frac{a_{3,1}(\tau)}{a_{3,0}(\tau)} - \frac{2a_{3,2}(\tau)}{a_{3,1}(\tau)}\right)\right]^{-\frac{1}{\left(\frac{2a_{3,2}(\tau)}{a_{3,1}(\tau)^2} a_{3,0}(\tau) - 1\right)}} \tag{8.6.5}$$

where

$$\begin{aligned}
a_{n,k}(\tau) &\cong \tau^{2n+1+4k} \cdot \\
&\cdot \left\{ \begin{array}{l} A_{n,k} e^{-i\alpha_{n,k}\nu\tau} \cdot \exp\left(-\frac{(\nu\tau)^2}{2(\gamma_{n,k})^2}\right) - \\ -iB_{n,k}(\nu\tau)^3 e^{-i\beta_{n,k}\nu\tau} \cdot \exp\left(-\frac{(\nu\tau)^2}{2(\eta_{n,k})^2}\right) \end{array} \right\}
\end{aligned} \tag{8.6.6}$$

Self-induced higher order generation 257

The coefficients $(A_n, B_n \ldots)$ are given in Table VIII.1a,b,c.
The inclusion of the effect of the energy spread in the dynamics of harmonic generation can be achieved by performing a convolution of equation 8.6.6 on a Gaussian energy distribution, thus obtaining $(\tau = 1)$, $n > 1$

$$a_\xi(\nu, \mu_\varepsilon) = A_\xi\, H(\nu, \mu_\varepsilon, \alpha_\xi, \gamma_\xi) - \quad (8.6.7)$$
$$-iB_\xi\, H(\nu, \mu_\varepsilon, \alpha_\xi, \gamma_\xi)\, \Phi\left(\nu, \mu_\varepsilon, \beta_\xi, \frac{(\eta_\xi)^2}{(\eta_\xi)^2 + \pi^2 \mu_\varepsilon^2}\right)$$
$$(\xi = n, k)$$

where

$$H(\nu, \mu_\varepsilon, \alpha, \gamma) = \frac{\gamma}{\sqrt{\gamma^2 + \pi^2 \mu_\varepsilon^2}} \exp\left\{-\frac{1}{2}\frac{(\alpha\gamma\pi\mu_\varepsilon)^2 + 2i\alpha\gamma^2\nu + \nu^2}{\gamma^2 + (\pi\mu_\varepsilon)^2}\right\} \quad (8.6.8)$$

$$\Phi(\nu, \mu_\varepsilon, \alpha, \gamma) = \gamma^2 \left(\nu - i\alpha(\pi\mu_\varepsilon)^2\right) \left[\gamma\left(\nu - i\alpha(\pi\mu_\varepsilon)^2\right)^2 + 3(\pi\mu_\varepsilon)^2\right]$$

The approximant 8.6.7 can be exploited to get an idea of the dependence of the intensity of the coherently generated harmonic on μ_ε.

Table VIII.1a - Coefficients of equation 8.6.6 $n=1$

s	$A_{1,s}$	$B_{1,s}$	$\alpha_{1,s}$	$\beta_{1,s}$	$\gamma_{1,s}$	$\eta_{1,s}$
0	0.083	$7.44 \cdot 10^{-6}$	0.5	$\alpha_{1,1}$	4.472	5.331
1	$-9.425 \cdot 10^{-4}$	$1.677 \cdot 10^{-6}$	0.868	0.628	3.372	$\gamma_{1,1}$
2	$1.19 \cdot 10^{-5}$	$-1.035 \cdot 10^{-8}$	1,228	1.013	2.655	$\gamma_{1,3}$

Table VIII.1b - Coefficients of equation 8.6.6 $n=3$

s	$A_{3,s}$	$B_{3,s}$	$\alpha_{3,s}$	$\beta_{3,s}$	$\gamma_{3,s}$	$\eta_{3,s}$
0	$7 \cdot 10^{-3}$	$1.762 \cdot 10^{-5}$	1.798	1.973	2.364	2.963
1	$-5.06 \cdot 10^{-4}$	$-1.143 \cdot 10^{-6}$	2.116	2.291	2.077	2.077
2	$2.174 \cdot 10^{-5}$	$4.157 \cdot 10^{-8}$	2.453	2.666	1.883	2.383

Table VIII.1c - Coefficients of equation 8.6.6 $n=5$

s	$A_{5,s}$	$B_{5,s}$	$\alpha_{5,s}$	$\beta_{5,s}$	$\gamma_{5,s}$	$\eta_{5,s}$
0	$1.127 \cdot 10^{-3}$	$6.912 \cdot 10^{-6}$	3.157	3.258	1.829	1.984
1	$-1.725 \cdot 10^{-4}$	$-9.636 \cdot 10^{-7}$	3.452	3.556	1.679	1.779
2	$1.408 \cdot 10^{-5}$	$6.1 \cdot 10^{-8}$	3.763	3.92	1.561	1.561

Roughly speaking the maximum emission occurs in the small signal regime, and a reference value for the attainable values is

$$|a_n|^2 \simeq 4\pi^2 g_n^2 a_0^{2n} \quad (8.6.9)$$

and since

$$|a_n|^2 \propto \frac{I_n}{I_{s,n}} \quad (8.6.10)$$

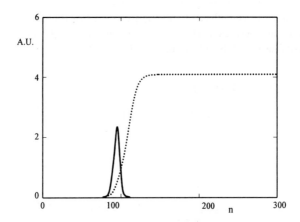

Figure 12 Intracavity harmonic generation vs the number of round trips. The dotted line is the intensity of fundamental harmonic and the continuous line is that of the coherently generated third harmonic (multiplied by $80 \cdot \pi^4$). $G = 0.2$, $g_0 = 0.2$, $\eta = 0.03$, $\nu_0 = 2.6$, $\mu_\varepsilon = 0.16$.

$$I_{s,n} = I_s \left(\frac{f_B(k)}{f_{B,n}(k)} \right)^2 \quad , \quad f_{B,n}(k) = J_{\frac{n-1}{2}}(n\xi) - J_{\frac{n+1}{2}}(n\xi)$$

we end up with

$$I_n \simeq 4 \cdot (0.8)^n \pi^{4n+2} g_n^2 I_{s,n} x^{2n} \tag{8.6.11}$$

It is therefore evident that the FEL can be exploited to get laser light at a given harmonic, and to generate a substantive amount of coherent power at higher harmonics.

We must also underline that in the case of a FEL oscillator, the SIHG mechanism can be exploited far from saturation, i.e. we may have a significant power emission in a small time interval before that the saturation causes a beam degradation which inhibits SIHG. An example is provided by fig.(12). Where we have reported the intracavity power evolution of a Linac based FEL along with the coherently generated power at the third harmonic.

When the system reaches the steady state regime, the quality of the beam is deteriorated enough to drastically reduce the power radiated at the third harmonic. A possible mechanism to enhance the coherently generated power is that of operating at reduced intracavity power (of the fundamental), by e.g. enhancing the cavity losses (see fig. 13) or by periodically dumping the cavity.

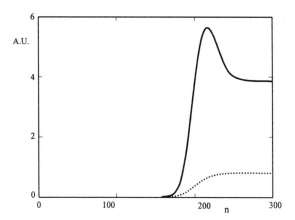

Figure 13 Same as fig.12 but with $G = 0.2$, $g_3 = 0.3$, $\eta = 0.1$, $\nu_0 = 2.6$, $\mu_\varepsilon = 0.16$.

In the case of fig.(13) larger stable power is coherently generated, because the larger cavity losses impose saturation at a lower power level of the fundamental and thus, less energy spread is induced on the e-beam.

It is evident that, if one were able to periodically dump the cavity, the higher harmonic coherently generated power could be periodically regenerated. To achieve these goals, two solutions have been considered:

1) energy modulation
2) cavity length modulation.

In both cases cavity dumping is achieved by means of a periodic reduction of the gain, due to a variation of the resonance a) or synchronism conditions b).

In the following we will describe the consequences of the two methods

a) SIHG and cavity length modulation

In chapter III we have discussed the dependence of the fundamental laser power on the cavity length mismatch parameter ϑ.

We can study the self-consistent evolution of the intracavity optical power and the consequent SIHG with the inclusion of the cavity length dependence, by using the following rate equation model

$$x_{r+1} = x_r + [(1 - \eta) G(\vartheta, g_0, \mu_c, x_r) - \eta] x_r \qquad (8.6.12)$$
$$a_n^{(r)} = 2\pi i g_n \left(0.8\,\pi^4 x_r\right)^{n/2} M_n \left[1 + \left(0.8\,\pi^4 x_r\right) L_n\right]^{P_n}$$

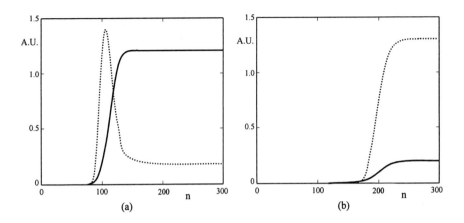

Figure 14 (a) Evolution vs the round-trip number n of the intracavity dimensionless power x (solid line) and of $|a_3|^2/g_3^2$ (dotted line). $g_0 = 0.3$, $\eta = 3\%$, $\mu_c = 1$, $\vartheta = \vartheta^* = 0.343$. (b) Same as (a) but $\vartheta \simeq 0.7$ (Recall that the maximum gain occurs at $\vartheta^* = \frac{\vartheta_s}{1+\frac{1}{3}\mu_c}$).

where r denotes the rth round trip and M_n, L_n and P_n are the functions appearing in equation 8.6.5 and $G(\vartheta, g_0, \mu_c, x_r)$ is the gain function containing the effect of slippage, pulse length and saturation.

Equations 8.6.12 describe the generation of the nth harmonic induced by the fundamental (recall that $|a|^2 = 0.8\,\pi^4\,x$) and, since the notation may be ambiguous, we dwell on the fact that x_r yields the first harmonic intracavity dimensionless power density at the rth round trip.

In fig.14 we have reported the evolution of the fundamental harmonic along with that of the SIHG power at the third harmonic. In this case too to lower intracavity power, corresponds larger stable SIHG power.

We can introduce in our model the effect of cavity modulation by assuming that the cavity mismatch varies with r as

$$\vartheta(r) = A \cdot \sin^2\left(\frac{\alpha\,r}{2}\right) \qquad (8.6.13)$$

To give an idea of the time scale involved in, we note that the single round trip time is linked to the cavity length by

$$\tau_r\,[\mu s] = \frac{2}{3} \cdot 10^{-2}\,L_c\,[m] \qquad (8.6.14)$$

while the rise-time signal is

Self-induced higher order generation   261

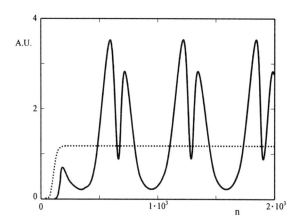

Figure 15 Dimensionless intracavity power vs n for $g_0 = 0.3$, $\vartheta = \vartheta^*$, $\eta = 0.03$, $\mu_c = 1$; stationary cavity (dotted line), modulated cavity (solide line), $A = \vartheta^*$, $\alpha = 10^{-2}$.

$$\tau_{rs}\,[\mu s] = 0.14 \frac{L_c\,[m]}{g} \qquad (8.6.15)$$

where g is the net gain. Denoting by $\tau_M\,[\mu s]$ the duration of the macropulse, we find that, after the on-set of the saturation, the system executes a number of round trips provided by

$$n_T = \frac{\tau_M - \tau_{rs}}{\tau_r} = \frac{\tau_M}{\tau_r} - \frac{21}{g} \qquad (8.6.16)$$

for large enough gain and τ_M of the order of hundreds of μs, we can take α between 10^{-2} and 10^{-3}. The oscillation period of the cavity length mismatch will be provided by

$$\tau_T\,[\mu s] = \frac{4\pi}{3} \cdot \frac{10^{-2}}{\alpha} L_c\,[m] \qquad (8.6.17)$$

and may range between few and tens of μs. The effect of the cavity modulation for a system having $A = \frac{\vartheta_s}{1+\mu_c/3}$, $\alpha = 10^{-2}$, $g_0 = 0.3$, $\mu_c = 1$ and $g_0 = 3\%$ is shown in fig.15,

which clearly displays that the cavity modulation induces an analogous modulation of the intracavity power whose amplitudes and periods can be controlled by varying either α and A (see figs.16,17).

The consequence of the cavity modulation on the SIHG are shown in figs.(18,19). As expected the harmonic generated power has a pulsed structure complementary to

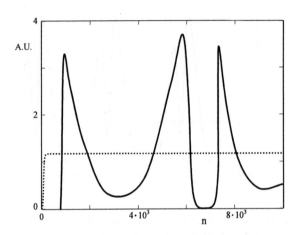

Figure 16 Dimensionless intracavity power vs n for $g_0 = 0.3$, $\vartheta = \vartheta^*$, $\eta = 0.03$, $\mu_c = 1$; stationary cavity (dotted line), modulated cavity (solide line), $A = \vartheta^*$, $\alpha = 10^{-3}$.

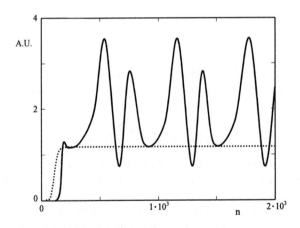

Figure 17 Same as fig.15 but $A = 0.5 \cdot \vartheta^*$.

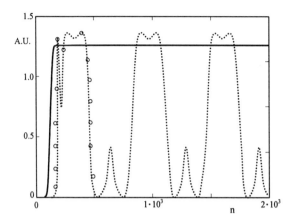

Figure 18 $|a_3|^2/g_3^2$ vs n. Same parameters for the fixed cavity (dotted line) and the modulated cavity (solide line), $A = \vartheta^*$, $\alpha = 10^{-3}$. The dots refer to a comparison with numerical computation.

that of the fundamental and it also follows that the power level is larger than that occurring for the fixed ϑ case, optimized at the maximum gain.

b) SIHG and e-beam energy modulation

A periodical modulation of the e-beam energy produces, in a FEL oscillator, effects analogous to those obtained with the cavity length modulation. This method offers the advantage of working for both pulsed and C.W. FELs.

This case does not present any conceptual difference with respect to the previously discussed mechanism. The only difference being that the gain variation is induced by the e-beam energy variation through the resonance parameter, which will be modulated according to

$$\nu_r = A\,\nu^* \cos^2\left(\frac{\alpha\,r}{2}\right) \qquad (8.6.18)$$

where ν^* is the value at which the maximum gain occurs and A and α are free parameters, which in the case of figs.(20,21) are 5 and $1 \cdot 10^{-3}$.

The consequences of the energy modulation are closely similar to those of the energy modulation. The cavity is indeed temporarily dumped, because the gain decreases below the losses threshold. The gain process and the consequent SIHG are therefore periodically restored.

Laser and harmonic power follow a complementary pulsed structure, whose periodicity can be controlled by controlling A and α in equation 8.6.18. The reader

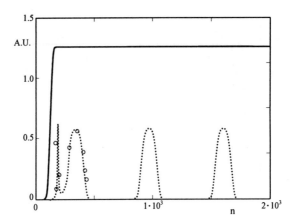

Figure 19 Same as fig.18 for $|a_5|^2/g_5^2$.

interested to the practical realization of the cavity and energy modulation systems is addressed to the bibliography at the end of the chapter.

8.7 Self-induced harmonic generation in Storage Rings

The SIHG process in FEL oscillators operating with $\mathcal{SR}$ deserves a separate treatment. To obtain a correct description, we should extend the rate equation model (equations 8.6.12), by including the effect of the dumping and of the induced bunch lengthening and energy spread. Since the higher harmonic gain coefficient g_n depends on the peak current the induced bunch lengthening will modify this quantity too. Our model equations will be provided by those of the $\mathcal{SR}$ FEL dynamics (see chapter III section 3.4), slightly modified to include the effect of the cavity detuning, namely

$$x_{r+1} = (1-\eta)\left[F\left(x_r, g_0, \sigma_\varepsilon(0), \sigma_r, \vartheta\right) + 1\right] x_r \tag{8.7.1}$$

$$\sigma_{r+1}^2 = \left(1 - \frac{2\tau}{\tau_s}\right)\left[\sigma_r^2 + \left(\frac{0.433}{N}\right)^2 \exp\left(-\frac{\beta x_r}{2}\right)\left(\frac{\beta x_r}{1-\exp(-\beta x_r)} - 1\right)\right]$$

where

$$F(x_r, g_0, \sigma_\varepsilon(0), \sigma_r, \vartheta) = -\frac{0.85\, g_0}{\sqrt{1+\left(\frac{\sigma_r}{\sigma_\varepsilon(0)}\right)^2}} \frac{\vartheta}{\vartheta_s}\left\{\ln\left[\frac{\vartheta}{\vartheta_s}\chi(x_r,\sigma_\varepsilon(0),\sigma_r)\right] - 1\right\}$$

$$\chi(x_r, \sigma_\varepsilon(0), \sigma_r) = \frac{1 + 1.7\,\mu_\varepsilon^2(0)\left(1+\frac{\sigma_r^2}{\sigma_\varepsilon^2(0)}\right)}{1 - e^{-\beta x_r}} \beta x_r \tag{8.7.2}$$

Self-induced harmonic generation in Storage Rings

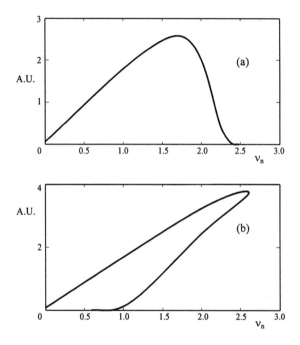

Figure 20 a) Dimensionless intracavity power vs ν_n for the first 800 round trips. b) Same for round trips 800 to 10^4.

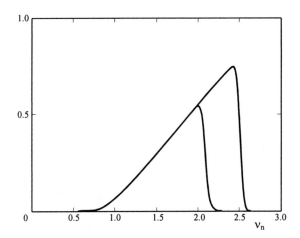

Figure 21 Dimensionless intracavity power vs n for $g_0 \cong 0.17$, $\eta \cong 0.08$, $\alpha = 10^{-3}$. The external loop is the stability diagram for the first 7160 round trips, the internal loop is the stability diagram for round trips up to 10^4.

Completed to these of harmonic generation

$$a_n^{(r)} = \frac{2\pi i\, g_n}{\sqrt{1 + \left(\frac{\sigma_r}{\sigma_\varepsilon(0)}\right)^2}} \left(0.8\,\pi^4\, x_r\right)^{n/2} M_n \left[1 + \left(0.8\,\pi^4\, x_r\right) L_n\right]^{P_n} \qquad (8.7.3)$$

which include the effect of induced bunch lengthening and energy spread.

The model function F takes into account the gain reduction due to natural and induced bunch properties and to the cavity length variations.

An example of the evolution of coherently generated power at 3rd and 5th harmonics, along with the evolution of the fundamental is provided by figs.22,23.

It is evident that the power generated at higher harmonics follows the pattern of the intracavity stored power. The maximum emission occurs just before that a substantial degradation of the beam occurs.

According to figs.24,25 the coherently generated power is relatively large during the first round trips, i.e. before that the system reaches the equilibrium with a substantively degraded e-beam.

By using the numerical parameter of Tables I-II and by inspecting fig. 25 we may conclude that the maximum output of the harmonically generated power is reached for round trip numbers ranging between 200-400. By keeping an average in this interval we get for the third harmonic a power of about $0.4\,W/cm^2$, to be

Self-induced harmonic generation in Storage Rings   267

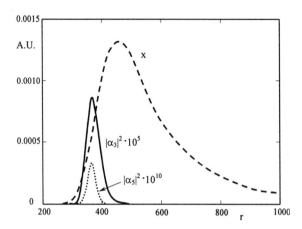

Figure 22   Intracavity dimensionless power x and harmonically generated power for $n = 3$ and $n = 5$ vs the first 10^3 round trips.(The third- and fifth-order sqare amplitudes have been multiplied by 10^5 and 10^{10}, respectively). We have assumed $g_3 = g_5 = 1$ and $\nu = 0$.

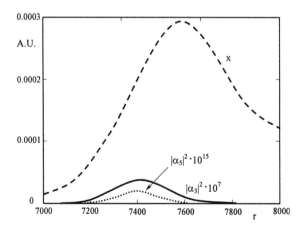

Figure 23   Same as fig.22 for round trips $7 \cdot 10^3$- $8 \cdot 10^3$.(The third- and fifth-order sqare amplitudes have been multiplied by 10^7 and 10^{15}, respectively).

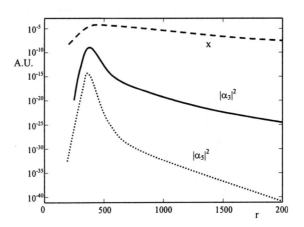

Figure 24 Same as fig.22 for the first $2 \cdot 10^3$ round trips. The vertical axis is given in logarithmic scale, $|a_{3,5}|^2$ are, therefore not multiplied by any magnifying factor.

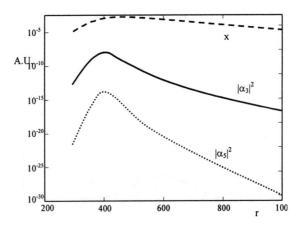

Figure 25 Same as fig.24 $\vartheta = 0.8$.

compared with an equilibrium power on the order of $1.4 \cdot 10^{-6} \, W/cm^2$. For the 5th harmonic we get much lower power. In the region of maximum emission we obtain $< 10^{-6} \, W/cm^2$ and an almost negligible power at equilibrium. This means that a Q-switched operation of the system could be convenient if a large amount of peak power is needed.

Table II- Numerical values used in the simulation

$$g_0 = 0.1$$
$$\gamma \cong 10^3$$
$$N = 40$$
$$k = 3$$
$$\lambda_u = 6\, cm$$
$$\tau_s = 1.5\, ms$$
$$T = 250\, ns$$
$$\sigma_1(0) \cong 6 \cdot 10^{-4}$$
$$\Delta = N \lambda_R$$
$$g_3 \cong 2 \cdot 10^{-2}$$
$$g_5 \cong 8.9 \cdot 10^{-3}$$

The results of this section seems to indicate that the mechanism of SIHG does not appear particularly convenient for FELs operating with $\mathcal{SR}$. This conclusion deserves an accurate experimental analysis and perhaps our model over-estimates the effect of induced energy spread and bunch lengthening. we have presented the problem in the above term to stimulate further consideration on these aspects, which we do not believe secondary, and in any case the reader is addressed to bibliography at the end of the chapter for further comments.

8.8 Storage-Ring FELs and longitudinal instabilities

In the previous chapter we have discussed the problems associated with longitudinal instabilities in $\mathcal{SR}$ as one of the limiting features of the e-beam brightness; in this section we will discuss their interplay with the FEL dynamics.

A semi empirical criterion fixes a threshold current above which this instability grows and manifests itself through an anomalous increase of the energy spread and bunch lengthening, according to Boussard we define the threshold value of the peak current as (for the notation see chapters. II,III,VI)

$$I_{th}[A] = 2\pi \cdot 10^9 \cdot a_c \frac{E\,[GeV]}{\left|\frac{Z_n}{n}\,[\Omega]\right|} \sigma_{e,n}^2. \qquad (8.8.1)$$

In the case of a Chasman-Green lattice one gets

$$a_c = \frac{\pi^2}{3 \cdot N_d^2} \cdot \frac{\rho_m}{C_R} \qquad (8.8.2)$$

with N_d being the number of achromats in the ring lattice, ρ_m is the bending radius and C_R is the circumference of the ring. By writing the natural energy spread as

$$\sigma_{\epsilon,n} \simeq 1.2 \cdot 10^{-3} \frac{E\,[GeV]}{\sqrt{\rho_m\,[m] \cdot J_s}} \qquad (8.8.3)$$

where J_s is the longitudinal partition number. By combining the above relations we get, for the threshold current

$$I_{th}\,[A] \simeq 2.976 \cdot 10^4 \frac{E\,[GeV]^3}{N_d^2 \cdot J_s \cdot C_R\,[m] \cdot \left|\frac{Z_n}{n}\,[\Omega]\right|}, \qquad (8.8.4)$$

and to give an idea of the numbers involved in, we note that by assuming $E = 1\,GeV$, $N_d = 20$, $J_s = 2$, $C_R = 100\,m$ and $|Z_n/n| = 0.5\,\Omega$ we find a threshold value of the current of $0.7\,A$.

For current values exceeding (8.8.2), the energy spread increases and link with the peak current is provided by

$$\sigma_\epsilon \simeq \frac{3.16}{\sqrt{2\pi}} 10^{-5} \left[\frac{\left|\frac{Z_n}{n}\,[\Omega]\right| \cdot I\,[A]}{\alpha_c E\,[GeV]}\right]^{1/2} \qquad (8.8.5)$$

which on account of eq.(8.8.2) yields

$$\delta^2 = \frac{I}{I_{th}} \qquad (8.8.6)$$

and note that if the Boussard criterion holds, we also have

$$\sigma_\epsilon^2 \simeq \delta^2 \cdot \sigma_{\epsilon,n}^2. \qquad (8.8.7)$$

As we have already remarked the natural energy spread is due to a balance mechanism between the damping and quantum diffusion effects. The energy spread induced by the microwave instability is due to a single bunch force, caused by the field generated by the electrons interacting with the vacuum pipe. An energy spread due to a diffusive effect, may combine quadratically with the natural energy spread, shift the threshold and eventually switch off the microwave instability. This is indeed the case of the FEL and according to eqs.(8.8.6-8.8.7) the threshold induced energy spread to counteract the instability is provided by

$$\sigma_i^2 = (\delta^2 - 1) \cdot \sigma_{\epsilon,n}^2. \qquad (8.8.8)$$

as already remarked the equilibrium energy spread induced in a $\mathcal{SR}$ FEL is provided by

$$\sigma_{i,\epsilon}^2 = \frac{7.47 \cdot 10^{-2}}{N^2} \frac{\tau_s}{T} \cdot x, \qquad (8.8.9)$$

Storage-Ring FELs and longitudinal instabilities

we remind that $x = \frac{I}{I_s}$, where I is the FEL intracavity power density. By exploiting the same procedure adopted to evaluate the output power in $\mathcal{SR}$ FEL's, we can derive the instability switching off power threshold and in fact from eqs.(8.8.8-8.8.9) we get

$$I^* \simeq 1.673 \frac{\delta^2 - 1}{g_0 \cdot \Sigma} \cdot \mu_\epsilon(0)^2 \left(\frac{1}{4N} P_s\right) \tag{8.8.10}$$

where P_s is the emitted in one machine turn and Σ is the e-beam cross section. The value given in eq.(8.8.10) can be compared to that corresponding to the $\mathcal{SR}$-FEL equilibrium power

$$I_e \simeq 1.422 \cdot \frac{\tilde{a}}{g_0 \cdot \Sigma} \cdot \mu_\epsilon(0)^2 \left(\frac{1}{4N} P_s\right), \tag{8.8.11}$$

where $\tilde{a}$ is given by

$$(1 + \tilde{a}) \cdot \left[1 + 1.7 \cdot \mu_\epsilon(0)^2 (1 + \tilde{a})\right]^2 = \frac{1}{r^2}. \tag{8.8.12}$$

According to the previous discussion the switching off of the instability requires

$$\frac{I_e}{I^*} \geq 1 \tag{8.8.13}$$

i.e.

$$\tilde{a} \geq 1.18 \cdot (\delta^2 - 1) \tag{8.8.14}$$

To give an example of how the above considerations apply to an actual experimental configuration we have shown in fig.26 the evolution of the r.m.s. bunch length of a $\mathcal{SR}$ e-beam affected by microwave instability, the evolution is very noisy and the bunch length (or equivalently the energy spread) is twice larger than the natural value.

When the FEL interaction is switched on, it may happen that if the power if the power is too large, large energy spread (and thus a correspondingly large bunch lengthening is induced). The evolution becomes however more regular. The bunch length does not increase if the amount of laser power is that corresponding to I^*. The level of noise, characteristic of the instability, has been, however, completely eliminated. In fig.(27) we have considered values below the power threshold and it is evident that the effect of the instability are not counteracted any more.

We have also mentioned the existence of a further instability of the microwave type the Saw-tooth Instability, which has been modelized by means of two coupled non linear differential equations, accounting for the evolution of the induced energy spread and of the instability growth rate. The interplay of the Saw-Tooth instability and FEL dynamics is easily understood by combining the $\mathcal{SR}$ FEL rate equation model with those of the instability evolution, thus getting

$$\frac{d}{dt} a = \left[\frac{A}{(1 + \sigma^2)^{\frac{1}{4}}} - B \cdot (1 + \sigma^2)^{\frac{1}{2}}\right] \cdot a, \tag{8.8.15}$$

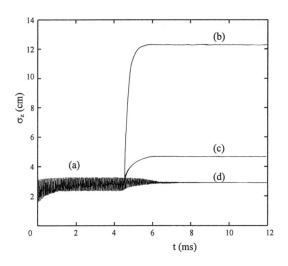

Figure 26 Bunch length evolution vs time. a) evolution dominated by M.I. only; b) the FEL interaction is switched on at $t = 4.5\,ms$ with $x \sim 100\,I^*/I_s$; c) same as b) with $x \sim 10\,I^*/I_s$; d) same as b) with $x \sim I^*/I_s$.

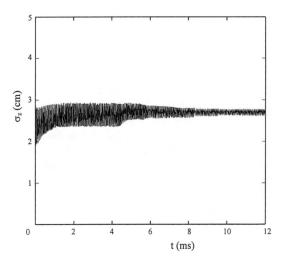

Figure 27 Bunch length evolution vs time with $x = 0.5\,I^*/I_s$. (values of $x = 0.1\,I^*/I_s$ leaves the evolution unaffected).

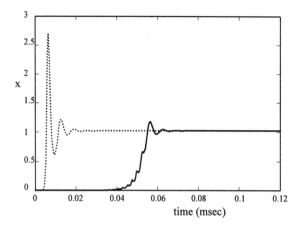

Figure 28 Evolution of the intracavity FEL dimensionless power (continuous line) in comparison with the case unaffected by the instability (dotted line).

$$\frac{d}{dt}\sigma^2 = \left(\alpha - \frac{2}{\tau_s}\right)\cdot\sigma^2 + \frac{2}{\tau_s}\cdot x,$$
$$\frac{d}{dt}x = D\cdot x\cdot\left[\frac{1}{\sqrt{1+\sigma^2}}\frac{1}{1+1.7\cdot\mu_\epsilon(0)^2(1+\sigma^2)} - r\right],$$
$$(D = \frac{0.85\cdot g_0}{T})$$

The second of equations (8.8.15) contains a further term with respect to the $\mathcal{SR}$ FEL equations discussed in chapter III, which is the FEL induced spread, the last equation is just that providing the laser power intracavity evolution. The evolution of the intracavity FEL dimensionless power, of the induced energy spread and of the instability growth rate is reported in Figs.28,29 and a comparison is done with the case unaffected by the instability.

It is evident that if the laser process start, the final power is independent of the presence of the instability, it is also interesting to note that the parameter controlling the laser evolution is r namely the ratio between the cavity losses and the FEL small signal maximum gain, this parameter can always be chosen in such a way that the laser power be just on the threshold to switch off the instability (see fig.30), this fact suggests that the FEL can be used as a passive element to provide a more stable operation.

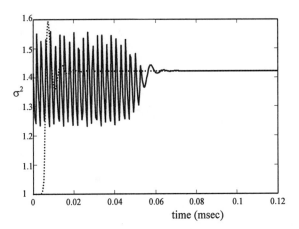

Figure 29 Evolution of the induced energy spread (continuous line) in comparison with the case unaffected by the instability (dotted line).

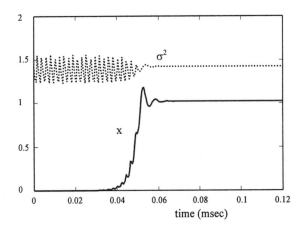

Figure 30 A comparison between the evolution of the intracavity FEL dimensionless power (continuous line) and the induced energy spread (dotted line).

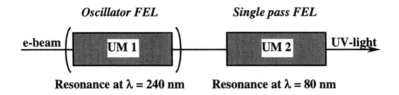

Figure 31   FEL-OA schematic layout.

8.9 Harmonic generation and FEL devices operating at short wave-lengths

In the previous section we have seen that a suitably prebunched e-beam may allow the generation and the growth of a laser signal without any input seed. This fact is an element of noticeable importance, since it may provide the lasing at short wavelength. However the way to achieve the bunching is by no means a secondary point. It is evident that the most efficient way to bunch an e-beam for FEL operation is the FEL itself.

This observation has been the starting point of the so called FEL-OA (oscillator-amplifier) devices. The concept underlying these systems is fairly straightforward. With reference to fig. 31, we note that the first oscillator may induce a suitable bunching at higher harmonics, the e-beam extracted from the oscillator can be injected into a second undulator tuned at e.g. the third harmonic of the first. It is evident that if the bunching is appropriate and the gain is enough, coherent radiation in the second undulator can be generated and driven up to the saturation.

Unfortunately along with bunching, energy spread too is induced and therefore when it is large enough, the laser in the second undulator may be switched off. To overcome this difficulty the first oscillator may be operated at very low stationary intracavity intensity.

Let us more quantitatively analyze the dynamical behaviour of a FEL-OA.

In the case in which we assume that the undulator of the amplifier-section has a period equal to $1/\zeta$ of the first, the coefficient responsible for the start up of the signal is b_ζ. We have already remarked that the prebunching can also be treated as an equivalent seed and that the growth of the signal may be specified by the simple relation

$$I\left[\frac{MW}{cm^2}\right] \simeq I_b\left[\frac{MW}{cm^2}\right] \cdot \exp\left[4\sqrt{3}\,\pi\,\rho\,\frac{z}{\lambda_u}\right] \qquad (8.9.1)$$

The equivalent input intensity may be written in terms of the bunching coefficient as

$$I_b \left[\frac{MW}{cm^2}\right] \cong 5,57 \cdot 10^{-7} \pi^2 \, |b_\zeta|^2 \, \gamma \rho \, \left|J\left[A/m^2\right]\right| \tag{8.9.2}$$

where

$$b_\zeta = \frac{1}{2\pi} \int_{-\infty}^{+\infty} b_\zeta(\nu) \, d\nu \tag{8.9.3}$$

By assuming that the e-beam is initially monoenergetic and by calculating 8.9.3 at $\tau = 1 + \delta$, $\delta = \frac{L_D}{L_{u,1}}$ we find

$$|b_\zeta| \cong \frac{9}{32\pi} (1+\delta)^\zeta \, |a|^{3/2} \tag{8.9.4}$$

where $|a|$ is the dimensionless intracavity intensity of the first oscillator.

Since the efficiency of a constant parameters FEL amplifier is just ρ, we can link saturation length z_s and bunching coefficient by imposing the condition

$$I_b \left[\frac{MW}{cm^2}\right] \cdot \exp\left[4\sqrt{3}\,\pi\rho \frac{z_s}{\lambda_u}\right] \cong \rho P_E \left[\frac{MW}{cm^2}\right] = 0.511 \cdot 10^{-4} \gamma \rho \left|J\left[A/m^2\right]\right| \tag{8.9.5}$$

which yields

$$|b_\zeta| \cong \zeta \, e^{-2\sqrt{3}\,\pi\rho N_s} \quad , \quad N_s = \frac{z_s}{\lambda_u}$$

The above relation states the link between the intracavity power of the oscillator and the number of periods of the second section undulator to reach saturation in the amplifier, namely

$$I\left[\frac{MW}{cm^2}\right] \simeq 1.5 \frac{I_s \left[\frac{MW}{cm^2}\right]}{1+\delta} e^{-\frac{2\sqrt{3}}{3}\pi\rho N_s} \tag{8.9.6}$$

By using the parameters given in Tab VIII.2 we find $\rho \simeq 2.27 \cdot 10^{-3}$ and thus a maximum attainable peak power of $2.44 \cdot 10^5 \frac{MW}{cm^2}$, by assuming that the number of periods of the second undulator is $N_2 = 300$ and $\delta \cong 0.59$ we find that this requires $\frac{I}{I_s} \sim 1.4 \cdot 10^{-3}$.

Table VIII.2

$E\,[MeV]$	=	215
$\hat{I}\,[A]$	=	200
σ_ε	$\leq$	1%
N_1	=	40
$\lambda_{u,1}\,[cm]$	=	4.25
k_1	=	$\sqrt{2}$
$\lambda_{u,2}\,[cm]$	=	$\frac{1}{3}\lambda_{u,1}\,[cm]$

The above simple considerations yields idea of what one can obtain with FEL-OA devices and how to combine oscillator and amplifier parameters to get reasonable predictions. In the following we will discuss a more realistic analysis, based on a deeper numerical treatment.

One might also argue that, to reduce the saturation length it would be sufficient to operate at large intracavity power. This solution is not practical because the induced energy spread may inhibit the on-set of the laser signal in the amplifier. We expect that when the intracavity power increases, larger bunching occurs along with larger induced energy spread so that the output power in the Amplificator is switched off.

An idea of the behaviour of the power at the end of the second undulator, as a function of $\frac{I}{I_s}$ of the oscillator is offered by fig. 32 (for the input data see Tab VII.1), which also displays the saturation of the gain in the bunching section. It is interesting to note that the maximum power is reached significantly before the on-set of the saturation in the oscillator.

In fig. 33 we report the final power vs. N_2, for different values of $\frac{I_0}{I_s}$, it is interesting to note that when $\frac{I_0}{I_s}$ increases (and thus for larger bunching coefficients) the first part of the signal growth is dominated by a quadratic behaviour and not by the exponential one.

For larger values undesired effects like "early" saturation arise. It is evident from the previous discussion that the evolution of the power in the amplifier, reflects, within certain limits, that of the oscillator.

The power grows inside the oscillator round-trip after round-trip and eventually reaches saturation which depends on the gain and cavity losses. The output power of the Amplifier grows up to a maximum and then reaches an equilibrium value, which does not coincide with the maximum attainable power. The reason is obvious. We have indeed seen that the maximum is reached for $\frac{I}{I_s} \ll 1$, while the intracavity equilibrium is obtained for $\frac{I}{I_s} \gtrsim 1$. An oscillator having an equilibrium intracavity power well below the saturation intensity can be conceived. It would require, however, such a small gain or such a large cavity losses that the rise time of the system would be very long and the operation intrinsically unstable. The problem would be therefore that of finding a compromise between large gain to reach large output power in the Amplifier and reasonable values of the cavity losses. To that the oscillator dynamics does not switch-off abruptly the Amplifier.

An idea of the above quoted dynamics is provided by fig.34. In fig.34(a) the oscillator cavity has been assumed to be $2.5\,m$ long and thus 60 round trips correspond to $1\,\mu s$. A remarkably negative feature of this configuration is that the start-up of the Amplifier occurs in a rather long time and that the output signal, averaged on the round-trip, remains low. This is presumably due to the fact that the intrinsic gain of the oscillator is only 20%. The situation significantly improves for the case of fig 34(b) (larger cavity, larger oscillator gain 50% with cavity losses 20%). in this case the Amplifier reaches the maximum value in about $2\,\mu s$.

According to the present discussion, it is evident that FEL-OA devices could be exploited to provide laser power in the VUV region below $100\,nm$.

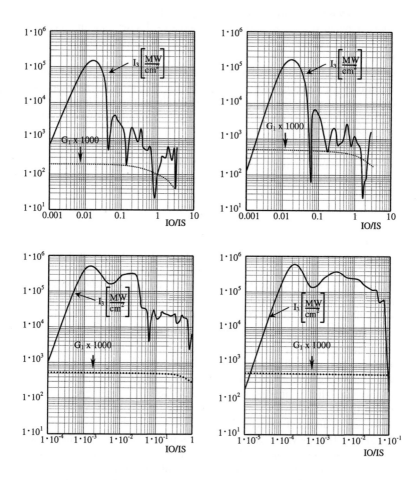

Figure 32 Output power vs dimensionless intracavity power for different number of periods of the amplifier, (a) 100, (b) 200, (c) 300, (d) 400.

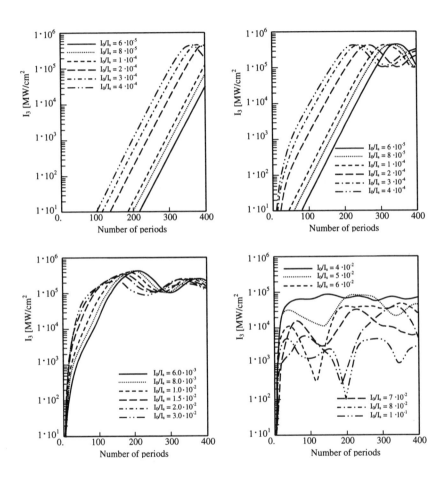

Figure 33 Output power vs the number of periods of the amplifier for different values of the intracavity power.

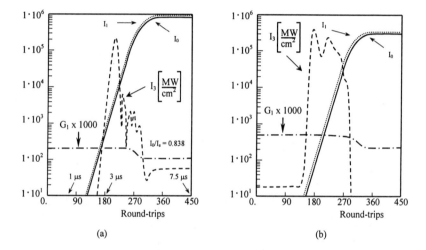

Figure 34 Intensity growth at the amplificator output vs the round-trip number. the other curves are relevant to the oscillator intracavity dynamics: I_1 is the power at the undulator output, after a cavity reflection, G_1 is the gain. (a) cavity length $2.5\,m$, $N_2 = 100$, $N_1 = 30$, $1\,\mu s = 60$ round-trips. (b) cavity length $3\,m$, $N_2 = 300$, $N_1 = 40$, $1\,\mu s = 50$ round-trips.

The problems associated to the oscillator dynamics, could be overcome by using an external conventional laser as buncher. This solution, even through useful, limits the region of operation to that of the available conventional lasers. In the cases discussed here, the lasing region is fixed by the availability of reflecting mirrors in the chosen region of operation.

The interested reader is addressed to the bibliography at the end of the chapter.

8.10 Fourth generation synchrotron radiation sources: Linacs or Storage Rings

In the previous sections we have seen that FELs operating with Linacs or Storage Rings may yield the possibility of providing high-brightness, short wavelength coherent radiation. A natural question is therefore which of the two solutions is more convenient to generate short wavelength laser light, a Storage Ring or a high energy Linac?

We have seen that in the case of VUV-X ray radiation, the e.m. field generated by an electron bunch cannot be trapped in an optical cavity. Coherence does not develop after many cavity round trips, in which, counterpropagating radiation, overlaps in the undulator a freshly injected electron bunch, induces energy and density modulation thus forcing the electrons to radiate coherently. For sufficiently high brightness e-beam and for sufficiently long undulators, the gain may be so high that spontaneous power emitted in the beginning part of the undulator could be amplified to an intense quasi-coherent radiation. As also remarked in the previous section, the use of a seed laser or of a suitable prebunching, playing the role of an equivalent seed, may be exploited to reduce the saturation length and to provide a preliminary spectral selection.

This type of mechanism has attracted significant attention during these last years because it seems the only capable of providing the "fourth generation" light sources yielding coherent radiation down to the X-ray wavelength region, with substantively larger brightness than that achievable with conventional sources.

We have also seen that various performances relevant to the SASE mechanism can be expressed in terms of ρ and we have remarked that the SASE saturates in about $\frac{1}{\rho}$ periods and the saturation power is about ρ times the e-beam power and the e-beam energy spread should be $\sigma_\varepsilon < \rho$ to avoid gain degradation effects. The further condition on the emittance $\varepsilon < \frac{\lambda}{4\pi}$, should be imposed to ensure spatial coherence and overlapping between electron and laser mode.

To be more quantitative we write the condition on the emittance as

$$\varepsilon\, [m \cdot rad] \leq 8 \cdot 10^{-12} \lambda \left[\overset{\circ}{A} \right] \qquad (8.10.1)$$

and manipulate the condition $I_L \sim \rho P_E$, to get (see equation 8.4.5)

$$I_L \left[\frac{MW}{cm^2}\right] \simeq 1.26 \cdot 10^3 \, (N\rho)^4 \, I_S \left[\frac{MW}{cm^2}\right] \quad (8.10.2)$$

if we now assume that $N\rho \sim 1$, we end up with the interesting conclusion that the laser power is uniquely specified by the "natural" saturation intensity of the device.

We can use the above relations to get a precise idea on the Linac beam parameters to operate e.g. at $1\,\text{Å}$. We first assume $\rho \sim 10^{-3}$, to limit the number of undulator periods and the e-beam energy around $15\,GeV$. From this last assumption it follows that we can choose an undulator with $\lambda_u = 6\,cm$ and $k = 2$, while from 8.10.1 we infer that the e-beam should have a normalized emittance of the order of $1.2 \cdot 10^{-7}$, with a current (see equation 8.10.2) of about $2\,kA^*$.

It is claimed that the present Linac technology is capable of providing the above requested performances. It seems indeed possible to use R.F. photocathode guns with the emittance correction technique, to produce $1\,nC$ electron bunches in $10\,ps$. After acceleration, pulse compression, further acceleration, e-beam with $15\,GeV$ energy, pulse length of $200\,ps$ and emittance $\varepsilon \sim 0.3 \cdot 10^{-10} mrad$ can be obtained. The above parameters seem to perfectly match the previous requirements.

Let us now go back to the Storage Rings. We have seen that the advantage offered by a Storage Ring is due to the fact that most of its characteristics are determined by the interplay between damping and quantum diffusion: so far the brightest electron beams are provided by Storage Rings designed for third generation synchrotron radiation sources. The strategy followed within this context is that of minimizing the effects due to the quantum excitation dilution by a proper choice of the magnetic lattice. As already remarked in chapter VI the best achieved performance yield Storage Ring emittances of the order $10^{-9}\,m \cdot rad$ for radial emittance and 10^{-11} for the vertical one. To fit the requirement of the $1\,\text{Å}$ operation the radial emittance should be improved by two orders of magnitudes. We have already noted that the problem of getting suitable emittance values for advanced synchrotron radiation facilities, is due to the fact that emittance scales with the square of the e-beam energy, while the optical beam emittance scales with the inverse of the square e-beam energy. as already remarked a lattice of low emittance $\mathcal{SR}$ has a large number of achromatic cells (see chapter VI). Limiting ourselves to the Chasman-Green lattice, we write the horizontal emittance in practical units as (see equations 6.5.6)

$$\varepsilon_x\,[m \cdot rad] \cong 7.7 \cdot 10^{-13} \frac{\gamma^3}{N_d^3} = 2.95 \cdot 10^{-6} \frac{E^2\,[GeV]}{N_d^3} \quad (8.10.3)$$

---

*To derive this value we have assumed that the transverse section of the electronbeam is

$$\Sigma_T = \pi\,\sigma_x^2 = \pi\,\beta_x\,\varepsilon_x = \pi\left(\frac{N\lambda_u}{4\pi} \cdot \frac{\lambda}{4\pi}\right)$$

for the operation at $1\,\text{Å}$, we require therefore $\sigma_x \sim 2\,\mu m$.

where N_d is the number of achromat cells in the ring lattice.
To better exploit equation 8.10.3 we cast it in the form

$$\varepsilon_x\,[m \cdot rad] \simeq 5.85 \cdot 10^{-3} \frac{\lambda_u\,[m]\left(1+\frac{k^2}{2}\right)}{N_d^3\,\lambda\,[\text{Å}]} \tag{8.10.4}$$

The use of 8.10.1 finally yields the condition

$$N_d^3 \underset{\sim}{>} 4.8 \cdot 10^6 \frac{\lambda_u\,[cm]}{\lambda^2\,[\text{Å}]}\left(1+\frac{k^2}{2}\right) \tag{8.10.5}$$

In the case of $\lambda = 1\,\text{Å}$, $\lambda_u = 6\,cm$, $k = 2$ we would end up with an almost unrealistic result $N_d \underset{\sim}{>} 500$, which can be reduced to $N_d \simeq 100$ for $\lambda \simeq 10\,\text{Å}$.

To better appreciate the interplay between the various design elements, we stress again that a very low momentum compaction characterizes $\mathcal{SR}$ operation with Chasman-Green lattices and with large N_d, in fact

$$\alpha_c = \frac{\pi^2}{3N_d^2} \cdot \frac{\rho_m}{C} \tag{8.10.6}$$

where ρ_m is the bending radius and C is the machine circumference. As also noted in chapter VI the maximum achievable peak current, without inducing instability effects of the microwave type, is fixed by

$$I\,[A] = 2\pi \cdot 10^9\,\alpha_c \frac{E\,[GeV]}{\frac{\mathcal{Z}_n[\Omega]}{n}}\sigma_\varepsilon^2 \tag{8.10.7}$$

where $\mathcal{Z}_n$ is the longitudinal coupling impedance at the nth harmonic of revolution frequency.

The $\mathcal{Z}_n/n$ is usually evaluated at the frequency σ_z/c for bunched beams and best achieved values are 0.1-0.5 Ω. It is therefore evident that the operation at short wavelengths is hampered by the achievable current and by the longitudinal instabilities. To be more realistic we could fix a lower limit of operation around 50 Å, thus finding in this case $N_d \simeq 40$ which can be considered an acceptable number. In conclusion, the present technology seems to suggest that high energy - high brightness Linacs may be good candidate to yield X-ray region around few Å, provided that one reaches the claimed emittances and is able to propagate the e-beam with low β and with a spot of few μm. This in turn implies strong constraints on the tolerances of the system. We have indeed seen that the SASE X-ray FELs require undulators about 100 m long with about 1000 periods. The undulator can also be segmented with interruptions between adjacent segments, without degrading the exponential gain in the undulator and without significantly reducing the output power. The interruption between the segments can be used for the e-beam focusing and for diagnostic purposes, but the alignments between all segments must be within the accuracy of microns.

The other problems are connected with the radio frequency photocathode gun and the bunch compression schemes. We have not touched these topics in the book, we only remind that to reach the required performances for few Å SASE operation, the already low emittance, from an R.F. photocathode, must be further reduced by an emittance compensation scheme. Furthermore since the bunch length is typically $10\,ps$ long, it needs to be compressed by a factor of 100, to increase the peak current to the level of kA. in addition emittance dilution must be avoided along the undulator. These tasks could be within the present level of technology. At the same time significant effort experimental and computational is going on to find schemes suppressing the various types of instability in $\mathcal{SR}$. Finally the SASE mechanism at short wavelengths has not been experimentally confirmed.

It is always difficult to make predictions even for the near future. These authors believe that at the moment (June 1999) it is quite difficult to decide in which direction synchrotron radiation sources will evolve, there are good arguments to say that the region down to 50 Å can be reached by using combined mechanisms of FEL and harmonic generation employing either Storage Rings or Linacs, but for the moment is perhaps better avoiding any further speculation.

Suggested bibliography

Most of the literature relevant to this chapter ha already been quoted. For single pass high gain FELs the reader is addressed to

1. Laser Handbook vol. 6 Ed. by W. B. Colson, C. Pellegrini and A. Renieri, North-Holland Amsterdam (1990),
   in particular the contributions by
   J. B. Murphy and C. Pellegrini, Introduction to the Physics of free electron lasers.
   W. B. Colson, Classical free electron laser theory.
   E. T. Scharlemann, Single pass free electron lasers.
   For a more accurate treatment of High gain FEL brightness see e. g.
2. E. L. Saldin, E. A. Schneidmiller and M. V. Yurkov, Phys Rep. 260, 187 (1995).
3. K. J. Kim, Critical review of high gain X- ray FEL experiments, Presented at the XVIIIth international Free Electron Laser conference, Rome (Italy) August 26-31 (1996).
4. G. Dattoli, P. L. Ottaviani and A. Renieri, Il Nuovo Cimento B (1999) to be published.
   For the treatment of emission by prebunched e- beam including self- induced coherent emission see e. g.
5. G. Dattoli, L. Giannessi and A. Torre J. Opt. Soc. Am. 10B, 2136 (1993).
6. A. Doria, L. Bartolini, J. Feinstein, G. P. Gallerano, R. H. Pantell, IEEE JQE 29, 1482 (1993).
7. G. Dattoli, L. Giannessi, P. L. Ottaviani and A. Segreto, J. Phys. I (France) 7, 1039 (1997)
8. G. Dattoli and P. L. Ottaviani, FEL small signal dynamics and electron beam prebunching, ENEA Frascati internal report RT/INN/93/10 (1993).
9. H. H. Weits and D. Oepts, IEEE-JQE 35, 15 (1999).
10. G. Dattoli, A. Dipace, L. Mezi and E. Sabia, J. Appl. Phys.,83, 5034 (1998).
    For Cavity length and energy modulation see
11. U. Bizzarri, G. Dattoli and P. L. Ottaviani, Rev. Sci. Instrum. 70, 1355 (1999).
12. G. Dattoli, L. Mezi, A. Renieri and A. Torre, IEEE-JQE 34, 1782 (1998) and references therein.
    For experimental proposals of seedless prebunched FEL operation see e. g.
13. I. Ben Zvi, A. Friedman, C. M. Hang, G. Ingold, S. Krinsky, K. M. Yang, L. H. Yu, I. Leherman, D. Weissenburg, Nucl. Instrum. Methods A318, 787 (1990).
14. F. Ciocci, G. Dattoli, A. De Angelis, B. Faatz, F. Garosi, L. Giannessi, P. L. Ottaviani and A. Torre, IEEE- JQE 31, 1242 (1995).
    For the interplay between FEL and storage ring instabilities see e. g.
15. G. Dattoli, L. Mezi, A. Renieri, M. Migliorati, M. E. Couprie, R. Roux, D. Nutarelli and M. Billardon, Phys. Rev. 58E, 6570 (1998).

16. G. Dattoli, A. Renieri and G. K. Voykov, Phys. Rev. 55E, 2056 (1997).
17. M. E. Couprie et. al. Nucl. Instrum. Methods, Phys. Res. A 407, 215 (1998).
    For the description of an High Gain FEL facility see e. g.
18. J. Rossbach, Nucl. Instrum. Methods 375A,269 (1996).

Chapter 9
SYNCHROTRON RADIATION BEAM LINES: X-RAY OPTICS

Summary

In this chapter, the X-ray optics are discussed as specific components of a beam lines, which collect the synchrotron radiation from the tangential points to the experimental stations. Thus, the working principle of x-ray (focusing) mirrors, multilayers and crystal monochromators are described. Their role in specific experimental layouts is discussed.

9.1 Introduction

Synchrotron radiation sources generally supports large number (typically 10-40) of experimental stations. These are located along beam lines which, as shown in fig. 1, collect the radiation emitted tangentially as the e-beam passes through the insertion device. This configuration realizes what is currently called *tangential ports*, through which synchrotron radiation leaves the vacuum enclosure and travels till sample in the experimental station. One or more of these ports on the machine are often split by optics, so that it serves simultaneously a number of users.

Optical devices as mirrors, gratings, monochromators and slits are used in beam-lines to achieve the suitable geometry for the experiment and to match the beam parameters to the experimental requests. Each experimental station uses a certain portion of the synchrotron radiation spectrum (IR, UV, VUV, X-ray, hard X-ray). Fig. 2 shows the distribution of experimental techniques and photon energy ranges exploited during the year 96-97 at the SRS at DARESBURY laboratory. It is evident that almost half of the beam time has been exploited for hard X-ray experiment. Usually a station performs one type of experiment, but multipurpose stations are available on many Storage-Rings. The time of a single measure may range from milliseconds to hours; similarly experiments may last from a few days to several months. Most synchrotron facilities operate as multi-user facilities for a broad range of scientific topics including physics, chemistry, material science, medicine, biology and so on.

As further example fig.3 shows the distribution of S.R.-based experiments between various disciplines at the SRS at DARESBURY laboratory.

Each insertion device is appropriate for a specific experiment. Some exper-

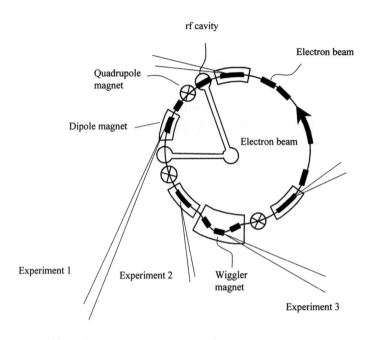

Figure 1 A schematic diagram of a Storage Ring with beam lines collecting radiation from the tangent points to the experimental stations.

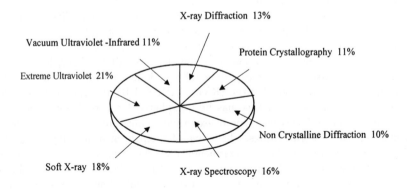

Figure 2 The distributed techniques and photon energy ranges exploited at the SRS at Daresbuty Laboratory in 1996/97.

Introduction

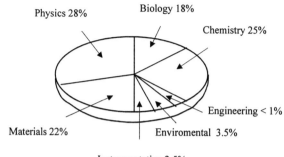

Figure 3 Scientific activities at the SRS at Daresbuty Laboratory in 1996/97.

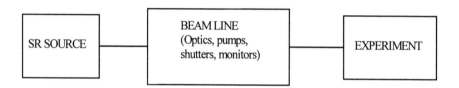

Figure 4 General configuration for SR-based experiments.

iments require e.g. high photon fluxes provided by multipole wigglers, other need high brightness undulator radiation to have beams with small cross sections and small divergences. The beam-lines represent the interface, between the source and the sample, delivering radiation from the tangent point to the Storage Ring to the experiments (Fig. 4).

The design of a beam-line involves problems of different nature in order to met the various requirements of a specific experiment. The first to be solved are those of geometrical nature, arising from the physical constraints imposed by the limited space where a large number of components have to be placed. The unique properties of synchrotron radiation have caused the evolution of specialized instrumentation, as we will see in the below with some details for X-ray optics. For further comments the reader is addressed to the bibliography at the end of the chapter.

A schematic example of a high-brightness beam-line, exploiting the high brilliance X-ray beam from undulator magnets is shown in Fig. 5. This type of beam-line provides high intensity parallel X-ray beams with a small cross-section and therefore useful for a wide range of X-ray scattering and diffraction experiments in structural biology, chemistry, material and surface science. Within this framework, a role of

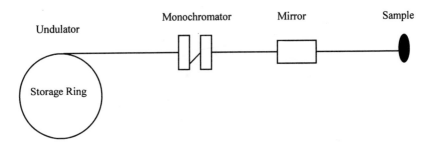

Figure 5   A schematic diagram of a high brilliance beam line.

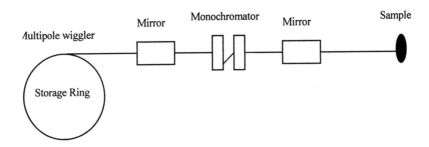

Figure 6   A schematic diagram of a variable wavelength beam line.

noticeable importance is played by the determination of the structure of large and complex protein assemblies, their dynamical behavior is of great relevance in medical science too, when proteins and their aggregates are associated to some pathology. Because of a combination of molecule's weak diffraction and large unit cell, diffraction based investigations require high-brightness X-rays.

The use of multipole wigglers, rather than undulators, yields a broader spectrum of wave-lengths, providing an ideal source for X-ray spectroscopy. A typical beam-line is sketched in Fig. 6, which shows a collimating pre-mirror and a double-crystal monochromator which gives a low wavelength spread at the sample. This is a typical facility for experiments requiring tunable X-rays, including anomalous diffraction and scattering from crystals and amorphous materials.

Insertion devices offer the further advantage of producing photon beams with fully selectable polarization. As already discussed, bending magnets produce radiation linearly polarized in the orbital plane and elliptically polarized below and above that plane. As a consequence, experiments requiring circularly polarized light exploit less brightness since only the periphery of the beam is used. On the other hand undu-

Introduction

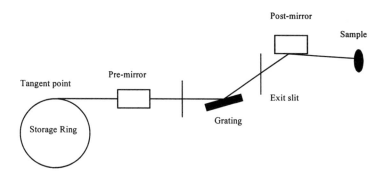

Figure 7   A schematic illustration of a soft x-ray beam line.

lators with variable polarization can be exploited to produce beams with horizontal, vertical and elliptical polarizations. The basic elements of a typical beam line for soft X-rays, as those emitted by helical undulators or multipole wigglers, are shown in Fig.7.

The selection of photon wavelengths is performed by means of gratings, whereas for hard X-rays, as in Figs. 5,6, crystal monochromators are more appropriate. Soft X-rays (0.3-1.5 KeV) represent an ideal tool for the study of surfaces, thin films, multilayers and gases because they allow a great flexibility of diffraction, absorption and spectroscopy techniques, like photoelectron diffraction, X-ray absorption and emission spectroscopy, magnetic X-ray circular dichroism and spin polarized spectroscopy. Such techniques may further benefit from the possibility of selecting the polarization of the photon-beam. To give specific examples, we note that angle resolved photoemission employing linearly or circularly polarized X-rays can reveal important informations about the magnetic moments of surfaces or thin films. Furthermore, circularly polarized beams are suitable to investigate the behavior of chiral molecules, both in the gas phase and adsorbed onto surfaces. Chiral compounds exist indeed in the left- and right-handed versions and often exhibit different chemical properties, thus interacting in a different way with photons having different circular polarizations.

The beam line lay-out of Figs. 5,7 show only optical devices i.e. slits, mirrors, monochromators and gratings. The optical elements are just a fraction of the instrumentation inherent to a synchrotron beam line, but they are its most important part since they are exploited to handle the photon beam extracted from the tangent point.

Beam optics for synchrotron beam-lines must be accurately considered to provide a usable beam with the required characteristics (spectral, angular and spatial) and to take full advantage of the unique properties of SR sources and the relevant design must take the proper account of radiation damage effects, thermo-mechanical problems and associated stability.

The presently well established applications of synchrotron radiation, as well those planned in correspondence with further improvements of both beam quality and power, are based on the efficient performance of the optical elements, which thus assess the final usefulness of such improvements: this concluding chapter is devoted to the illustration of the optical elements commonly used in synchrotron beam-lines. Since, as already stressed, most experiments use hard X-rays, the discussion will be focused on mirrors, multilayers and crystal monochromators. While a detailed bibliography on soft X-ray optics (gratings and Fresnel lenses) is given at the end of the chapter.

Table IX.1 lists the various optical devices currently used in the X-ray beam conditioning. The table contains indication on the photon energy ranges where the element is used and the beam parameters they define (angular divergence ψ, spatial resolution Δx, energy spread ΔE, polarization P.

The plane of the chapter is given below. In section 2 we briefly comment on the role of X-ray optics and the problems underlying the design of optical devices for SR beam lines, which ideally should transfer the full brightness and low emittance of the source to the experimental stations. Section 3, 4 are devoted to illustrate the basic principles of X-ray mirrors in flat and focusing geometry. Multilayer structures are treated in section 5, whereas section 6 deals with monochromators. Finally, Section 8 contains a schematic description of three of the principal SR based measurements along with a list of specific experiment.

Reflecting optics for X-rays exploits the total external reflection occurring at the mirror surface at grazing incidence. Multilayer structures, synthesized on flat or figured surfaces, stand out as non grazing incidence reflecting systems, the exhibited high reflectivities is a by-product of constructive interference processes. Perfect or mosaic crystal based monochromators take advantage of the characteristic Bragg-reflection effect of crystal periodicity. In the following we will see that at a given glancing angle, mirrors reflect X-ray of all energies smaller than the critical energy characteristic of a given material, on the other side single crystals and multilayers select sets of narrower or wider energy ranges corresponding to the Bragg reflection and to the relevant harmonics. According to the experimental requests one or more of these devices can be used in appropriate geometries. Their combined effect on the incoming beam and hence their optimum sequence can be determined by means of accurate codes, based on ray-tracing, or graphic schemes, as the *Du Monde diagrams*.

Introduction

Table IX.1 X-ray optical devices presently existing or under development

	Devices	Soft X-rays 0.3-3 KeV (40-4 Å)		X-rays 3-30 KeV (4-0.4 Å)		Hard X-rays 30-300 KeV (0.4-0.04 Å)		Beam parameters
		R	T	R	T	R	T	
Beam Shapers	Pinholes, diaphrams		f		f		f	$\psi, \Delta x$
	Soller slits		o		f		f	ψ
Total reflectors	Mirrors	f		f				$\psi, \Delta x, E, \Delta E$
	Guide Tubes		o		o			$\psi, \Delta x, E, \Delta E$
Linear and Planar Microstructures	Gratings	f	m	o				$\psi, \Delta x, E, \Delta E$
	Fresnel plates	m	m	o				$\psi, \Delta x, E, \Delta E$
Bragg Optics	Multilayers	f	o	m	o		f	$\psi, \Delta x, E, \Delta E$
	Single Crystals	o		f	m	o		$\psi, \Delta x, E, \Delta E, P$
Combined Systems	Multilayer Gratings	o		o(p)			o(p)	$\psi, \Delta x, E, \Delta E$
	Bragg-fresnell Optics	o(p)		o(p)			o(p)	$\psi, \Delta x, E, \Delta E, P$

9.2 Role of X-ray optics: preservation of the source brightness and emittance.

The specific needs of the various types of S.R. X-ray experiments, as to spatial and spectral resolution of the beam at the sample, are quite different. For instance, a beam cross section of $1\mu m^2$ or less is required by an X-ray microprobe, whereas $10\, cm^2$ are adequate for topography. The angular resolutions range from $1\mu rad$ or even less, for plane wave topography, to typical values of $10\, mrad$ for a microprobe. Similarly, the energy resolution is between 10^{-7} (phonon scattering) and 10^{-1} (microscopy). These wide ranges of parameters is associated to a wide range of optical performances, requested by the experimental conditions. The research work relevant to geometrical shape, crystalline structure, material composition and so on, aimed at producing *ideal* optical elements, is practically never resting. An optical element, transforming the beam parameters to match the experimental conditions, is said *ideal* when it is able to transport both the low emittance and the high brightness of the source downstream to the sample.

The beam transformation and the relevant transport are governed by the Liouville theorem. Therefore, in absence of dissipative effects, phase-space volume and density associated with the source and thus with the beam should be constant. According to the discussion of the previous chapters the optical beam emittance is linked to the phase space area and the brilliance to the corresponding phase-space density. A degradation of the emittance corresponds to an increase of the phase space area and this amounts unavoidably to a deterioration of the brightness, which may also occur owing to the losses associated with the limited efficiency of the optical devices. The problem of emittance and brilliance preservation, within the established tolerances, is therefore one of the challenging features of the design.

Although the problem should be addressed for any specific experimental configuration, we can say, in general terms, that quality and intensity losses are associated either to fundamental and technological limits.

As to the former aspect, the problems are not different from those faced with e-beam transport, with the further complication that diffraction must be taken into account too. Beam shapers, pinholes, slits and aberration terms occurring in focusing schemes may contribute to emittance dilution. The cylindrical and toroidal mirrors commonly used as focusing devices, because cheaper and easier to manufacture than the ideal aspherical surfaces, introduce spherical aberrations responsible for the degradation of the beam quality. A correct design of the optical element should therefore be inspired to a minimization of aberrations, taking into account all kinds of effects arising from the finite source size, small beam divergence and grazing incidence.

The technological limits depend on the quality of the single crystal and on the surface and interface imperfections (roughness and slope errors) of mirrors and multilayer structures respectively and the stability of the materials when exposed to very high power density beams. The present day technology allows to contain the roughness and slope errors of mirror surfaces and thin-film interfaces within a few tenth of nm and $1\mu rad$ respectively. Similarly, highly perfect silicon and germanium single crystals can be artificially grown, where variation of the lattice plane orientation do not exceed $1\mu rad$ and 10^{-6}, respectively.

The most severe technological problem for optical beam quality and quantity conservation arises from the high power density produced by the existing insertion device, in particular by wigglers. To give an example of power levels, we note that power densities up to $150 W/mm^2$ must be handled in the first elements of the beam-line. Windows and filters allow to reduce the power density at the mirror (or crystal) surface by a maximum of two orders of magnitudes. Notwithstanding the power loading can still produce thermal stresses across the surfaces illuminated by the beam, with consequent deformations and/or deteriorations of the optical surfaces. Technical solutions have been adopted to avoid the above quoted problems. These include passive cooling by radiation and conduction, active cooling by water or liquid nitrogen, use of pre-monochromators, practically grazing incidence multilayers and special design of monochromators. As we will see in the following, mirrors are also used as high harmonic rejector, and hence placed before the monochromator in the beam-line reduce the heat loading on it.

The field is in continuous progress and suggests continuous technological improvements. Within this context the researches on new materials with high radiation pressure, low thermal expansion, high thermal conductivity and low X-ray absorption, have been strongly stimulated. Furthermore new techniques like cryogenic cooling of single crystals and adaptive optics have been developed to cure or prevent the *thermal disease*.

9.3 X-ray mirrors

As visible light, the macroscopic interaction of x-rays with matter can be described in terms of a complex refractive index $\tilde{n}$

$$\tilde{n} = n - i\beta = 1 - \delta - i\beta \qquad (9.3.1a)$$

The optical constants δ, β denoting the refractive index decrement and the absorption or extinction coefficient respectively, account for the dispersive and absorptive properties of the sample material. They can be expressed in terms of the sample material parameters and of the X-ray wavelength as,

$$\delta = \frac{r_0}{2\pi} \cdot N \cdot \lambda^2 \qquad (9.3.2a)$$

and

$$\beta = \frac{\mu}{2\pi}\lambda \qquad (9.3.3a)$$

where r_0 is the classical electron radius, μ the linear absorption coefficient and N is the electron density of the material. By expressing this last quantity in terms of the atomic parameters, i.e. the atomic number Z and the atomic weight $A\,(gr/mole)$, the δ coefficient can be rewritten as

$$\delta = \frac{r_0}{2\pi} \cdot \lambda^2 \cdot N_A \cdot \frac{\rho \cdot Z}{A} \qquad (9.3.4a)$$

where N_A is the Avogadro number ($N_A = 6.022 \cdot 10^{23} \cdot mole^{-1}$) and $\rho\,(\frac{gr}{cm^3})$ is the mass density of the material. More accurate expressions give δ and β in terms of the real and imaginary parts of the atomic scattering factor $f = f_1 + if_2$, which describe the overall scattering and absorption of X-ray with the atom. Explicitly, one has

$$\delta = K \cdot f_1, \qquad (9.3.5a)$$
$$\beta = K \cdot f_2$$

where

$$K = \frac{r_0 \cdot \lambda^2}{2\pi} \frac{N_A}{A} \rho \qquad (9.3.6a)$$

The scattering factors $f_{1,2}$ must be calculated from the relativistic quantum dispersion theory. The relevant general expression are omitted here, because out of the purposes of this book. We note however that, limiting ourselves to the Thomson scattering of X-rays by the single atomic electrons, considered as free, f_1 is equivalent to the atomic number Z, thus yielding the expression (9.3.4a) for δ.

X-ray optics is strongly influenced by the values of δ and β, which are both real and typically $|\delta| \simeq 10^{-2} - 10^{-5}$ and $\beta \simeq 10^{-2} - 10^{-6}$. The small values of δ mean that the real part of the refractive index $n = 1 - \delta$ is close to unity. as a consequence, the reflectivity at normal incidence is extremely small, being proportional to the square of the change of this index across a single interface. At angles far from normal incidence (grazing), much higher reflectivity can be achieved. In this regime, usually referred as total external reflection, X-rays are reflected in a specular way if absorption is ignored. When absorption is included, a substantial fraction of the incident beam is still reflected , so that X-rays reflected optics can be realized. X-ray mirrors are, indeed, grazing-incidence reflection based devices. When inserted in S.R. beam-lines, they act as high-energy cut-off filters and/or as focusing devices. X-ray mirrors are also widely exploited as higher order harmonic rejectors to reduce the power load (typically by a factor of two) on the more delicate single crystal monochromators, following downstream along the beam-line. As focusing devices they offer the advantage that the heat and radiation is spread over a much larger surface than in the case of crystals because the reflection angle is very small, usually of the order of some $mrad$, as we will see in the following.

The above qualitative assertions can be made more quantitative by employing the Fresnel equations. The possibility of treating the interaction of an X-ray with a mirror surface as a boundary-value problem for the reflection and refraction of a plane electro-magnetic wave at an absorbing medium, is well supported from the experimental point of view. Such a model yields the well known Fresnel equations for the reflection and transmission coefficients.

The smallness of the grazing-incidence angle and of the optical constants δ and β, in the cases we are considering, allows to work out a simplified expression of the surface reflectivity, which can be expressed in terms of a step function which falls off to zero with increasing angle.

In the following we will derive the Fresnel formulae in the general case of an e.m. wave reflected by a boundary of two materials with different refractive indices. The general expressions are then specialized for the X-ray region and for the usual experimental arrangement in which the first medium containing the incident beam is the vacuum.

We consider an infinite planar interface between two media, described by permeabilities $\epsilon_{1,2}$ and by permittivities $\mu_{1,2}$. The relevant refractive indices are therefore defined as $n_\alpha = \sqrt{\epsilon_\alpha \cdot \mu_\alpha}$. A plane e.m. wave with wave number $k_I = n_1 \cdot k_0$ ($k_0 = \frac{2\pi}{\lambda}$ is the vacuum wave number) moves in medium 1 at some angle φ_I of incidence. As the field must satisfy certain boundary conditions at the interface, taken as the $z = 0$ plane, it is in general necessary to have a second wave in medium 1 (the reflected wave) and another in medium 2 (the transmitted wave). The existence of boundary conditions at the interface plane, which must be satisfied at all points on the plane and at all times, implies the same spatial periodicity for the three waves. The three waves, incident, transmitted and reflected have therefore the same frequency and satisfy at the interface the condition

$$\vec{k}_I \cdot \vec{r} = \vec{k}_R \cdot \vec{r} = \vec{k}_T \cdot \vec{r} \qquad (9.3.7a)$$

where $\vec{k}$ denotes the propagation vector of the wave, the subscripts I, R and T refers to the three fields and $\vec{r}$ is any vector lying in the interface plane. According to eq. (9.3.7a) the directions of the three waves must be coplanar and the vectors $\vec{k}_{I,R,T}$ lye in the so called plane of incidence specified by the direction of the incident wave and by normal to the interface. The identity (9.3.7a) can be exploited to derive the well known reflection law (see fig.8)

$$\phi_I = \phi_R = \phi \qquad (9.3.8a)$$

for incident and reflected waves, for which $\vec{k}_I = \vec{k}_R$. The Snell law of refraction can be derived in a similar way. By recalling, indeed, that $k_I = n_1 \cdot k_0$ and $k_R = n_2 \cdot k_0$ we can conclude

$$n_1 \sin \phi_I = n_2 \sin \phi_R \qquad (9.3.9)$$

The above relations reflect kinematical properties and follow from the wave nature of the phenomenon and the fact that there are boundary conditions to be satisfied. They are however independent of the detailed nature of the waves or the boundary

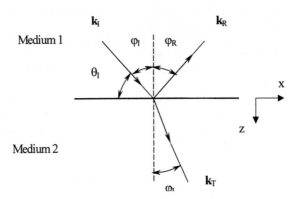

Figure 8   Geometry of the reflection and refraction of an e.m. wave at the interface between two media.

conditions, which determine the so called dynamical relations, relevant to the relative amplitudes of the three waves.

The wave fields can be cast in the general form

$$\vec{E}_\alpha = \vec{E}_{0_\alpha} \exp(i\vec{k} \cdot \vec{r} - \omega t), \quad (9.3.10)$$
$$\vec{B}_\alpha = \frac{n_j}{k_\alpha} \vec{k}_\alpha \times \vec{E}_\alpha,$$
$$(\alpha = I, R, T \text{ and } j = 1, 2)$$

The propagation vectors relevant to incident, reflected and transmitted waves are given by

$$\vec{k}_I = n_1 k_0 \left(\hat{i}_x \sin \phi_I + \hat{i}_z \cos \phi_I \right), \quad (9.3.11)$$
$$\vec{k}_R = n_1 k_0 (\hat{i}_x \sin \phi_I - \hat{i}_z \cos \phi_I),$$
$$k_T = n_2 k_0 (\hat{i}_x \sin \phi_T + \hat{i}_z \cos \phi_T).$$

where $\hat{i}_{x,z}$ denote unit vectors along x and z and the plane x, z is taken as the incident plane (see fig. 8).

The wave polarization must be taken into account too, however since any polarization can be obtained as a superposition of two plane-polarized waves, we will discuss the plane polarized case only. The principal orientations are designed as σ or π- polarization according to whether the electric vector is perpendicular or parallel to the plane of incidence.

X-ray mirrors

The Fresnel equations for the reflection and transmission coefficients r and t, defined as the ratios of the incident electric field amplitude to the reflected and transmitted counterparts, are obtained by requiring the continuity of the tangential (to the boundary surface) components of the electric vector $\vec{E}$ and of the magnetic intensity vector $\vec{H} = \frac{1}{\mu}\vec{B}$.

In the following, we report the expressions for r and t in the case of typical arrangement of an X-ray experiment, where the first medium is a vacuum and the magnetic properties of the second medium have negligible effect on the X-ray interaction so that $\mu_{1,2}$ are equal to the magnetic permittivity of the vacuum.

The explicit expressions for r and t read (see the bibliography at the end of the chapter for further details)

$$r_\sigma = \frac{\sin(\theta_I) - \sqrt{n^2 - \cos(\theta_I)^2}}{\sin(\theta_I) + \sqrt{n^2 - \cos(\theta_I)^2}}, \qquad (9.3.12)$$

$$t_\sigma = \frac{2\sin(\theta_I)}{\sin(\theta_I) + \sqrt{n^2 - \cos(\theta_I)^2}}$$

and

$$r_\pi = \frac{n^2 \sin\theta_I - \sqrt{n^2 - \cos^2\theta_I}}{n^2 \sin\theta_I + \sqrt{n^2 - \cos^2\theta_I}}, \qquad (9.3.13)$$

$$t_\pi = \frac{2n \sin\theta_I}{n^2 \sin\theta_I + \sqrt{n^2 - \cos^2\theta_I}}$$

For the definition of the angle θ_I see fig. 8.

The general formulae for the amplitude reflectivity and transmissivity can be obtained from the previous relations by replacing n with $\frac{n_2}{n_1}\frac{\mu_1}{\mu_2}$ (see problem 1). An example of behavior of r and $t-$ coefficients is provided in the case of an air-to-glass interface (fig. 9), for which $\frac{n_2}{n_1} = 1.5$ and $\mu_1 = \mu_2$. It is interesting to note that r_σ and t_π (as well as r_π after crossing the axis) take negative values as a consequence of a 180-degree phase-shift of the σ-polarized reflected and π-polarized refracted (reflected) wave (Problem 2).

Equations (9.3.12-9.3.13) hold also in the hypothesis, of interest to the X-rays, of complex refractive index $\tilde{n}$. In this case, n^2 is replaced by $\tilde{n}^2$,which on account of eq. (9.3.1a) reads

$$\tilde{n}^2 = (1-\delta)^2 - \beta^2 - 2i\beta(1-\delta) \qquad (9.3.14)$$

The intensity reflectivity R (usually referred as the reflectivity) is just provided by the square modulus of (9.3.12) and (9.3.13), namely

$$R_{\sigma,\pi} = |r_{\sigma,\pi}|^2 \qquad (9.3.15)$$

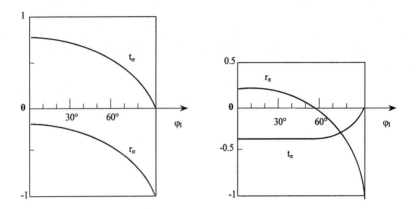

Figure 9 Fresnel coefficients for a σ- and π-polarized wave at an interface $\frac{n_2}{n_1} = 1.5$.

The Fresnel's formulae strictly apply to reflection at plane surfaces; however, they can also be used for curved surfaces provided that the curvature radius is much greater than the wavelength, as it always happens for X-rays. If the glancing angle of incidence θ_I takes *critical* value θ_C defined as

$$\cos\theta_C = n \quad (9.3.16)$$

the amplitude reflectivities reduce to unity and hence the incident wave is totally reflected. According to the Snell's law, it comes out that, for $\theta_I = \theta_C$, $\theta_T = 0$ so that no radiation is enters the medium 2. For angles $\theta_I < \theta_C$ the wave is totally reflected as well, since θ_T takes non physical values being $\cos\theta_T > 1$, and radiation cannot propagate in the second medium. The reflectivity R should exhibit a cut-off, falling off to zero for angles $\theta > \theta_C$. Absorption produces a broadening of the sharp cut-off, as clearly displayed in fig. 10, where the theoretical reflectivity is plotted for different values of the ratio $\frac{\beta}{\delta}$.

In the general case of complex refractive index $\tilde{n}$ the critical angle θ_C is defined according to (9.3.16), with n being the real part of $\tilde{n}$, and therefore on account of eq. (9.3.1a)

$$\cos\theta_C = 1 - \delta \quad (9.3.17)$$

As already remarked, the values of δ are typically of the order $10^{-2} - 10^{-5}$ and hence θ_C is very small, usually of the order of $mrad$. In particular by expanding the cosine up to θ_C^2, we get

$$\theta_C = \sqrt{2\delta} \quad (9.3.18)$$

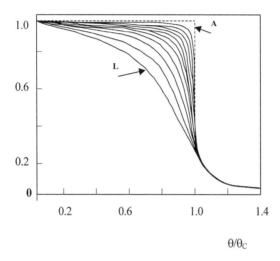

Figure 10 Theoretical reflectivity vs ϑ/ϑ_c for a perfectly smooth surface for selected values of the ratio $\frac{\beta}{\delta}$. the curves must be labelled from A to L corresponding to the values of $\frac{\beta}{\delta}$. A: 0.0, B: $5 \cdot 10^{-3}$, C: 10^{-2}, D: $15 \cdot 10^{-3}$, E: $2 \cdot 10^{-2}$, F: $25 \cdot 10^{-3}$, G: $3 \cdot 10^{-2}$, H: $4 \cdot 10^{-2}$, I: $6 \cdot 10^{-2}$, J: $8 \cdot 10^{-2}$, K: 10^{-1}, L: $14 \cdot 10^{-2}$.

The use of eqs. (9.5.5-9.3.4a) yields in practical units

$$\theta_C \, [mrad] \simeq 2.3 \cdot 10^{-11} \sqrt{N} \lambda(\mathring{A}) = 2.33 \sqrt{\frac{Z\rho}{A}} \lambda(\mathring{A}) \qquad (9.3.19)$$

In table II, the critical angles θ_C for a representative of mirror materials are listed.

An equivalent expression for θ_C can be worked out in terms of the photon energy corresponding to the wavelength λ, by recalling indeed that the photon energy can be written as

$$E\,[KeV] = \frac{12.4}{\lambda\left[\mathring{A}\right]} \qquad (9.3.20)$$

one obtains

$$\theta_C \, [mrad] \simeq \frac{29}{E\,[KeV]} \sqrt{\frac{Z\rho}{A}} \qquad (9.3.21)$$

For low Z-materials, a good approximation is given by setting $\frac{Z}{A} = \frac{1}{2}$, thus getting

$$\theta_C \, [mrad] \simeq 1.6 \sqrt{\rho} \lambda(\mathring{A}) \simeq \frac{20}{E\,[KeV]} \sqrt{\rho} \qquad (9.3.22)$$

The following two approximated (within 10%) relations, can be exploited for low and high Z-materials respectively

$$\theta_C \, [mrad] \simeq \frac{33}{E\,[KeV]}, \; \text{(low Z)} \qquad (9.3.23)$$

$$\theta_C \, [mrad] \simeq \frac{77}{E\,[KeV]}, \; \text{(high Z)}$$

The dependence of θ_C on $\sqrt{N}$ implies that coating the mirror surface with an heavier element one can increase the critical angle, as shown in fig.11a), where the reflectivity $R(\theta)$ at $\lambda = 1.54 \mathring{A}$, is plotted as a function of the glancing angle θ for quartz and gold-coated quartz mirrors. In the latter the critical angle is larger (more than a factor two), but the cut-off is broadened by the larger absorption and consequently the reflectivity at θ_C is smaller than that of the coated mirrors. Let us also note that from eq. (9.3.19) one can derive a critical X-ray wavelength and a critical energy according to

$$\lambda_C \left[\mathring{A}\right] = \frac{0.3 \cdot 10^{10}}{\sqrt{N}} \theta_C \, [mrad], \qquad (9.3.24)$$

$$E_C \, [KeV] = 37.2 \cdot 10^{-11} \frac{\sqrt{N}}{\theta_C \, [mrad]}$$

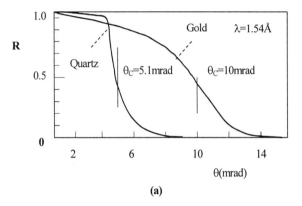

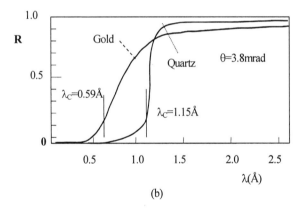

Figure 11 Reflectivities of uncoated and gold-coated quartz mirrors calculated as function of (a) the glancing angle ϑ for $\lambda = 1.54\,\text{Å}$ and (b) the x-ray wavelength λ for $\vartheta = 3.8\,mrad$.

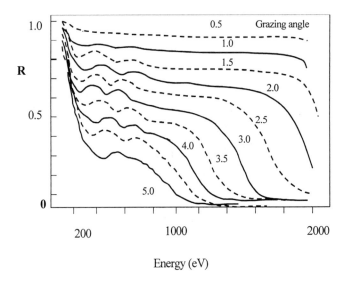

Figure 12   Reflectivity vs x-ray energy for platinum at different angles of incidence.

Thus, for a range of glancing angles of incidence ($\theta \leq \theta_C$) X-rays with energies up to the critical value are reflected efficiently, whereas more energetic X-rays are not reflected. This is the basic mechanism allowing the use of mirrors as efficient high-frequency filters for harmonic rejection when exploited with a crystal monochromator. In fig. 11b) the reflectivity $R(\lambda)$ of the quartz and gold-coated quartz mirrors of fig. 11b) is reported as function of the wavelength for a grazing angle $\theta \simeq 3.8$. It is evident that, for a given glancing angle, a high-Z surface reflects X-rays with shorter wavelengths than a low Z-material, having a sharper cut-off than high Z-surfaces. In fig. 12 reflectivity profiles for platinum, which is widely used in X-ray mirror coatings, are plotted vs. the photon energy from 0.2 to 2 KeV for a wide range of grazing angles.

The further practical motivation of increasing the critical angle is motivated by the possibility of reducing the length of the X-ray mirror, when used in S.R. instrumentation. For technical reasons, the optical elements in a S.R. beam cannot be located too close to the source point, typical distances are of the order of 10-20 m. Being the horizontal and vertical opening angles, ψ_1 and ψ_2, on the order of 10 mrad and 0.2-0.4 mrad respectively, long mirrors are needed to collect the vertically dispersing beam. With the typical arrangement of fig.13 it is easy to conclude the

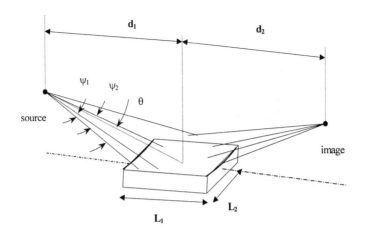

Figure 13   Typical arrangement of a mirror with respect to a SR beam.

following minimum value for the mirror dimensions

$$L_1 \geq \frac{d_1 \psi_2}{\theta}, L_2 \geq d_1 \psi_1 \qquad (9.3.25)$$

for small angles $\psi_{1,2}$ and θ. For instance, at a distance $d_1 = 15m$ from the source a quartz mirror at $\lambda = 1.2\text{Å}$ ($\theta_C = 3mrad$) should have dimensions $L_1 \simeq 1 - 2m$ and $L_2 = 1.5cm$. The use of heavy metals (platinum or gold) coating of the quartz surface, allows an increase of the critical angle by more than a factor of two, and thus a reduction of the mirror length by the same amount. Long mirrors are expensive; composed systems, consisting of several aligned short segments are widely used. In contrast, they demand for more precise mechanical holders, being the focusing achieved by adjusting the heights of the segments until the reflected beams overlap at the focus position.

Before closing this section, we derive an analytical expression for the mirror reflectivity $R(\theta)$ as a function of the grazing angle θ, valid in the approximation allowed by the smallness of both θ_C and the optical constants in the case of X-rays. Approximating the circular functions in eqs. (9.3.12) up to the θ^2 and keeping first order only in the quantities δ and β, we end up with

$$R_\sigma \simeq \frac{h - \frac{\theta}{\theta_C}\sqrt{2(h-1)}}{h + \frac{\theta}{\theta_C}\sqrt{2(h-1)}} \qquad (9.3.26)$$

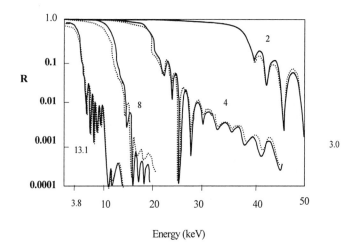

Figure 14 Computed (full line) and measured (dotted line) reflectivity vs energy for a platinum-coated mirror at different grazing angle (in $mrad$). For both profiles, a 29 Å rms surface roughness for the air-platinum interface and a 7 Å rms for the platinum-glass interface have been used.

with

$$h \simeq \left(\frac{\theta}{\theta_C}\right)^2 + \sqrt{\left[1 - \left(\frac{\theta}{\theta_C}\right)^2\right]^2 + \left(\frac{\beta}{\delta}\right)^2} \qquad (9.3.27)$$

As previously stressed, in absence of absorption the reflectivity is the unity for $0 < \frac{\theta}{\theta_C} < 1$ and then drops abruptly for $\frac{\theta}{\theta_C} > 1$. The step function is more and more smeared with increasing absorption, as shown in fig.10, where the function is plotted for selected values of $\frac{\beta}{\delta}$. As further remark it is worth stressing that the analytical expression (9.3.26), and hence the theoretical reflectivity profiles, holds for perfectly smooth homogeneous mirror surfaces. Unfortunately, in practice an accurate knowledge of the reflectivity is unlikely to be available for a practical surface which, to some extent, is rough and impure and may degrade on exposure to the S.R. beam. Specular reflection from real surfaces is based on a modified Fresnel theory, that takes into account surface roughness and slope errors. For an accurate discussion of this topic the reader is addressed to the bibliography at the end of chapter. In fig. 14, we report the computed and measured reflectivity vs. photon energy of a platinum coated mirror. The oscillatory modulation are an effect of the coating thickness, whilst the surface roughness causes a steeper descent of R as θ approaches θ_C.

9.4 X-ray focusing mirrors

An other important application of mirrors is to focus an X-ray beam onto the sample. Optical arrangements for synchrotron radiation experiments usually demand for focusing optics because, due to the large source to experiment distance, one needs to exploit the extreme brilliance of the source. On the other hand, diffraction and scattering experiments at hard X-ray energies require a focussed spot either at the sample or at the detector. Similarly, for VUV experiments or soft X-ray spectroscopy, the beam is focussed twice, at the entrance slits and exit slits of the grating monochromators (fig.7). Although focussing of X-rays and UV radiation can in principle be performed by using crystals or diffraction gratings, in practice one uses prefocusing mirrors, working also, as previously stressed, as high-energy cut-off filters.

The efficiency of the focusing is determined by aberrations and by mirror imperfections such as surface roughness and slope errors. The best conditions are achieved by shaping the mirror surface in elliptical or parabolic profiles.

In fact, it is well known in optics that elliptic and parabolic mirrors are stigmatic imaging systems *. It is easy to prove that in an ellipsoidal mirror the two foci (fig.15a)) represent a stigmatic pair, since the radiation diverging from a source, located at one of them, let us say S, for instance, is focussed into the other, and viceversa. Similarly, in a paraboloidal mirror rays parallel to the axis are all brought to a focus at the same point, denoted as I in fig.15b), just the focus of the parabola.

More precisely, in a cylindrical mirror whose section is a segment of an elliptic profile (see fig.15a)), the object point at the focus S is imaged into a line through the other focus I, with the parameters being linked as

$$a^2 = \frac{1}{2}\left[d + \sqrt{d^2 - \frac{1}{4}(d_1^2 - d_2^2)}\right] \quad (9.4.1)$$

$$b^2 = a^2 - \frac{1}{4}(d_1^2 + d_2^2)$$

where

$$d = \frac{1}{2}(d_1^2 + d_2^2) + 4\left(\frac{d_1 d_2}{d_1 + d_2}\right)^2 \vartheta^2 \quad (9.4.2)$$

The above specify the semiaxis a and b of the elliptic section, whilst d_1 and d_2 represent the distances of the incidence point from the object and image points, respectively, measured along the major semiaxis (Problem 3).

If indeed the mirror surface is a segment of a parabolic profile, the object point at infinity is imaged into the line through the parabola focus I. In that case, one has

---
*Let us recall that a reflector forms a geometrically perfect image of the source if all rays emerging from the *object point* are reflected through the same point, the *image point*, independent of the initial ray direction. The two point are called a *stigmatic pair*.

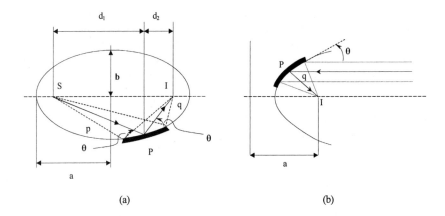

Figure 15 Stigmatic X-ray focusing mirrors, machined from an ellipsoidal (a) and paraboloidal (b) surface.

$$a = \frac{d_1}{2}\left[\sqrt{1+\vartheta^2} - 1\right] \quad (9.4.3)$$

where a is the distance of the focus I from the parabola generatrix, d_1 denotes the incidence point to focus distance, taken along the axis, and ϑ as before is the grazing incidence angle.

However, a true ellipsoidal figure is difficult to produce and also with a real (i.e. extended rather than point) source the advantage of the ideal ellipsoidal geometry is rapidly lost. Spheres and cylinders with circular cross sections can be much easier produced and hence are most frequently used. In fact, on account of the angular extent of the S.R. beam, which is much wider than it is high, and on account of the small grazing angle of incidence, a section of a cylindrical mirror with a circular cross section of radius R_S (fig.16) can be an effective focusing system. The minimum mirror dimensions are dictated by the opening angles, horizontal ψ_1 and vertical ψ_2, of the beam and the parameters d_1, R_S and ϑ of the reflection geometry, as

$$L_1 \geq \frac{d_1\psi_2}{\vartheta} + \frac{L_2^2}{8\vartheta R_S}; \quad L_2 \geq d_1\psi_1 \quad (9.4.4)$$

where the first term reproduces eq. (8.3.25), whilst the second specifies the additional length, required by the curved intersection of the flat beam with the curved mirror surface.

It is evident that a great improvement in the focusing effectiveness of the mirror can be achieved by curving the mirror in the *long* (we recall that L_1 and

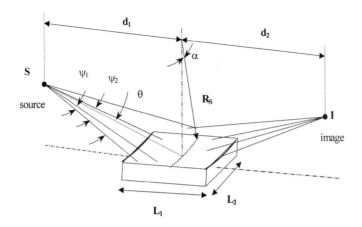

Figure 16   SR beam reflector geometry using a cylindrical mirror.

L_2 differ typically by a factor ten) direction, with radius of curvature $R_T < \infty$. Actually, toroidal mirrors are widely used as X-ray reflectors instead of spherical mirrors. Indeed, the latter are affected by undesired effects as astigmatism, spherical aberrations and come, arising from the grazing incidence, required to have high X-ray reflectivities.

We briefly describe the image formation of a concave spherical mirror, stressing the drawbacks, associated with grazing incidence, and their minimization.

Consider a concave spherical mirror with radius of curvature R and a source point at S. According to the usual nomenclature, the *tangential line* connects the object point S to the centre of curvature C of the mirror and the *tangential* (or *meridian*) *plane* is each plane passing through the tangential line. The tangential plane intersects the mirror in a *meridian* circle. The *sagittal* planes are orthogonal to the meridian ones.

To deduce the focusing conditions for a spherical mirror, we look at fig.17, where a meridian circle is depicted along with the source point S, from which radiation is incident at the point P on the mirror surface at a glancing angle ϑ, and the point I, where an image is formed. The angular divergence of the radiation at S is δ, whilst the convergence angle at the image point is denoted by γ. Both δ and γ are assumed to be small enough to assure that the length $\mathcal{L}$ of the mirror surface, which the radiation is spread over, is much smaller than R, the curvature radius of the mirror: $\mathcal{L} \ll R$. Furthermore, denoting by α the angle subtended by $\mathcal{L}$, we also have $\mathcal{L} = R\alpha$. Let us denote by p and q the source to mirror and mirror to image distances, respectively, taken along some chosen rays; in fig.17, for instance, $p \equiv SP$ and $q \equiv PI$.

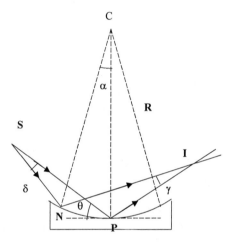

Figure 17 Image formation in the meridian plane by a spherical mirror.

Exploiting the geometrical relations between the triangles associated to the angles α, δ and γ, and the law of reflection, we can deduce that

$$\delta + \gamma = 2\alpha \qquad (9.4.5)$$

Then, applying the sine rule to the triangles SNP and INP, we can express the angles δ and γ in terms of the mirror radius R and the parameters, p, q, ϑ and α of the reflection geometry; explicitly, one obtains

$$\delta = \frac{R\alpha \sin(\vartheta - \frac{\alpha}{2})}{p - R\alpha \cos(\vartheta - \frac{\alpha}{2})}, \qquad (9.4.6)$$
$$\gamma = \frac{R\alpha \sin(\vartheta + \frac{\alpha}{2})}{q + R\alpha \cos(\vartheta + \frac{\alpha}{2})}$$

which, once inserted into (9.4.5), for very small angle α provide the relation

$$\frac{2}{R \sin \vartheta} = \frac{1}{p} + \frac{1}{q} \qquad (9.4.7)$$

for the image formation in the tangential plane.

The above is usually rewritten in terms of the so called tangential (or meridian) radius ρ_M, or equivalently the tangential focal length f_M, given as

$$\rho_M = R \sin \vartheta, \qquad f_M = \frac{\rho_M}{2} \qquad (9.4.8)$$

Accordingly, the focusing condition (9.4.7) in the meridian plane also writes as

$$\frac{1}{f_M} = \frac{1}{p} + \frac{1}{q} \qquad (9.4.9)$$

A similar relation, specifying the imaging condition in the sagittal plane can be obtained as

$$\frac{1}{f_S} = \frac{1}{p} + \frac{1}{q} \qquad (9.4.10)$$

where the sagittal focal length f_S is defined as

$$f_S = \frac{R}{2 \sin \vartheta} \equiv \frac{\rho_S}{2} = \frac{f_M}{\sin^2(\vartheta)} \qquad (9.4.11)$$

ρ_S being the corresponding sagittal radius.

Equations (9.4.9) and (9.4.10) resemble the well known Descartes law for the image formation by a curved mirror in the paraxial approximation. In that case, being the incidence angle almost normal: $\vartheta \simeq 90°$, $f_M = f_S \simeq \frac{R}{2}$ and hence the tangential and sagittal images come to coincide. The formation of two images is therefore a direct consequence of the grazing incidence. The term *astigmatism* is adopted to denote this effect, arising when the incident rays make an appreciable angle φ with the mirror axis. The result is that, instead of a point image, two mutually perpendicular line images are formed in the tangential and sagittal planes. The images are severely astigmatic with smaller glancing angle ϑ, as confirmed by the ratio $\frac{f_S}{f_M} = \frac{1}{\sin^2(\vartheta)}$, which for small angles may become quite large. At $\lambda = 1.5$ Å, for instance, for a quartz surface used at the critical angle $\vartheta_C = 5.1\,mrad$, one has $\frac{f_S}{f_M} = 4 \times 10^4$, whereas for a gold-coated quartz surface at the critical angle $\vartheta_C = 10\,mrad$ the above ratio turns to be $\frac{f_S}{f_M} = 10^4$.

Then, even with dense materials as gold and platinum (see Table II), the astigmatism cannot be reduced to acceptable levels. However, it can be removed at a fixed glancing angle of incidence, shaping the mirror surface at a toroid with radii of curvature R_M and R_S in the tangential and sagittal planes such that the corresponding focal lengths f_M and f_S be equal, thus yielding the relation

$$R_S = R_M \sin^2(\vartheta) \qquad (9.4.12)$$

Typical values for R_M and R_S follow directly from (9.4.9) and (9.4.12). With $p = q = 15\,m$ and $\vartheta = 10^{-2}\,rad$, for instance, one gets $R_M = 2\frac{pq}{p+q}\sin\vartheta = 1.5\,km$ and $R_S = 15\,cm$.

As previously noted, double focusing toroidal mirrors are manufactured by grinding a cylinder of radius R_S into a substrate, which is subsequently bent to a radius R_M, by applying some mechanical couples at the ends. This technique allows to tune the mirror curvature radius R_M, when changing the incidence angle by simply altering the applied forces, and also to correct the deformation of the mirror surface arising from gravitation (typically, the natural deflection due to gravity of a segment of $\sim 50\,cm$ length and 3-4 cm thickness is $5\mu m$) and thermal load. On the other hand, as the deflection at the centre of the mirror is $\alpha \simeq \frac{L_1^2}{8R_M}$, and hence, for the typical values of L_1 and R_M, $\alpha \simeq 20 \div 25\mu m$, high precision is needed for mirror preparation techniques and supports. As a consequence, double focusing single mirrors are often replaced by two single focusing cylindrical mirrors, orthogonally placed, thus forming the so called *compound systems*.

Compound systems are widely used to reduce the spherical aberrations and coma, which, as previously stressed, are the unavoidable defects affecting the image formation at grazing incidence by a spherical mirror. As astigmatism, spherical aberrations and coma [†] are strictly associated to the imaging by a spherical mirror of an

---

[†]Spherical aberrations, coma and astigmatism relate respectively to the first, second and third Seidel sum of the third-order aberration theory of geometrical optics, for which the reader is addressed to the specific text-books. Here, we briefly recall that spherical aberrations arise from the non paraxiality of rays emerging from an on-axis object point. All rays from the object point are not reflected into the common paraxial image point; the image formed by the envelope of the reflected rays extends along, and transversely to, the mirror axis. Similarly, coma is generated by non

X-ray focusing mirrors

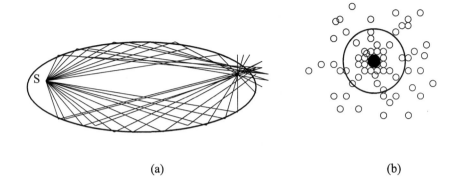

(a) (b)

Figure 18 Imaging of a off-axis object point by an ellipsoidal reflecting surface (a). The reflected rays do not intersect the focal plane at a common point (b). The circle is the intersection with the ellipsoid.

extended source with angular extent.

Spherical aberrations may be removed by using elliptical or paraboloidal mirrors, which however cannot image an extended object. As shown in fig.18, the rays from an off-axis source point are not all reflected into a single image point in the plane of the second focus, but spread over distances that are larger than the distance of the object from the axis. In other words, the magnification (i.e. the ratio between the heights from the mirror axis of the image and source points) changes along the surface of the ellipsoid. In fig. 18 rays that are reflected near the source point on the left side produce a highly magnified image, whereas those reflected on the right produce a demagnified image, a magnification of 1 being obtained for rays reflected from the center of the ellipse.

In fact, there is a very little change in the magnification for mirrors used near normal (i.e. paraxial) incidence; conversely, for reflecting surfaces used at small grazing angles, the magnification changes linearly with position and hence only very short sections of the surface can be used to produce good images. In geometrical optics, the requirement for constant magnification is expressed by the Abbe sine condition

$$\frac{\sin \varphi}{\sin \varphi'} = M = const. \qquad (9.4.13)$$

where φ and φ' are the angles, with respect to mirror axis, of rays from the object and

---

paraxial incidence, which causes the magnification to be different for different parts of the mirror surface. Thus, instead of a point image, a point object produces an image pattern with the typical cometlike appearance, from which this kind of aberration takes the name.

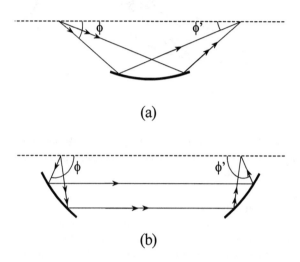

Figure 19   The Abbe sine condition cannot be satisfied for a single reflection in two dimensions (a); two reflections are needed (b).

to the image, respectively, and M is the magnification. It is easy to recognize that the above condition cannot be satisfied by any ray from the source for a single reflection in two dimension. As graphically depicted in fig.19a) the ratio $\frac{\sin\varphi}{\sin\varphi'}$ cannot be kept constant for any ray if only one reflecting surface is involved, because for any surface shape the angle φ' decreases when increasing φ; conversely, using two reflectors, the angles φ and φ' can be made to be equal for any rays from the object point, as schematically depicted in fig.19b), thus satisfying the sine condition (9.4.13). For a three-dimensional systems four reflections are needed to satisfy the sine condition, two in each plane, meridian and sagittal.

For these reasons, compound systems are used involving at least two grazing incidence reflecting surfaces. There are two main types of such compound systems: Kirkpatrick-Baez and Wolter mirror systems.

The simplest implementation of Kirkpatrick-Baez systems uses two cylindrical mirrors with equal radii of curvature, placed at right angle to each other (fig.20). In

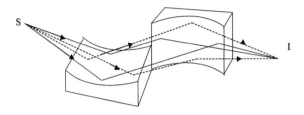

Figure 20   A two-component Kirkpatrick- Baez mirror system.

practice, to significantly reduce spherical aberrations and coma, more than two reflectors are used, the relevant parameters being chosen in order to design an aberration-free reflection geometry.

Wolter optical systems exploit the focusing properties of conicoidal surfaces, where the foci are imaged into each other. Accordingly, optical systems composed by two confocal conical reflecting surfaces can be designed to minimize the aberrations. Examples of such systems are shown in fig.21.

The system of fig.21a), working as a microscope, consists of section of a hyperboloid and ellipsoid, which have a common focus at the point F_1. Rays emerging from the other focus of the hyperboloid (at O) are first reflected from the surface of the hyperboloid, hit the elliptical surfaces as if coming from F_1 and are then reflected to the second focal point I.

Similarly, in the system of fig.21b), which works as a telescope with the object being at infinity, the ellipsoid is replaced by a paraboloid.

In both systems of fig.21, astigmatism and spherical aberrations are eliminated; and with an appropriate choice of the conical surface parameters coma can also be made small.

The above discussed mirror arrangements belong to the so called Wolter type I systems, which involve reflections from two internal surfaces. They are most common since the two reflecting surfaces can be machined from a single block, making easier the subsequent alignment. For completeness' sake, we note that Wolter type I optics involve an internal followed by an external reflection from conicoidal surface sections, whereas the inverted sequence (external-internal) is actuated in Wolter type II optics.

It is needless to say that many other compound systems have been envisaged

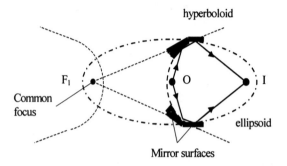

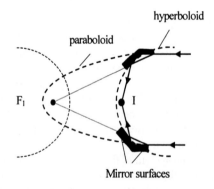

Figure 21 Wolter type I optics: (a) a microscope and (b) a telescope.

as possible X-ray reflectors having high reflectivity and good image quality. For a more exhaustive discussion of the topic, the reader is addressed to the bibliography.

In this connection, let us briefly stress that increasing the mirror components in the reflection geometry the performances of the system especially for off-axis imaging can hardly be analyzed by means of analytical techniques. Design of optics, indeed, is normally performed using ray-tracing computer programs. As input to such programs, one has to provide the geometrical description of the mirror system and to specify the optical properties of the surface materials and the X-ray source location. Incident photons of specified energy enter the mirror system at random points and are followed through the system, using calculated reflectivities, until they reach the desired image plane or are lost. Then, as output from such programs one obtains informations about the imaging properties of the system, accounting also for the effects of aberrations and surface roughness and shape errors as well as misalignments of the system components.

The above discussion about compound systems has been dictated by sake of completeness; Wolter type I telescopes, indeed, are used for astronomical X-ray source studies, as Kirkpatrick-Baez and Wolter type II microscopes are used in X-ray imaging of laser-fusion plasmas. Actually, S.R. beam-line optics require single reflectors. Typically, at least two mirrors are used; the first usually flat, which rather frequently is also the first optical element in the beam line, is used to remove X-rays of higher energy than required. Subsequent mirrors, usually spherical or cylindrical, are used to focus the beam and to deflect it.

Finally, let us note that, as already stressed in section 2, presently, the most severe problem connected to the utilization of the high-power S.R. beam is related to radiation damage and thermal gradients, produced by the intense photon beam hitting on the mirror surface, especially when the mirror is the first optical element along the beamline, as it usually is. We recall that high-power wigglers yield a total power of several kW and power densities up to a few $100 \, W/mm^2$ perpendicular to the beam. A substantial part of radiation is absorbed by the mirror, thus producing a deterioration of the coating with consequent reduction of reflectivity. In addition, thermal gradients parallel and perpendicular to the surface create strains which deform the surface and degrade the mirror performance even if they are cooled.

Deterioration of mirror surfaces exposed to S.R. beam is object of intensive studies in order to envisage effective methods and suitable materials to reduce the effects of the thermal load, the optimum strategy being dictated by the type of experiment, the energy range of interest and also by the source. We mention for instance the so called adaptive mirror technology, that is able to change the shape of the reflecting surface to dynamically compensate for thermal deformations and for some other effects (e.g. gravitational forces).

Recently suggested and still under development at ESRF, this technique takes advantage of the well-established experience in the astronomy field, where adaptive optics are required to correct for varying wavefront due to both atmospheric turbulences and gravitational effects and thermal variation of the telescope.

The thermal deformation of the mirror surface is compensated by a mechanical deformation, which is produced by means of piezo-electric actuators attached to the back of the mirror body, whose configuration, i.e. the number of actuators, their forces and positions, is to be optimized according to the parameters of the system, as mirror size and function, source and beam parameters. An optical system and a feedback loop are implemented, the former to inspect continuously the surface shape, the latter to correct for changes in the surface shape, which is then maintained in the desired figure (with a fixed error) even if the perturbations change with time.

As stressed in the relevant references in the bibliography, the adaptive optics technology may be convenient for other applications, as for instance stabilization against external perturbations of various origins or variation of the focusing properties of the mirror, by varying its curvature.

9.5 Multilayers

In the previous sections, we have shown as the design and imaging characteristics of X-ray optic systems are strongly affected by the magnitude of the refractive indices of solids for X-rays, being nearly equal to but slightly less than unity. A layered structure in which the refractive index varies periodically throughout its depth offers considerable potential in overcoming optical and thermal problems, associated with the X-ray reflection from a single surface. Multilayers are artificial crystal synthesized on flat or figured surfaces by alternatively depositing coatings of high and low atomic number (i.e. of different refractive indices). With appropriately chosen layer thicknesses (typically of the order of tens of angstroms: $\geq 15 \text{Å}$) and relative refractive indices, such man-made structures with several hundred layers can work as soft X-ray diffractors, selectively reflecting particular wavelengths at Bragg angles greater than 10. In fact, multilayers have been recognized as possible candidate to developing normal incidence X-ray imaging systems. Thereby, many of the optical aberrations encountered with grazing incidence mirrors, as discussed before, should be considerably reduced. In addition, due to the refractory nature of the coating, multilayered structure can be synthesized to support the high-power load when inserted in S.R. beam-line.

The potential of multilayer structures was recognized soon after the discovery of X-ray diffraction by crystals, when it was proposed that such synthetic structures would extend the range of this phenomenon to longer wavelengths. Earliest attempts to synthesize multilayer materials failed primarily as a result of technological limitations. In fact, optical coating technology came into existence rather later (around 1940). The field was expanded so that presently the multilayer structures have the further attractive advantage of a very flexible fabrication. They can be grown on curved surfaces to produce focusing elements, their thickness can be graded in depth or/and laterally, and the constituting materials can be varied over a wide range to achieve optimum performance for a given application. Here, we do not go into detail, describing the processes for formation of multilayer structures, which are basically three: evaporation, sputtering, and molecular beam epitaxy. It is however worth stressing that in order to obtain a good and stable multilayer structure interdiffusion between the layers must be avoided and the two materials must be chosen chemically compatible and capable to withstand the high-power of X-ray synchrotron beams. Presently, tungsten and carbon are the most used materials producing stable structures. Furthermore, particular attention must be devoted, when manufacturing such structures, to the interface roughness as well as the layer thicknesses, which can affect the reflectivity, especially when working with X-ray of higher energies, demanding indeed for smaller coating thicknesses.

We deduce the expression for the multilayer reflectivity, confirming the potentiality of such structures as nongrazing incidence X-ray imaging systems.

In this connection, two approaches can be adopted: the diffraction approach, according to which the theory of X-ray diffraction by periodic structures is applied to a layered material, and an optic approach which utilizes the optical multilayer theory based on the Fresnel equations. These two approaches are equivalent, the only difference being in the formal complexity of the analytical tools involved in the two formalisms; the optic approach treats each layer in the structure as a continuum, while the diffraction approach includes the discrete atomic nature of layers .

In the following, we will adopt the optic approach using the expressions (8.3.15) for the amplitude reflectivity; we address the reader to the bibliography for a diffraction theory based analysis.

In its simplest arrangement, a layered structure is made up of two materials, A and B, deposited alternatively in layers of thickness d_A and d_B, respectively, as sketched in fig.22, where the coordinate axes are as in fig.8, y coordinate extending out of the plane of the figure.

As the reflectivity at a layer interface is relatively large when there is a large difference in the refractive indices, the two materials have to be chosen with high (i.e. small δ) and low (i.e. large δ) refractive indices. The structure is thus periodic in the direction perpendicular to the coating planes with a period $D = d_A + d_B$; however, the thicknesses d_A and d_B need not to be constant as function of z, the depth into the multilayer, so long as D remains constant.

We suppose that the structure consists of L such layer pairs, so that the total number of media involved, including vacuum and substrate, is $N = 2L+2$. neglecting absorption, at near normal incidence typical values of the amplitude reflectivity are $10^{-3} \div 10^{-2}$, then structures of $10^2 \div 10^3$ layers, producing in phase reflections, exhibit a significant enhancement of reflectivities at non-grazing incidence angles. Thus, L, the number of layer pairs, is typically of the order of 10^2 and more.

X-rays of wavelength λ, incident at a glancing angle ϑ on the planes, are scattered by each layer at angle ϑ_ℓ given by a Bragg-like relation accounting for refraction. In fact, as displayed in fig.23a), in the case where the propagation between

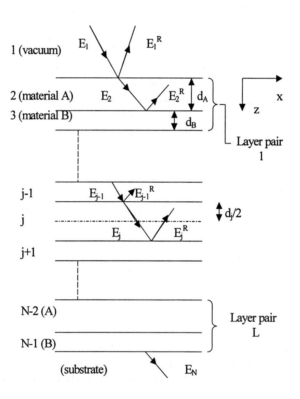

Figure 22 Layered structure made up of N layer pairs of materials A and B.

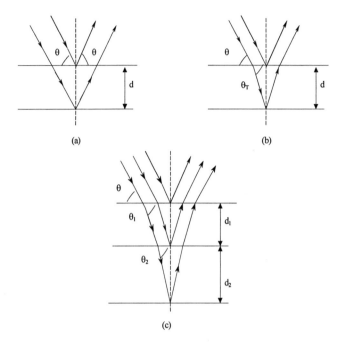

Figure 23 Bragg diffraction from (a) two reflecting surfaces, (b) two refracting surfaces and (c) three refracting surfaces.

two reflecting planes does not suffer of refraction, the reflections at angle ϑ will add in phase if

$$2d\sin\vartheta = m\lambda \tag{9.5.1}$$

where d is the plane-to-plane distance and the integer m represents the order of the reflection.

On account of refraction, the Bragg relation (9.5.1) modifies into (see fig.23b))

$$2nd\sin\vartheta_T = 2d\sin\vartheta\sqrt{1 - \frac{2\delta}{\sin^2\vartheta}} = m\lambda \tag{9.5.2}$$

according to the Snell's law and the relation (8.3.1) for $\tilde{n}^2$, in which quadratic terms in δ and terms in β have been neglected.

Due to the smallness of δ, the above relation can be further simplified as

$$2d\sin\vartheta\left(1 - \frac{\delta}{\sin^2\vartheta}\right) = m\lambda \qquad (9.5.3)$$

The above can be easily extended to the layer pair of fig.23b), according to

$$2D\sin\vartheta\left(1 - \frac{\overline{\delta}}{\sin^2\vartheta}\right) = m\lambda \qquad (9.5.4)$$

where D= $d_A + d_B$ and the refractive index decrement $\overline{\delta}$ is the weighted average index of the two materials, A and B, namely

$$\overline{\delta} = \frac{d_A\delta_A + d_B\delta_B}{d_A + d_B} \qquad (9.5.5)$$

Solving for $\sin\vartheta$ in (9.5.4), we obtain the grazing angles for the maximum of reflectivity as

$$\sin\vartheta = \frac{m\lambda}{2D}\left(1 + \frac{\overline{\delta}D^2}{m^2\lambda^2}\right) \qquad (9.5.6)$$

The reflectivity of a structure, as that depicted in fig.22, can be calculated recursively, applying the boundary conditions at each interface and using the expressions (8.3.12)-(8.3.13) for the amplitude reflectivities for σ- and π-polarized waves. In particular, the Fresnel coefficients for the reflection at the interface between the j-th and (j+1)-th layers may be written as

$$r^{(\sigma)}_{j,j+1} = \frac{g_j - g_{j+1}}{g_j + g_{j+1}} \qquad (9.5.7)$$

$$r^{(\pi)}_{j,j+1} = \frac{\frac{g_j}{\tilde{n}_j^2} - \frac{g_{j+1}}{\tilde{n}_{j+1}^2}}{\frac{g_j}{\tilde{n}_j^2} + \frac{g_{j+1}}{\tilde{n}_{j+1}^2}} \qquad (9.5.8)$$

In the above, $\tilde{n}_j$ and $\tilde{n}_{j+1}$ represent the refractive indices of the two materials; obviously, $\tilde{n}_1 = 1$ since medium 1 is the vacuum. Furthermore,

$$g_j = \sqrt{\tilde{n}_j^2 - \cos^2\vartheta} \qquad (9.5.9)$$

where ϑ is the glancing incidence angle, to which the refraction angle ϑ_j in the j-th layer relates according to the Snell's law as

$$\tilde{n}_j\cos\vartheta_j = \cos\vartheta \qquad (9.5.10)$$

We denote as $E_{j(j+1)}$ and $E^R_{j(j+1)}$ the incident and reflected amplitudes of the electric vector midway through the j-th ((j+1)-th) layer. Then, the continuity of the tangential components of the electric and magnetic vectors at the interface between the j-th and (j+1)-th layers for the σ-component of polarization, may be respectively expressed as (see sect. 3)

$$a_j E_j + \frac{E_j^R}{a_j} = \frac{E_{j+1}}{a_{j+1}} + a_{j+1} E_{j+1}^R \qquad (9.5.11)$$

$$g_j \left(a_j E_j - \frac{E_j^R}{a_j} \right) = g_{j+1} \left(\frac{E_{j+1}}{a_{j+1}} - a_{j+1} E_{j+1}^R \right) \qquad (9.5.12)$$

where the amplitude factor a_j (a_{j+1}) accounts for the phase shift through half the thickness d_j (d_{j+1}) of the layer, then

$$a_j = \exp\left[-i\frac{\pi}{\lambda} g_j d_j\right] \qquad (9.5.13)$$

We define

$$\mathfrak{R}_{j,j+1} = a_j^2 \frac{E_j^R}{E_j} \qquad (9.5.14)$$

which is indeed the amplitude reflectivity at the interface. Dividing eqs. (9.5.11)-(9.5.12) one by the other, a recursive relation for $\mathfrak{R}_{j,j+1}$ can be developed, namely

$$\mathfrak{R}_{j,j+1} = a_j^4 \frac{\mathfrak{R}_{j+1,j+2} + r_{j,j+1}}{\mathfrak{R}_{j+1,j+2} r_{j,j+1} + 1} \qquad (9.5.15)$$

The suffix σ (or π) has been omitted from the Fresnel coefficient $r_{j,j+1}$ in the above expression as it holds for both the perpendicular ($r^\sigma_{j,j+1}$) and parallel ($r^\pi_{j,j+1}$) components of polarization, as it can be easily verified writing down the counterpart of relations (9.5.11)-(9.5.12) for the π-polarized fields (Problem 5).

The computational scheme starts at the substrate, the N-th layer, which is much thicker than the layers, and hence $\mathfrak{R}_{N,N+1} = 0$ is a good approximation to a starting value, particularly for a large number of layer pairs for which the substrate will have little influence on the overall reflectivity. Then, applying eq. (9.5.15) successively at consecutive interfaces backward to the first surface, where $a_1 = 1$ and $g_1 = \sin\vartheta$, the value of $\mathfrak{R}_{1,2} = \frac{E_1}{E_1^R}$ is obtained. This is related to the intensity reflectivity R through the relation

$$R(\vartheta) = \frac{I(\vartheta)}{I_0} = |\mathfrak{R}_{1,2}|^2 \qquad (9.5.16)$$

where I_0 is the incident intensity ($\propto |E_1|^2$) and $I(\vartheta)$ is the intensity reflected at ϑ ($\propto \left|E_1^R\right|^2$).

Calculated reflectivities of molybdenum/silicon multilayer mirror are plotted in fig.24 as a function of the layer pairs at $\lambda = 23.6\,nm$ and as a function of wavelength, for different values of the ratio $\Delta = \frac{d_{Mo}}{d_{Si}}$ of the layer thicknesses. It is evident that there

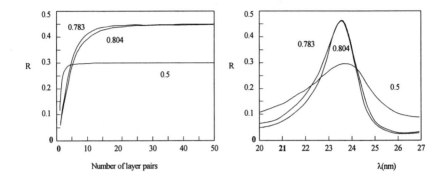

Figure 24 Reflectivities of Mo/Si multilayer mirror as a function of (a) the number of layer pairs at $\lambda = 23.6\,nm$ and (b) the wavelength, for a variety of the ratio $\Delta = \frac{d_{Mo}}{d_{Si}}$ of the layer thicknesses.

exists an optimum value of Δ, for which the reflectivity is maximum: $\Delta_{opt} = 0.783$. In general, the optimum value of Δ can be found differentiating eq. (9.5.15) with respect to Δ, equating the resulting expression to zero and then solving for Δ.

Finally, fig.25 shows the reflectivity curves computed for a multilayer mirror consisting of 101 layer pairs of tungsten and carbon ($d_W = d_C = 10\,\text{Å}$) for a variety of X-ray wavelengths. The value of the angle ϑ_1, shown in the figure, corresponds to the first-order peak, as predicted by the Bragg equation with $2d = 40\,\text{Å}$. The large difference between ϑ_1 and the peak in the reflectivity is due to the refraction (see eq. (9.5.6)).

9.6 Crystal monochromator

As already stressed, the continuous spectrum of S.R. X-rays offers the advantage of a facility that can satisfy the requirements of different experiments. On the other hand, to make this X-radiation usable some form of monochromator device is needed to extract from the primary beam the component with the wavelength and bandwidth suitable to particular experiments. The wavelength range, covered in the experiments so far performed, is quite large going indeed from $0.2 \div 0.3\,\text{Å}$ to about $15 \div 20\,\text{Å}$; similarly, the required bandwidth $\frac{\Delta\lambda}{\lambda}$ varies over a range of six orders of magnitude, from $\sim 10^{-1}$ to $\sim 10^{-7}$. For instance, in fluorescence analysis or small angle X-ray scattering a $\frac{\Delta\lambda}{\lambda} \sim 10^{-1}$ is adequate. In contrast, for inelastic X-ray scattering spectroscopy a high resolution $\sim 10^{-7}$ in $\frac{\Delta\lambda}{\lambda}$ is necessary to detect the energy transfer between X-ray photons and crystalline phonons. In Table III the wavelength resolutions required in some experiments are reported.

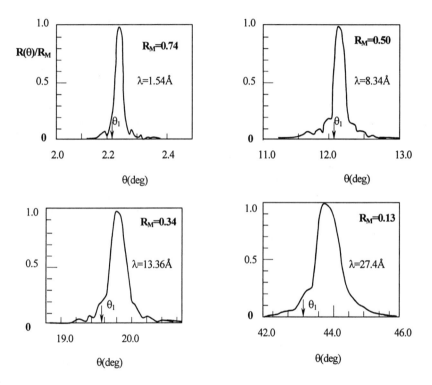

Figure 25 Reflectivity curves of a multilayer consisting of 101 layer pairs of W and C ($d_W = d_C = 10$ Å) at four wavelengths. ϑ_1 denotes the Bragg angle relevant to the first-order reflection. R_M is the peak reflectivity.

Table III. Monochromator energy resolutions
required in various experiments.

Experiment	$\frac{\Delta\lambda}{\lambda}$
X-ray microscopy	$10^{-1} \div 10^{-2}$
Structure analysis	$10^{-2} \div 10^{-3}$
Compton scattering	$10^{-2} \div 10^{-4}$
EXAFS	$10^{-3} \div 10^{-4}$
Anomalous scattering	$\sim 10^{-4}$
Interferometry	$\sim 10^{-4}$

It is needless to say that no single monochromator could handle the whole wavelength range, where two limiting regimes can be roughly drawn, corresponding to the VUV and soft X-rays ($\lambda < 30nm$) and X-rays ($\lambda > 30nm$). The monochromators used in the two regimes differ in that reflection and transmission gratings are used in the VUV to soft X-ray experiments, which furthermore are carried out in vacuum, while crystals are suitable for hard X-ray experiments, usually conducted at atmospheric pressure.

In the following we will limit ourselves to consider crystal monochromators; the discussion will not exhaust the whole subject, which indeed is constantly and rapidly progressing. We recommend the reader to the bibliography for a more detailed illustration of specific monochromator designs.

As is well known, the X-ray diffraction in crystalline materials provides an efficient mechanism for wavelength selection. This is clearly expressed by Bragg's law

$$2d\sin\vartheta = m\lambda \qquad (9.6.1)$$

which relates wavelengths and incident beam directions to have intense peaks in the intensity of the radiation scattered by the lattice planes, separated by the distance d. The positive integer m specifies the order of the Bragg reflection.

Equation (9.6.1) applies to any of the infinite sets of parallel, equally spaced lattice planes, into which the crystal can be sectioned, each family of planes being characterized by its own interplanar spacing d (more appropriately, one should write d_{hkl}, where the integers (hkl) are the Miller indices of the planes). It is assumed that the radiation incident at an angle ϑ on the lattice planes of spacing d, is specularly reflected by any one plane. Then, the scattered amplitude will be maximum if the reflected rays from successive planes interfere constructively and hence if the relevant path difference is an integral multiple of wavelengths. Equation (9.6.1) is just the formal writing of this constructive interference condition.

Since $\sin\vartheta$ should be less than unity, a set of lattice planes spaced by d, do not reflect wavelengths longer than $2d$. This indicates that we need crystals with long interplanar spacings to monochromatize soft X-rays. On the other hand, an excessively long d spacing causes the Bragg angle ϑ to be inconveniently small. As noted in the previous section, multilayer structures, for which the modified Bragg relation (8.5.2) holds, have been devised as soft X-ray monochromator devices, in which the interplanar distances can be artificially produced, thus allowing for a no-grazing incidence selective reflection. Indeed, the naturally occurring crystals, whose interplanar distances are on the order of a few angstroms, are suitable to monochromatize (hard) X-rays.

The crystals most commonly used for S.R. X-rays are silicon and germanium, for which the planes (220) and (111) are appropriate as Bragg planes. The corresponding 2d spacings for Si are respectively 3.84 and 6.27Å, and 4.00 and 6.53Å for Ge. Table IV lists the most commonly used Bragg planes for Be, diamond, Si and Ge crystals at $\lambda = 1.54$Å ($E = 8.05 KeV$) along with other significant properties.

Table IV. Some crystal reflection data for Be, diamond, Si and Ge monochromators at $\lambda = 1.54$Å.

Material (hkl)	d_h[Å]	ω_S[μrad]	$\varepsilon_S(10^{-6})$	t_e[μm]	t_a[μm]
Be(002)	1.7916	10.7	22.8	5.0	1200
Be(110)	1.1428	6.49	7.1	10.3	1874
C(111)	2.0589	23.3	57.8	2.27	250
C(220)	1.2609	15.4	19.9	4.03	408
Si(111)	3.1355	34.5	136	1.53	8.7
Si(220)	1.9201	23.5	53.7	2.27	14.2
Ge(111)	3.2664	74.8	308	0.68	2.94
Ge(220)	2.0002	57.2	137	0.93	4.79

Since synchrotron radiation is white, the simplest monochromator configuration allows for a tune of the reflected wavelength by rotating the crystal to vary the incidence angle ϑ on the Bragg planes. Crystal monochromators are typically arranged in Bragg (or reflection) geometry, where the diffracted wave emerges from the same face illuminated by the incident beam, and in Laue (or transmission) geometry, where the diffracted wave leaves the crystal from the rear face (fig.26). Laue configurations are rather rare at wavelengths longer than about 2.5Å, because of the high photoelectric absorption in the crystal, which causes a substantial intensity loss. Furthermore, reflections are said symmetric or asymmetric according to whether the angle α between the reflecting planes used and the entrance crystal surface is equal to or different from zero (fig.26c). The usefulness of asymmetric crystals as to the monochromator performance will be clarified later. In all cases except the symmetric Laue case the reflection takes place at angles slightly different from that expected according to the Bragg's law because of refraction.

Before proceeding to describe some properties of the monochromator devices, we stress that the relation (9.6.1) expresses the condition for diffraction in the direct lattice space. An equivalent relation can be worked out in the reciprocal lattice space

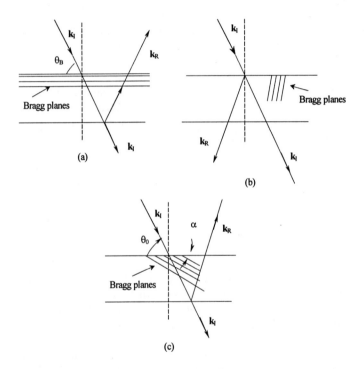

Figure 26 Sketch of X-ray diffraction in (a) symmetric Bragg geometry, (b) Laue or transmission geometry and (c) asymmetric (i.e. $\alpha \neq 0$) Bragg geometry.

‡ in the form

$$\vec{k_R} - \vec{k_I} = \vec{K} \qquad (9.6.2)$$

where $\vec{K}$ is a reciprocal lattice vector, $\vec{k_I}$ and $\vec{k_R}$ are the incident and reflected wave vectors and, being the scattering elastic, $k_I = k_R = \frac{2\pi}{\lambda}$. The above relation represents the Laue formulation of X-ray diffraction by a crystal. The equivalence with the Bragg picture (9.6.1) is a straightforward consequence of the properties of the reciprocal lattice (see footnote). However, we will no longer comment on the relation (9.6.2), which is of some relevance in the present context as it allows a geometrical view of the diffraction condition in the reciprocal lattice by means of the Ewald's construction (fig.27), widely used in the graphic approach to X-ray optics.

As shown in fig.27, one marks the incidence and reflected wave-vectors, $\vec{k_I}$ and $\vec{k_R}$, in magnitude and direction, in the reciprocal lattice, from a common origin and with $\vec{k_I}$ ending at a lattice point O. The condition (9.6.1) or (9.6.2) requires that the reflected wave-vector $\vec{k_R}$ ends at a lattice point P as well, thus ensuring the difference $\vec{K} = \vec{k_R} - \vec{k_I}$ be a reciprocal lattice vector.

An important property of a monochromator is the wavelength or energy resolution. Differentiating the Bragg equation, we obtain

$$\frac{\Delta\lambda}{\lambda} = \frac{\Delta E}{E} = \Delta\vartheta \cot\vartheta_B \qquad (9.6.3)$$

---

‡We briefly recall that the reciprocal lattice associated to a Bravais lattice is formed by all the vectors $\vec{K}$ that yield plane waves with the periodicity of the Bravais lattice. In formal terms, representing the Bravais lattice as the set of vectors $\left\{ \vec{R} = n_1\vec{a_1} + n_2\vec{a_2} + n_3\vec{a_3} \right\}$ for given primitive vectors $\vec{a_1}, \vec{a_2}, \vec{a_3}$ and any possible choice of the integers n_1, n_2, n_3, the corresponding reciprocal lattice can be thought as the set of vectors $\left\{ \vec{K} = l_1\vec{b_1} + l_2\vec{b_2} + l_3\vec{b_3} \right\}$ for arbitrary integers l_1, l_2, l_3. The vectors $\vec{b_1}, \vec{b_2}, \vec{b_3}$ are determined by the $\vec{a}$'s of the direct lattice, according to explicit formulae, that we do not report here (see the specific textbooks).

Here, we simply report the basic property

$$\vec{a_i} \cdot \vec{b_j} = 2\pi \delta_{ij} \qquad i,j = 1,2,3$$

or equivalently

$$\exp\left(i\vec{K} \cdot \vec{R}\right) = 1$$

for all vectors $\vec{K}$ and $\vec{R}$.

Finally, in the present context it is worth mentioning the further useful property of the reciprocal lattice vectors. For any family of lattice planes, there are reciprocal lattice vectors perpendicular to the planes (and hence parallel to each other), the shortest of which is represented as $\vec{K} = h\vec{b_1} + k\vec{b_2} + l\vec{b_3}$, where (h,k,l) are the Miller indices of the planes and its modulus is just the inverse of the interplanar distance d_{hkl}, namely $K = \frac{2\pi}{d_{hkl}}$. It is needless to say that the inverse proposition holds as well.

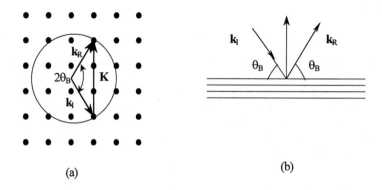

Figure 27 (a) Ewald's construction, in reciprocal space, as the condition for diffraction $\vec{K}$ to be excited. (b) Corresponding Bragg picture of the reflection process in direct space.

Then, for a given Bragg angle ϑ_B, the energy resolution is determined by the angular spread $\Delta\vartheta$. Two factors contribute to $\Delta\vartheta$: the angular spread of the incident beam and the intrinsic reflection width of the monochromator (i.e. the width of the reflectivity curve). The former depends on the geometry of the experiment and is determined by the angular width of the source and/or by the preceding optics and distance to the source. For the latter the perfection of the crystal lattice is of primary concern. In this connection, let us say that in treating the diffraction of X-rays from a crystal two theoretical approaches may be adopted, corresponding to different crystal modellization. In the kinematical approximation, one assumes that the magnitude of the incident wave amplitude is the same at all points in the crystal, with only the phase changing from point to point. This implies that the diffraction amplitude is small enough so that it can be neglected with respect to the incident amplitude. On the contrary, the dynamical theory deals in a self-consistent way with the propagation of waves in the periodic structure, where indeed the incident wave produces a diffracted wave, which in turn can be rediffracted back into the incident wave direction. There is of course a progressive reduction in amplitude of the forward wave due to subsequent reflections as it travels into the crystal. This can be quantified in terms of a sort of penetration depth of the X-rays into the crystal, called *extinction length* t_e. Usually, this length is much smaller than the absorption length t_a: $t_e \ll t_a$ (see Table IV).

The dynamical treatment accounts for diffraction by large perfect crystals, while the kinematical theory is more appropriate to thin or mosaic crystals. The kinematical behaviour of thin (i.e. the crystal thickness t is smaller than the extinction length: $t < t_e$) ideal crystals is justified by that there is so little material that a diffracted wave has no chance of being rediffracted into the incident direction, and that the amplitude diffracted out of the incident beam is negligible.

Similarly, kinematical diffraction occurs in mosaic crystals, which consists of a large number of small perfect blocks slightly misoriented around a mean direction and diffracting the incident wave independently. If the mosaic blocks have linear dimensions $\ll t_e$, the extinction of the primary beam can be neglected and the incident amplitude assumes nearly the same value on each scattering centre. Mosaic crystals provide a handy model to study the behaviour of imperfect crystals, whose complicate defect structure is well accounted for by the mosaic picture.

As to the materials in Table IV, it is worth mentioning that Si and Ge can be obtained as highly perfect crystals, whose lattice spacing and orientation, as shown experimentally, are defined to within 10^{-8}. Diamond is less perfect but deformations produced by growth defects can be as small as $5\mu rad$. On the other hand, Be is a mosaic crystal; the best presently available Be crystals have a mosaic spread (i.e. the width of the angular distribution of the mosaic blocks) of about $180\mu rad$ and block size of about a μm.

Intermediate theories of the kinematical approach and dynamical have been developed. The formal treatment of the X-ray diffraction by crystals is out of the purposes of the present exposition; the interested reader is addressed to the bibliography.

Here, we briefly comment on some useful formulae for understanding the behaviour of crystal monochromators. In this connection, the crystal's *rocking curve* is of basic relevance; it should be understood as the curve of the diffracted intensity scaled to the incident intensity of a monochromator and parallel X-ray beam, versus the deviation of the incident wave from the exact Bragg angle [§]. In practical measures, the crystalline sample is rotated with respect to the incident beam, and the reflected intensities recorded as function of the crystal rotation angle. Two quantities are directly linked to the crystal rocking curve: the *reflectivity*, represented by the height of the rocking curve, and the *integral reflecting power*, given by the area under the curve. The total intensity reflected by the crystal is determined by the latter when the angular spread of the incident beam is much larger than the rocking curve width. Both the quantities are affected by crystal perfection: in fact, in the range of the crystallographic wavelengths, mosaic crystals have usually low reflectivity while high integral reflecting power. In contrast, perfect crystals exhibit high reflectivities reaching 100% and low integral resolving power.

To fix mind, we write down the expression of integral reflecting power $I_{\vec{K}}$ for the reflection $\vec{K}$ by an ideally mosaic crystal

$$I_{\vec{K}} = \frac{1}{2\mu} \frac{r_0^2 \lambda^3}{V_C} \frac{C^2}{\sin 2\vartheta_B} \left|F_{\vec{K}}\right|^2 e^{-W} \tag{9.6.4}$$

[§] By *exact* Bragg angle we mean the angle ϑ_B given by the Bragg law of geometrical theory : $2d \sin \vartheta_B = \lambda$, d being the interplanar spacing of the reflecting planes used.

In the above, r_0 is the classical electron radius, μ the linear absorption coefficient, V_C the unit cell volume. The $\frac{1}{\sin 2\vartheta_B}$ term is called the Lorentz factor, whereas C is the polarization factor, which specializes into $C_\sigma = 1$ and $C_\pi = \cos 2\vartheta_B$ for σ and π-polarization. For unpolarized incident beam, one has $C^2 = \frac{1+\cos^2 2\vartheta_B}{2}$. Furthermore, $F_{\vec{K}}$ is the crystal structure factor ¶ relevant to the reflection $\vec{K}$ (see eq. (9.6.2)) and the exponential e^{-W} is the Debye-Waller factor, which accounts for the thermal motion of the atoms around their equilibrium position in the primitive cell.

On the other hand, fig.28a) shows the diffraction patterns obtained in the symmetric Bragg geometry with a σ-polarized X-ray beam and a *thick* $(t \gg t_e)$ perfect silicon crystal. The curves correspond to four reflection orders from the (220) planes of the silicon crystal, the relevant first-order wavelength being $\lambda = 1.54\text{Å}$; the broken lines reproduce the reflectivity profiles in the absence of absorption. The general shape of the diffraction patterns in both Bragg and Laue symmetric geometry for a perfect infinitely thick crystal in the zero-absorption limit are shown in fig.28b)$^\|$. Apart the angular shift of the Bragg peaks, it is evident that there is a region, ω_S wide in angle, for which there is total reflection and the reflectivity is 1. On either side, reflection is partial; then the reflectivity decreases steadily with increasing deviation from Bragg angle $\Delta\vartheta \equiv \vartheta_0 - \vartheta_B$, according to the inverse square $\Delta\vartheta^{-2}$, as in the case of specular reflection by mirror surfaces. The Laue rocking curve is basically different showing a typical Lorentzian behaviour with an half-width of $2 \sec arc$ and a maximum reflectivity of 0.5.

Explicitly, we have that the angular shift $\Delta\vartheta_S$ of the Bragg peak, taken at the middle of the region where the reflection curve is flat, is given for the symmetric Bragg case by

$$\Delta\vartheta_S = \frac{\omega_S F_{\vec{O}r}}{2C \left|F_{\vec{K}r}\right| e^{-W}} \qquad (9.6.5)$$

where $F_{\vec{h}r}$ (with $\vec{h} = \vec{O}, \vec{K}$) denotes the real part of the crystal structure factor relevant to the reflection $\vec{h}$; in particular $F_{\vec{O}r}$ is the real part of the structure factor

---

¶We recall that in the X-ray diffractionists' usage the structure factor is a number, the ratio of the amplitude diffracted by one unit cell to that diffracted by one free electron placed at the origin.

The standard expression, obtained from the Fourier transform $\tilde{\rho}(\vec{K})$ of the electron density $\rho(\vec{r})$ in the unit cell, is

$$F_{\vec{K}} \equiv \tilde{\rho}(\vec{K})V_C = \sum_j f_j \exp\left[i\vec{K} \cdot \vec{R}_j\right]$$

where the summation runs over all the atoms in the unit cell, f_j being the form factor of the j-th atom in the unit cell localized by the position vector $\vec{R}_j$.

$^\|$The smooth rocking curves of fig. 28b) are obtained by averaging over the fast oscillations, arising from the interference effects between the wave-fields overlapping through the crystal. Usually, such oscillations cannot be resolved experimentally and hence the measured rocking curves exhibit the smooth contours as in the figure.

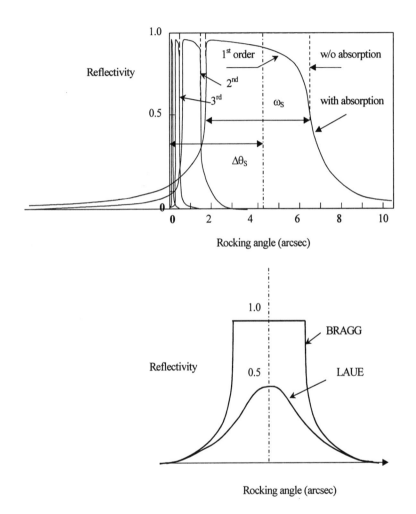

Figure 28 (a) Intrinsic perfect crystal diffraction profiles (Si (220)) for a first-order wavelength of $\lambda = 1.54 \text{Å}$. (b) General shape of the Bragg and Laue rocking curves, averaged over fast oscillations for a perfect absorbing crystal.

in forward direction. As in (9.6.4), C and e^{-W} are the polarization and Debye-Waller factors. Furthermore, ω_S is the width of the rocking curve (usually known as *Darwin width*), for which the following expression can be given

$$\omega_S = \frac{2\lambda^2 r_0 C \left|F_{\vec{K}r}\right| e^{-W}}{\pi V_C \sin 2\vartheta_B} \quad (9.6.6)$$

with the meaning of the symbols being as before. Exploiting the Bragg law $2d_{\vec{K}} \sin \vartheta_B = \lambda$, $d_{\vec{K}}$ being the spacing between the crystal planes involved in the reflection $\vec{K}$, one can rewrite the above expression in the form

$$\omega_S = \varepsilon_S \tan \vartheta_B \quad (9.6.7)$$

with

$$\varepsilon_S = \frac{4}{\pi V_C} d_{\vec{K}}^2 r_0 C \left|F_{\vec{K}r}\right| e^{-W} = \omega_S \cot \vartheta_B \quad (9.6.8)$$

As (9.6.7) resembles the relation (9.6.3), ε_S can be understood as the intrinsic crystal energy resolution which is to first order independent of energy but varies with the square of the d-spacing and is proportional to the structure factor, that in turn depends on $\frac{\sin \vartheta_B}{\lambda} = \frac{1}{2d_{\vec{K}}}$. If the crystal is put in a white incident beam of angular divergence ψ_0 (*white* means an energy bandwidth wider than that accepted by the crystal) the convoluted energy resolution, accounting for both the intrinsic width ω_S of the Bragg reflection and the angular intensity distribution of the incident beam, is given by

$$\frac{\Delta E}{E} = \sqrt{\omega_S^2 + \psi_0^2} \cot \vartheta_B \quad (9.6.9)$$

Clearly, the total reflected intensity is obtained by integrating over the rocking curve and hence, for maximum intensity, the Darwin width ω_S should be as large as possible, in which case Ge(111) is preferred to Si(111). On the other hand, for the highest resolution, ω_S should be as small as possible, so Si(220) is better than Si(111), although the reflected intensity is less. The best compromise between intensity and energy resolution is often achieved if the white beam divergence equals the rocking curve width of the crystal: $\psi_0 \sim \omega_S$. In this connection, we recall that for an energy of $8KeV$ the beam divergence is about $30\mu rad$ for undulator sources and ten times higher for bending magnets and wiggler radiation; compare with the values of ω_S listed in Tab. IV for some crystals.

Finally, for completeness'sake, we report the expression of the extinction thickness, t_e, i.e. the penetration length perpendicular to the lattice planes, that is

$$t_e = \frac{\pi V_C}{2d_{\vec{K}} r_0 C \left|F_{\vec{K}r}\right| e^{-W}} \quad (9.6.10)$$

which reveals the further interesting relation to the intrinsic energy resolution ε_S, namely

$$\varepsilon_S = \frac{2d_{\vec{K}}}{t_e} = \frac{2}{\pi N_p} \qquad (9.6.11)$$

expressed also in terms of the number N_p of lattice planes participating in the diffraction process.

Now, we turn again to fig.28 to pointing out the effect of absorption on the reflection curves. Within this respect, it is easy to recognize another similarity to the mirror case: in fact, the absorption causes a smearing of the step function representing the reflectivity of a non-absorbing crystal. Thus, the rocking curves lose their characteristic symmetric shape; the asymmetry is due to the fact that the maximum of the standing wave amplitudes inside the crystal are located between the atomic layers (thus yielding minimum absorption) on the left side of the curve and at the atomic layers (that produces maximum absorption) on the right side.

An explicit expression of the reflectivity as function of the deviation angle $\Delta\vartheta \equiv \vartheta_0 - \vartheta_B$, accounting also for the absorption, can be worked out. Limiting ourselves to the Bragg case, which is indeed the most interesting when dealing with X-ray monochromator, we report the expression of the reflectivity in the general case of asymmetric diffraction. The geometry is displayed in fig.29. As previously stressed, the entrance crystal surface is not parallel to the diffracting planes; they form indeed an angle α, which is considered positive when the angle between the incidence direction and the crystal surface is smaller than the Bragg angle ϑ_B. An asymmetry factor is defined as

$$b = \frac{\sin(\vartheta_B - \alpha)}{\sin(\vartheta_B + \alpha)} \qquad (9.6.12)$$

which turns to 1 for $\alpha = 0$.

The diffraction pattern relevant to the reflection $\vec{K}$ is given by (I_0 is the incident intensity)

$$R_{\vec{K}}(L) = \frac{I_{\vec{K}}}{I_0} = L - \sqrt{L^2 - 1} \qquad (9.6.13)$$

where L depends in a complicated way on the crystal structure factor, absorption, polarization and asymmetric factor. Explicitly, one has

$$L = \frac{1}{1+\mathfrak{F}^2}\sqrt{(y^2 - g^2 - 1 + \mathfrak{F}^2)^2 + 4(yg - \mathfrak{F})^2} \qquad (9.6.14)$$

where $\mathfrak{F}$ is the ratio of the imaginary part (accounting for absorption) of the structure factor $F_{\vec{K}}$ to the real part:

$$\mathfrak{F} = \frac{F_{\vec{K}i}}{F_{\vec{K}r}} \qquad (9.6.15)$$

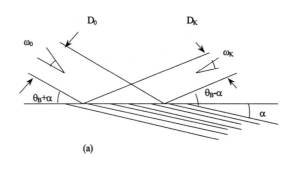

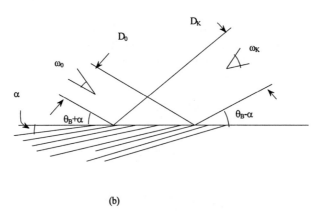

Figure 29 Bragg reflection from asymmetrically cut crystals: (a) $\alpha < 0$, $|b| < 1$ and (b) $\alpha > 0$, $|b| > 1$.

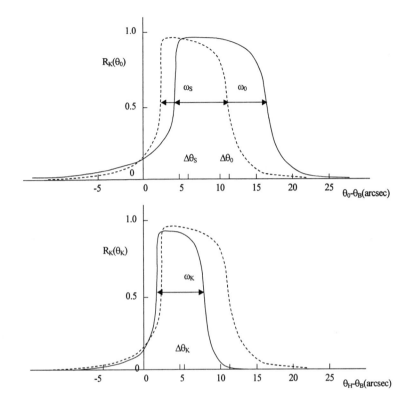

Figure 30 Acceptance $R_K(\vartheta_0)$ and emergence $R_K(\vartheta_K)$ curves for the (111) reflection of a perfect silicon crystal at 1.6Å.

and

$$g = -\frac{1+b}{2C\sqrt{|b|}} \frac{F_{\vec{O}i}}{F_{\vec{K}r}} \tag{9.6.16}$$

Finally, y is defined as

$$y = \frac{1}{2C\sqrt{|b|}F_{\vec{K}r}} \left[\frac{2\pi V_C}{r_0 \lambda^2}(\vartheta_0 - \vartheta_B)\sin 2\vartheta_B - (1+b)F_{\vec{O}r} \right] \tag{9.6.17}$$

where ϑ_0 denotes the angle between the direction of incidence and the diffracting planes.

Figure 30 shows the acceptance $R_K(\vartheta_0)$ and emergence $R_K(\vartheta_K)$ curves for the (111) reflection of a perfect silicon crystal at 1.6Å. $R_K(\vartheta_0)$ and $R_K(\vartheta_K)$ are the reflectivity for an incident plane wave, recorded as a function of the incidence ϑ_0 and reflection ϑ_K angles, respectively. We stress that ϑ_K, the angle between the diffracted beam and the Bragg planes, is related to the incidence angle ϑ_0 by

$$\vartheta_K - \vartheta_B = b(\vartheta_0 - \vartheta_B) \tag{9.6.18}$$

Then, $R_K(\vartheta_K)$ is given by the expression (9.6.13) as well, with ϑ_0 replaced by ϑ_K and b by $\frac{1}{b}$ in the pertinent relations (9.6.14, 9.6.16, 9.6.17). In the symmetric case ($b = 1.0$), it is obviously $R_K(\vartheta_0) = R_K(\vartheta_K)$, as displayed by the broken lines in the figure. For $b \neq 1$, the angular ranges over which the two functions have an appreciable value (~ 1) differ from the Darwin width ω_S in (9.6.5). Precisely, the acceptance angular width ω_0 is given to a good approximation by

$$\omega_0 = \frac{1}{\sqrt{|b|}} \omega_S \tag{9.6.19}$$

while the angular width of the emergence curve $R_K(\vartheta_K)$ is

$$\omega_K = \sqrt{|b|} \omega_S \tag{9.6.20}$$

Moreover, the center of $R_K(\vartheta_0)$, where $y = 0$, is

$$\Delta\vartheta_0 = \frac{1}{2}(1 + \frac{1}{b})\Delta\vartheta_S \tag{9.6.21}$$

whereas the center of $R_K(\vartheta_K)$ is

$$\Delta\vartheta_K = \frac{1}{2}(1 + b)\Delta\vartheta_S \tag{9.6.22}$$

$\Delta\vartheta_S$ being the angular shift from the Bragg angle in the case of symmetrical diffraction (eq. (9.6.1)). From eqs. (9.6.19) and (9.6.20) one can see that it is possible to change the angular divergence of a beam; in fact, one has

$$\omega_K = |b|\omega_0 \tag{9.6.23}$$

Associated with the angular transformation is the change in spatial cross-section of the beam. Let D_0 and D_K denote the spatial widths of the incident and diffracted beams, respectively. Being according to the Liouville's theorem

$$S_0\omega_0 = S_K\omega_K \tag{9.6.24}$$

one gets

$$S_K = \frac{S_0}{|b|} \tag{9.6.25}$$

Therefore, if for instance $|b| < 1$ the angular dispersion of the emergent beam is $|b|$ times smaller than that of the incident beam, whilst its width is $\frac{1}{|b|}$ times greater (fig.29b)

Finally, for comparison with the expression (9.6.4) relevant to mosaic crystals, we report the expression of the integral reflecting power $I_{\vec{\kappa}}$ for a weakly absorbing perfect crystal, obtained by integrating the Darwin curve in the zero-absorption. Explicitly, one has

$$I_{\vec{\kappa}} = \frac{1}{\sqrt{b}} \frac{8}{3\sin 2\vartheta_B} \frac{r_0 \lambda^2}{\pi V_C} C \left|F_{\vec{\kappa}}\right| e^{-W} \qquad (9.6.26)$$

which accounts for asymmetric reflection as well.

After dealing with the theoretical aspects of monochromator systems, we turn our attention to practical designs, which basically involve flat and curved crystals in single or multiple reflection.

Since S.R. X-rays are linearly polarized in the plane of electron orbit with the electric vector oriented in this plane, monochromator crystals should be oriented so that the dispersing plane be at right angle to the orbital plane. In fact, due to the polarization factor $C_\pi = \cos 2\vartheta_B$ in (9.6.4) and (9.6.26), in Bragg reflections dispersing in the orbital plane the reflectivity and the angular width of the emerging beam can be very small when $2\vartheta_B$ approaches $90°$. On the other hand, Bragg reflections dispersing in the vertical plane create serious difficulties in the layout of experiments. Then, configurations involving double or multiple reflections crystals are frequently used to reflect the beam back into the horizontal plane. Figure 31a) shows a typical two-crystal arrangement. Both crystals are aligned to within the Darwin width and inclined at an angle ϑ to the incident beam, thus producing a monochromatic beam parallel to the incident one. By rotating the crystal pair about a horizontal axis, the monochromator is rapidly and continuously tunable. If D is the gap between the two crystals, the diffracted beam is vertically displaced by $h = 2D \cos \vartheta_B$. Typically, $D \sim 10mm$ and h can be significant for large ϑ_B values. This effect can be compensated translating the second crystal with respect to the first, thus producing a constant exit beam as monochromator is scanned.

Double or multiple-Bragg reflection systems are the commonly used monochromator configurations. Usually asymmetric reflections are preferred; in fact, according to the discussion above, the oblique cut of the crystal reflecting planes at an angle α to the crystal surface produces a compression of the diffracted beam width compared with the incident beam width by the factor $\frac{1}{|b|}$ ($|b| > 1$); see eq. (9.6.23) and fig. 29a). Also, the angular acceptance of the crystal is reduced by $\frac{1}{\sqrt{|b|}}$ compared with that of the symmetric cut crystal (eq. (9.6.6)). Thus, the rocking curve width contributes to the spread of wavelengths $\frac{\Delta\lambda}{\lambda} = \Delta\vartheta \cot \vartheta_B$ through a negligible amount with respect to the total spectral spread, which is thereby determined by the source size and other geometrical factors, as the crystal curvature, for instance (see below). Values for the asymmetry factor between 10^{-1} and 10^2 are typically produced. For instance, the Darwin width for symmetric reflection from a Ge(111) at $\lambda = 1.5\text{Å}$ is $\omega_S \sim 16''$ and correspondingly $\frac{\Delta\lambda}{\lambda} = \omega_S \cot\vartheta_B = 3.2 \times 10^{-4}$ ($\vartheta_B = 0.23 rad$ for reflection from Ge(111) at $\lambda = 1.5\text{Å}$). For asymmetrically cut crystal at $\alpha = -10.5°$** the angular acceptance reduces to $\omega_0 \sim 6''$ and the wavelength spread becomes $\frac{\Delta\lambda}{\lambda} \sim 1.2 \times 10^{-4}$. Furthermore, the accepted section of the incident beam increases by the amount $\frac{\sin(\vartheta_B - \alpha)}{\sin\vartheta}$. The latter occurrence may be useful if the width of synchrotron beam available is greater than $L \sin\vartheta_B$, L being the length of the monochromator, as it usually happens in the horizontal direction.

Thus, utilizing multiple asymmetric reflections the angular and wavelength spreads of the emergent beam from the monochromator can be controlled to high accuracy: multiple-reflection collimators have been realized, for instance, producing beams of angular divergence as small as $5 \times 10^{-8} rad$.

We devote some comments to curved crystal monochromators, frequently used in synchrotron radiation experiments with the further task to collect source photons, which propagating from the tangent point to the experiment location spread over an area ($100 \div 200 \times 10 mm^2$ at a $10m$ distance) usually a good deal larger than needed, and to focus them to a smaller area. A typical arrangement using a curved crystal monochromator is sketched in fig.31b); a triangular shaped perfect crystal is bent to an arc of appropriate radius of curvature and rotated about a vertical axis for tuning. Once more asymmetrically cut crystals can be used thus improving the focussing property of the system. Then, we turn to deduce the imaging relation for a curved obliquely cut crystal, which within many aspects is similar to a curved mirror.

We refer to the geometry in fig.32, where a crystal bent to a radius R in the scattering plane is shown. The rays from the source point S are focused after Bragg reflection to the image point I. As in fig.17, β and γ denote respectively the opening and convergence angles of the incident and diffracted beams. Also p and q represent the source-to-crystal surface and crystal surface-to-image point distances, respectively, measured along a mean ray in the beam, incident on the crystal surface at the point P. The glancing angle of incidence of that ray is $\vartheta_B + \alpha$, the emergence angle of the corresponding diffracted ray is $\vartheta_B - \alpha$.

Retracing the analysis of section 4, we end up with the imaging equation

$$\frac{2}{R} = \frac{\sin(\vartheta_B + \alpha)}{p} + \frac{\sin(\vartheta_B - \alpha)}{q} \qquad (9.6.27)$$

In passing, we stress that as for a curved mirror, the sum of the angles β and γ is just equal to twice the angle subtended by the length L of the crystal involved in diffraction:

**Recall that, according to the usual convention, the case $\alpha < 0$ corresponds to the geometry of fig.29a).

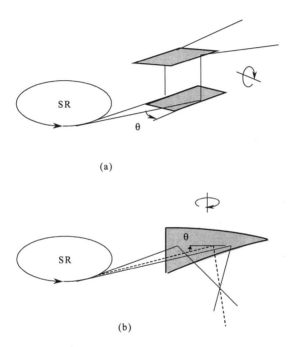

Figure 31 Two typical configurations for crystal monochromators in SR beam line: (a) two-crystal monochromator steering the beam vertically and (b) triangular shaped crystal bent to focus the beam.

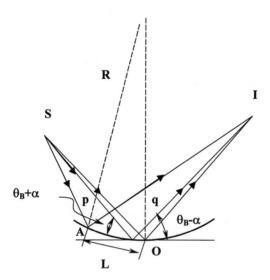

Figure 32 Focusing geometry monochromator: X-rays coming from the source point S are focused to the image point I.

$$\beta + \gamma = 2\frac{L}{R} \tag{9.6.28}$$

whilst the difference angle $\gamma - \beta$ provides the variation $\Delta\vartheta$ of the angle of incidence along the crystal length L:

$$\Delta\vartheta = |\gamma - \beta| = \frac{L}{2}\left|\frac{\sin(\vartheta_B - \alpha)}{q} - \frac{\sin(\vartheta_B + \alpha)}{p}\right| \tag{9.6.29}$$

Since most of the considerations developed in section 4 as to the aberrations arising in image formation by curved mirrors at grazing incidence apply to Bragg reflection from curved crystal as well, we will not repeat them here, addressing the reader to that section.

The spread of wavelengths, $\frac{\Delta\lambda}{\lambda}$, in the reflected beam is the convolution of the crystal rocking curve contribution with the geometrical factors of source size and variation of the incident angle along the curved monochromator surface. Then, if h is the linear extension of the source in the incidence plane, the energy spread of the diffracted beam is calculated as

$$\frac{\Delta\lambda}{\lambda} = \frac{\Delta E}{E} \simeq \cot\vartheta_B \sqrt{\omega_0^2 + \left(\frac{h}{p}\right)^2 + \Delta\vartheta^2} \tag{9.6.30}$$

In S.R. beamlines, horizontal focusing of the beam is usually required, and hence h is the horizontal extension of the source ($\sim 100 \div 200 mm$).

As previously stressed, with oblique cut of the crystal surface, the contribution to (9.6.30) from the crystal rocking curve can be made negligible: $\omega_0 \ll \Delta\vartheta, \frac{h}{p}$. On the other hand, geometry can be arranged to make $\Delta\vartheta$ nearly vanishing. The so called *Guinier condition*, according to which $\Delta\vartheta = 0$, requires that

$$q_G = p\frac{\sin(\vartheta_B - \alpha)}{\sin(\vartheta_B + \alpha)} \tag{9.6.31}$$

At the Guinier condition (fig.33), rays from a point on the source make equal angles with the monochromator surface, but rays from different points on the source subtend different angles at every point of the monochromator. The result is a photon energy gradient across the focus with total energy spread $\frac{\Delta E}{E} \simeq \frac{h}{p}\cot\vartheta_B$ according to (9.6.30). Thus, using a focus slit, a narrow $\frac{\Delta E}{E}$ can be selected.

The width of the focus line depends on several factors, such as the angular width of the diffracted beam ω_e, the magnification $M = \frac{q}{p}$ and the various aberrations.

For a symmetric cut crystal ($\alpha = 0$), the Guinier condition turns into $q = p$, thus yielding the so called *Johann focusing condition*

$$q = p = R\sin\vartheta_B = 2f_M \tag{9.6.32}$$

recovering the notation of section 4.

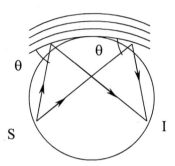

Figure 33   A curved crystal monochromator for which the Guinier condition is satisfied: $\Delta\vartheta = 0$.

As in S.R. experiments the pertinent p-distances are quite large, the above condition demands for large focal lengths; moreover, with $q = p$ symmetrical focussing would extend the experiment a further p ($\sim 10 \div 20 m$) distance. Asymmetric cuts allow for shorter focal lengths and image distances; the Guinier condition is indeed satisfied by

$$p = R\sin(\vartheta_B + \alpha)$$
$$q = R\sin(\vartheta_B - \alpha) = pb \qquad (9.6.33)$$

and q can be made smaller than p, if $|b| < 1$.

The Johann geometry monochromator is shown in fig.34a); a curved crystal of the Johann geometry is built by simply bending a crystal plate to a radius of curvature R centered at O. In the figure it is also shown the *Rowland circle*, i.e. the focal circle of radius $\frac{R}{2}$ tangent to the plate central point C. The source and the focal caustics are both circles with centre in O and radius $R\cos(\vartheta_B + \alpha)$ and $R\cos(\vartheta_B - \alpha)$ respectively. This arrangement is easy to prepare but suffers from geometrical aberrations.

A monochromator free of geometrical aberrations can be obtained if the so called *Johansson geometry* is used, as shown in fig.34b). A crystal, whose surface is ground to a radius $\frac{R}{2}$ is bent so that the Bragg planes lie on a circular cylinder of radius R. In this configuration the crystal surface is tangent to the Rowland circle and a line focus F is obtained from a point source S.

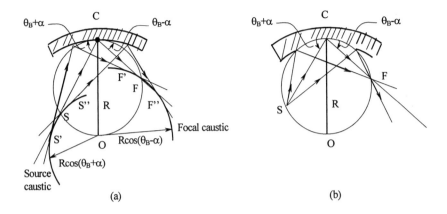

Figure 34 Curved crystal monochromator in the Johann and Johansson geometries.

We close here our discussion on crystal monochromators. For reasons of space, most of the formal aspects of the relevant theory have been only touched on and the graphic approach, which provides a powerful tool in designing crystal monochromators, has been completely skipped; then, we address the reader to the bibliography.

9.7 X-rays from synchrotron radiation sources: some applications.

Before closing this chapter we describe the experimental layouts relevant to some synchrotron radiation experiments.

Basically there are two geometries used for synchrotron radiation experiments, involving vertically or horizontally diffracting monochromators, as schematically depicted in fig.31. In the configuration of fig.31a), the white beam from the storage ring is vertically diffracted by a double-reflection monochromator, thus yielding a monochromatic beam parallel to the incoming beam. On the contrary, diffraction of the incoming beam from the storage ring in the horizontal plane is usually actuated by a single bent crystal monochromator, as shown in fig.31b).

The arrangement of fig.31a) results in the best energy resolution because the vertical emittance of the storage ring is usually superior to the horizontal emittance, whereas monochromators involving a single deflection (fig.31b)) usually benefit from greaterlight gathering power and result in the most brilliant monochromatic beams.

We turn now to briefly comment on two of the principal synchrotron radiation experiments, which use the above-described geometries: X-ray absorption spectroscopy and X-ray diffraction.

X-ray absorption spectroscopy experiments demand for both high intensity and excellent energy resolution. The typical experimental layout is shown in fig.35.

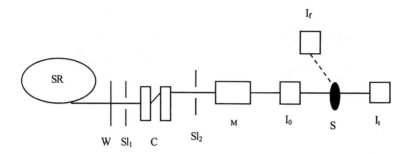

Figure 35 Layout for X-ray absorption spectroscopy experiments. W Be window; $Sl_{1,2}$, slits; C, 2-crystal monochromator; M, toroidal mirror; I_o, I_t, I_f, ion chambers; S, specimen.

A two-crystal, vertically dispersing wide aperture monochromator C is used; unless soft X-rays are required a Be window, W, is inserted in the beam line, operating as harmonic rejector. The monochromatic beam emerging from the monochromator can be focussed with a toroidal mirror M at the specimen S, after being registered by a semitransparent reference ion-chamber I_o in order to correct the absorption spectra for the decaying storage ring beam current and the instrument function of the monochromator/mirror combination.

If the element whose absorption is being measured is sufficiently concentrated, an ion chamber I_t, placed behind the sample is sufficient to measure the transmitted beam. The X-ray absorbance of dilute systems is recorded from the characteristic X-ray emission using a scintillator or solid state photoconductive detector (SSD) I_f, placed at right angles to the monochromatic beam, the sample facing the incident radiation at an angle of 45° (as shown by the dashed lines in the figure). Alternative arrangements can be used as well. Polished specimens for instance can be inclined at grazing incidence to the incoming X-rays, in which case the rear ion chamber measures the reflected beam or the X-ray fluorescence is collected above the surface.

X-ray absorption spectroscopy represents one of the most effective tools to produce a microscopic knowledge of a system at the atomic and/or electronic scale.

It is well known that correlation between atomic order and electronic properties exists and is subsequently object of investigations. Electromagnetic waves, i.e. photons, interact with electrons and therefore have a natural ability to probe electronic properties. In addition, X-ray have a wavelength of the order of the interatomic distance in condensed matter; this is the crucial property for X-ray scattering that allows determination of structures.

In the region of 1-40 KeV, which is pertinent to atomic order investigations, the X-ray photon is absorbed in the atom and its energy E is transmitted to an electron

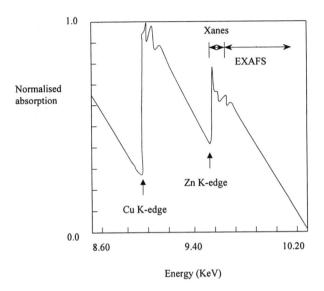

Figure 36 X-ray absorption spectrum of Cu, Zn-metallothionein.

in the K-shell, L-shell, etc. with binding energy E_K, E_L, etc. The electron is expelled from the atom with an energy $E - E_K$, $E - E_L$ etc., leaving the atom in an excited state with a hole in one of the shells. Subsequently, this hole is filled by an electron from one of the higher lying shells with concomitant fluorescent radiation. When the photon energy exceeds the threshold for expelling an electron from a certain shell, another channel for absorption is opened up and the absorption cross section exhibits a discontinuous jump at the threshold energy E_K, E_L etc. These discontinuities are called *absorption edges*.

Actually, near a threshold energy the cross section of a selected atom embedded in condensed matter has more structure than just a discontinuous jump. In fact, it shows (fig.36):
- near edge features (XANES = X-ray Absorption Near Edge Structure) and
- extended oscillations above the edge (EXAFS = Extended X-ray Absorption Fine Structure) which contain informations about the oxidation state, the symmetry of the local environment, the partial density of unoccupied states, the coordination number and the distances of the neighbours.

In physical terms, the photoabsorption cross section is proportional to the transition probability which is the squared matrix element between the initial state and the final state. The latter involves the expelled free electron with wave-number k_e

given by $\frac{(\hbar k_e)^2}{2m_e} = E - E_{\text{shell}}$, where m_e is the electron effective mass in the structure we are considering and E_{shell} denotes E_K, E_L, etc. The electron wave is scattered by the electron clouds of neighboring atoms and interference phenomena (oscillations) occur versus k_e and thereby versus the X-ray energy E.

The primary objective of EXAFS studies is to determine the local atomic environment of the excited atom by analyzing the measured oscillatory structure. In fact, the interference reflects the total phase of the backscattered wave, which is the product of the photoelectron wave-vector and the distance travelled but contains also contributions from the scattering process and from the passage of the electron out and back through the potential of the excited atom. The amplitude of the oscillations depends on the number and electron scattering strength of the scattering atoms. Consequently, by measuring the EXAFS spectrum one can determine the distance (roughly by the period of oscillations) as well as the number and chemical type of the near neighbours of the excited, or central, atom. Also, since the EXAFS spectrum is measured on a known absorption edge, due to an atom of a known chemical type, the technique is chemically specific, giving the coordination of a known type of atom. The EXAFS depends only on the local atomic environment; in fact, only elastically scattered electrons can contribute to the interference and the elastic mean free path of the electrons is short. Thereby, the technique is particularly useful to determine the structure of non-crystalline solids (as glasses), for which the local atomic environment is of main interest. EXAFS spectra typically contain information on atoms up to about 5Å from the central atom.

EXAFS technique is presently used to resolve a wide variety of structural problems, as those inherent, for instance, to a group of materials that can be classified together as *ionically conducting solids*. This classification requires some clarification as the compounds are not simple, normal ionic crystals (e.g. alkali halides) where the structures can be determined by diffraction methods. The ionically conducting solids have the characteristic feature of being unusually good ionic conductors of electricity. In general, a high ionic conductivity in crystals is associated with a highly defective lattice, due either to intrinsic disorder or the effect of added impurities; thus there are complex problems concerning the local structures around the ions. In this respect, the EXAFS technique is ideally matched to study the relation between the structure and ionic transport of crystalline ionic materials.

Also, EXAFS is uniquely appropriate technique for establishing structural parameters of metal catalysts, where the active species is generally a metal centre present in low concentration in a disordered medium.

Surface sensitive spectroscopy based on X-ray photons benefits from the unique features of synchrotron radiation. Let us mention, for example, the surface extended X-ray absorption fine structure (SEXAFS) spectroscopy, which produces interatomic spacings for absorbate systems with more precision than any other methods; and the related technique of near-edge X-ray absorption fine structure (NEXAFS) spectroscopy, which provides a direct means of determining the orientation of molecular absorbates with respect to the surface plane. These techniques demand for the tunability over a wide energy range, the brightness and the polarization characteristics, that presently are unique to synchrotron radiation.

Finally, EXAFS technique is ideally suited to investigate the local environment of atoms, particularly the d-transition metal atoms, in biological systems. In fact, being the technique not limited by the physical state of the sample, it allows to probe the nature of a catalytic site of a metallo-enzyme, directly and selectively. Since EXAFS signals are small in comparison to the atomic absorption from excitation of the core electron, good quality EXAFS data require an intense and stable X-ray source. In this regard, the spectral purity, high brightness and stability of the beam from synchrotron radiation sources are essential for EXAFS (and XANES) investigations of biological systems.

X-ray diffraction is one of the most standard laboratory techniques. Powder diffraction has been used in a routine manned for phase identification, lattice parameter determination and for solving simple crystal structures, while single crystal techniques have been employed for many decades to provide accurate crystal structure. The availability of synchrotron radiation allowed to greatly expand the range and power of *classical* powder diffraction methods as well as to conceive entirely new types of experiment.

The experimental layout for X-ray diffraction from powders and crystallites is typically similar to the EXAFS layout (fig.31a)) for *angle dispersive* mode of data collection, a monochromatised X-ray beam being used [††].

When biological systems are concerned, the experimental configuration is illustrated in fig.37. The scheme is essentially that of fig.31b), involving single deflection monochromators, which usually benefit from greater light gathering power and result in the most brilliant monochromatic beams. With some small modifications, the geometry of fig.37 is suitable both for small angle X-ray scattering (SAXS) and protein crystallography.

Typically, a Be window W is used to separate the experiment from the storage ring ultra high vacuum. Vertical focussing is achieved by means of a cylindrical mirror M, which also acts as a high energy X-ray rejector. Horizontal focussing and monochromatisation is obtained with a bent triangular-shaped crystal C. The beam is focussed at the position sensitive detector in SAXS experiments and at the specimen S, in the case of protein crystallography. The incident beam intensity is monitored with an ion chamber I_o, placed behind a fine collimator. As shown in the figure, one- or two-dimensional position sensitive detectors can be used to measure the scattered or diffracted X-rays.

[††] *Energy dispersive* experiments demand for very simple configurations using a white beam and a fixed detector, usually sited at a low scattering angle. The scattered radiation is energy analysed.

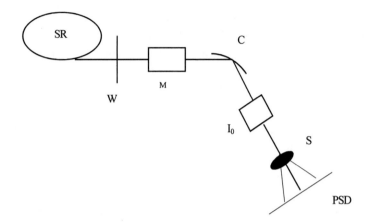

Figure 37 Layout for small angle X-ray scattering and protein crystallography experiments. W, Be window; M, cylindrical mirror; C, bent single crystal monochromator; I_o, reference ion chamber plus collimator; S, specimen; PSD, position sensitive. detector.

9.8 Conclusions

In this chapter we have briefly illustrated the optics inherent to S.R. beam lines. Thus, the working principle of X-ray mirrors and their role as focussing devices and/or high-energy rejectors are discussed. The advantages of multilayer structures, which are multiple Bragg-diffractors, are commented. Finally, the features of crystal monochromators, which represent a basic component of X-ray beam-lines, allowing to extract from the primary beam the component with the wavelength and bandwidth suitable to the specific experiment, are briefly illustrated.

As already stressed, the discussion here does not cover all the topic and has a cursory character; for a complete analysis the reader is addressed to the bibliography.

9.9 Problems

1. Deduce the general form of the Fresnel formulae, imposing the continuity of the tangential component of $\vec{E}$ and $\vec{H}$ and equivalently of the normal component of $\vec{D} = \epsilon \vec{E}$ and $\vec{B}$. Then, using the Snell's law and the relation $\vartheta_I + \varphi_I = \frac{\pi}{2}$, turn the obtained relations into the formulae (9.3.12-9.3.13).

2. Find the condition under which $r_\sigma \leq 0$ and $r_\pi = 0$. (Hint: using the Snell's law, express the general expression of r_π, obtained in Problem 1 in the form

$$r_\pi = \frac{\tan(\varphi_I - \varphi_T)}{\tan(\varphi_I + \varphi_T)},$$

to find the Brewster angle φ_B as

$$\tan \varphi_B = \frac{n_2}{n_1}.$$

3. Deduce the relation (9.4.1) for the geometry of fig. 15a) and the relation (9.4.3) for the geometry of fig. 15b)

4. Find the magnification for the Wolter type systems of fig. 21. Verify that it can made nearly equal to 1.

5. Deduce the expression of the multilayer reflectivity for the π-polarization.

Suggested bibliography

As to the mirror optics we suggest the books

1. A.G. Michette, "Optical Systems for Soft X-rays" (Plenum Press, New York, 1986).
2. E. Spiller, "Soft X-Ray Optics" (SPIE Press, Bellingham, 1994).
3. "X-ray Science and Technology", eds. A.G. Michette and C.J. Buckley (Inst. Phys. Publ., Bristol 1993).
   As to the crystal monochromators the reader is invited to read the review papers
4. B.W. Batterman and D.H. Bilberback, in Handbook on Synchrotron Radiation, Vol. 3 ed. G. Brown and D.E. Moncton, pp.105-153 (Elsevier,1991).
5. R. Caciuffo, S. Melone, F. Rustichelli and A. Boeuf, "Monochromators for X-ray Synchrotron Radiation", Phys. Rep. **152**, 1 (1987).
   Finally, the applications and relevant experimental layouts are discussed in
6. "Applications of Synchrotron Radiation", eds. C.R.A. Catlow and G.N. Greaves (Blackie and Son, Glasgow, 1990).
7. "Neutron and Synchrotron Radiation for Condensed Matter Studies", eds. J. Baruchel, J.L. Hodeau, M.S. Lehmann, J.R. Regnard and C. Schlenker, Vol.1 (Springer-Verlag and Les Editions de Physique, Les Ulis, 1993).

Index

A
Abbe sine condition, 327
angular distribution, 89
astigmatism, 326
asymmetry factor, 349

B
bandwidth, 85, 194, 257
beam
    heating, 116
    life-time, 215
    matching conditions, 124
    spot size, 178
beam-gas interaction, 215
bending
    magnet, 51, 63, 206
    radius, 65
    magnet, 22
betatron
    oscillations, 38
    phase-advance, 208
    wave number, 124
Boussard's criterion, 223, 283–284
Bragg
    diffraction, 335
    geometry, 341
    law, 340, 348
    planes, 341
Bravais lattice, 343
brightness, 251, 265
    beam emittance dominated, 211
    function, 167, 171
    of a high gain FEL device, 261
bunch lengthening, 117
bunching coefficient, 268

C
Cauchy repeated integrals, 108
cavity
    desynchronism, 125
    losses, 113
central emission frequency, 97
charged particle
    acceleration, 17
    momentum, 20
Chasman-Green lattice, 206, 283
chromaticity, 39
circular accelerator
    ideal orbit, 25
    transverse motion, 25
circular path, 31
cohercitivity magnetic strength, 234
coherent
    collective effects, 214
    emission, 100, 262
    harmonic generation, 251
    radiation, 251
coma, 326
complex
    curvature radius, 178
    optical field amplitude, 104
concave spherical mirror, 323
Courant-Snyder
    form, 38
    invariant, 78
critical frequency, 200
cross-section, 215–216, 218, 220
crystal
    monochromator, 338
    rocking curve, 345
    structure factor, 346

cylindrical mirror, 321–322

D
damping, 74, 116–117
  time, 79
Darwin width, 348
detuning parameter, 125
diffraction
  gratings, 321
diffraction-limited
  brightness, 196
  operating FEL, 252
  regime, 189
dipole magnet (see bending), 22
direct lattice space, 341
directional power spectrum, 166
dispersion
  function, 31, 34, 206
  term, 41
distribution function, 80

E
easy axis, 234
efficiency function, 258
electron emittance dominated regime, 189
ellipsoidal mirror, 321
emittance, 28, 205
  minimization, 206, 211
emitter time, 62
energy
  acceptance, 110
  emitted per unit frequency, 71
  radiated per unit solid angle, 68
  spread, 99, 284
ensemble
  average, expectation value, 163
equilibrium equation, 258
equilibrium intracavity
  brightness, 253
  dimensionless intensity, 256
  power, 255
ESRF, 195

Evershed's criterion, 233
evolution
  matrix, 29
  operator method, 32
EXAFS, 361
exponential growth, 268
extinction length, 344

F
FEL
  "active-medium", 100
  amplifier, 268
  energy acceptance, 116
  gain, 105
  gain regimes, 264
  high gain devices, 261
FEL-OA, 289
first generation sources, 195
flux, 87, 196, 198, 200
focusing conditions, 323
Fokker-Planck equation, 80–81
fourth generation light sources, 251
Free Electron Lasers, 97
free-space section, 148
frequency detuning parameter, 98
Fresnel
  coefficients, 336
  equations, 313

G
gain
  function, 101
  peak, 102
Gaussian
  beam, 178
  beam emittance, 179
  optics, 140, 144
geometrical optics, 136, 153
grazing
  angles, 336
  incidence angle, 311
Guinier condition, 357

H

Halbach configuration, 83
Hall probe technique, 239
Hamiltonian formalism, 56
Hamiltonian of ray optics, 136
Helmholtz coil technique, 239
high gain
    FEL, 261
    growth-rate parameter, 110
    regime, 265
higher harmonic
    dimensionless field, 270
    generation, 262
homogeneously broadened regime, 107
horizontal emittance, 78

I

imaging condition, 325
incoherent collective effects, 214
induced energy spread, 117, 259, 263
inhomogeneous broadening, 99, 107, 117, 123
insertion device, 211, 213, 231, 252
instability growth rate, 287

J

Jacobian matrix, 140
Johann
    focusing condition, 357
    geometry monochromator, 358
Johansson geometry, 358
Jones vectors, 202

K

Kirkpatrick-Baez systems, 328

L

Landau damping, 226
lattice, 22, 38
    Chasman-Green, 206
    DFA, 206
    FODO, 208
    isomagnetic, 65

TAL, 207
TBA, 208
Laue geometry, 341
lens-like medium, 151
Lie
    algebra, 141
    matrices, 147
    operator, 56, 147
Lienard-Wiechert potentials, 66
Linac, 251
linear
    absorption coefficient, 310
    optics, 140, 142
Liouville
    equation, 270
    phase-space evolution, 26
    theorem, 140
logistic
    equation, 255, 261
    function, 114
longitudinal damping time, 75
Lorentz force, 21

M

Möller formula, 218
machine duty-cycle, 255
macropulse duration, 255
magnetic
    lenses, 22, 51
    moment, 237
    rigidity, 20–21
    scalar potential, 52
magnetic field
    integrals, 244
    mapping, 243
    measuremets, 243
    strength, 236
magnifier, 151
Master equation, 80
meridional rays, 143
microwave instability, 223, 285
mirrors, 309

mode-locking, 125
modified Bessel function, 198
momentum compaction, 74, 297
monochromator energy resolution, 343
mosaic crystals, 345
motion stability condition, 37
multilayers, 332
multipoles, 36
multiturn FEL interaction, 116

N

natural emittance, 79
non-synchronous particle, 74
normalized emittance, 124

O

observer time, 62
off-momentum particles, 212
optical
    axis, 138
    constants δ, β, 309
    direction cosines, 137
    Hamiltonian, 143
    phase space, 136
    reciprocity property, 159
    transfer function, 153
Optical Klistron, 259
Optical Klystron, 119
optical pulse, 256
ordinary betatron contribution, 41
output laser brightness, 254

P

Padé approximant, 112–113
paraboloidal mirror, 321
paraxial
    approximation, 142
    propagation, 135
charged particle
    radius of curvature, 51
particles loss rate, 220
peak
    brillance, 194

laser brightness, 252
periodicity conditions, 36
phase advance, 37
phase-front radius of curvature, 178
phase-space
    distribution, 187, 270
    ellipse, 171
    evolution, 33
    Liouville evolution, 26
    longitudinal evolution, 75
    representation, 136
    fluctuation tensor, 29
    locus, 27
    particles distribution, 27
photon
    emittance, 196, 210
    flux, 194
point-spread function, 155
Pointing vector, 68
Poisson brackets, 141
polarization, 200
    circular, 201
    directions, 205
    linear, 201
    properties, 202
    rate, 202
positional power spectrum, 166
prebunched e-beam, 262
pulsed wire measurement system, 246

Q

quadratic form, 28
quadrupole
    matrix representation, 30
    scalar potential, 52
    magnet, 22
quantum fluctuactions, 75
quantum mechanical expectation value, 164

R

radial coupling factor, 211
radius of curvature, 22

rare hearth permanent magnets, 233

S
sagittal focal length, 325
SASE, 251, 260, 269
saturation, 261, 268
   brightness, 252
   intensity, 122, 124, 252, 259, 261
Saw-tooth instability, 224, 285
SAXS, 363
schimming technique, 242
self- induced harmonic generation (SIHG), 263
sextupole
   magnets, 34
   compensation, 42
   Hamiltonian, 35
   Liouville equation, 35
   strength, 35
SIHG, 273, 277–278
   mechanism, 270, 272
   number of round trips, 275
slippage distance, 125
small signal
   condition regime, 111
   gain, 113, 120
   gain coefficient, 104
small signal
   gain coefficient, 119, 261
spectral brightness, 181, 194
spherical aberrations, 326
stability criterion, 217
Storage Ring, 214, 251
   critical energy, 66
   dynamical aperture, 211
   longitudinal coupling impedance, 223
Storage Ring FEL, 259, 278, 284
   brightness, 258
   equilibrium power, 285
   output power, 118, 257
stretched wire measurement system, 246
symplectic
   group, 150
   matrix, 37
   transformation, 139
   unit matrix, 139
symplecticity condition, 140
synchronous particles, 74
synchrotron magnet, 31
   hamiltonian, 32
   Liouville equation, 32
synchrotron radiation, 17, 61
   angular properties, 61
   brightness, 193
   critical frequency, 63
   from magnetic undulators, 82
   integrals, 79
   intensity, 73, 204
   intensity dependence, 61
   quality factors, 194
   spectral broadness, 83
   spectral properties, 69
   spectral width, 61

T
tangential
   focal length, 325
   ports, 301
   radius, 325
TE field, 100
thin crystals, 344
thin lens, 149
   approximation, 55, 57
time-squeezing, 63
Touschek effect, 218, 222
transfer operator, 37
transmission coefficient, 313
transmissivity, 313
transport system, 26
   defocusing lens, 38
   drift section, 38
   focusing lens, 38
transverse motion coupling, 42
trapping process, 217

tune, 38
tuning studs, 242
Twiss parameters, 27, 29, 79, 124, 194
two-crystal monochromator, 353

U
undulator, 82, 231
    brightness, 189, 196
    crossed, 203
    emission linewidth, 98
    hybrid configuration, 242
    variably polarization, 203

V
vertical emittance, 79
Volterra type integro-differential equation, 104

W
wave response, 169
weak focusing, 32
wiggler, 88
Wigner
    distribution function, 171
    optical distribution function, 165, 170
    representation, 164
Wolter mirror systems, 328

X
X-ray
    absorption spectroscopy, 359
    compound systems, 326
    critical energy, 316
    critical wavelength, 316
    focusing mirrors, 321
XANES, 361